PRENTICE-HALL INTERNATIONAL SERIES
IN THE PHYSICAL AND CHEMICAL ENGINEERING SCIENCES

AMUNDSON *Mathematical Methods in Chemical Engineering*
ARIS *Elementary Chemical Reactor Analysis*
ARIS *Introduction to the Analysis of Chemical Reactors*
ARIS *Vectors, Tensors, and the Basic Equations of Fluid Mechanics*
BALZHISER, SAMUELS, AND ELIASSEN *Chemical Engineering Thermodynamics*
BERAN AND PARRENT *Theory of Partial Coherence*
BOUDART *Kinetics of Chemical Processes*
CROWE, HAMIELEC, HOFFMAN, JOHNSON, SHANNON, AND WOODS *Chemical Plant Simulation*
FREDRICKSON *Principles and Applications of Rheology*
HAPPEL AND BRENNER *Low Reynolds Number Hydrodynamics*
HIMMELBLAU *Basic Principles and Calculations in Chemical Engineering, 2nd ed.*
HOLLAND *Multicomponent Distillation*
HOLLAND *Unsteady State Processes with Applications in Multicomponent Distillation*
KOPPEL *Introduction to Control Theory*
LEVICH *Physicochemical Hydrodynamics*
MEISSNER *Processes and Systems in Industrial Chemistry*
PETERSEN *Chemical Reaction Analysis*
PRAUSNITZ *Molecular Thermodynamics of Fluid-Phase Equilibria*
PRAUSNITZ AND CHUEH *Computer Calculations for High-Pressure Vapor-Liquid Equilibria*
PRAUSNITZ, ECKERT, ORYE, O'CONNELL *Computer Calculations for Multicomponent Vapor-Liquid Equilibria*
WHITAKER *Introduction to Fluid Mechanics*
WILDE *Optimum Seeking Methods*
WU AND OHMURA *The Theory of Scattering*

PRENTICE-HALL, INC.
PRENTICE-HALL INTERNATIONAL, INC., UNITED KINGDOM AND EIRE
PRENTICE-HALL OF CANADA, LTD., CANADA

Chemical Plant Simulation

An Introduction to Computer-Aided Steady-State Process Analysis

C. M. CROWE
A. E. HAMIELEC
T. W. HOFFMAN
A. I. JOHNSON
D. R. WOODS

McMaster University

P. T. SHANNON

Digital Systems Corp.

PRENTICE-HALL, INC.

Englewood Cliffs, N. J.

PRENTICE-HALL INTERNATIONAL, INC., *London*
PRENTICE-HALL OF AUSTRALIA, PTY. LTD., *Sydney*
PRENTICE-HALL OF CANADA, LTD., *Toronto*
PRENTICE-HALL OF INDIA PRIVATE LIMITED, *New Delhi*
PRENTICE-HALL OF JAPAN, INC., *Tokyo*

Current printing (last digit):
10 9 8 7 6 5 4 3 2 1

Library of Congress Catalog Card Number 70-151822

13-128686-2

Printed in the United States of America

Preface

With the advent of digital computers and their ability to handle large quantities of information, engineering horizons have been dramatically extended. Formerly, only small sets of equations could be solved; currently, solving sets of hundreds and thousands of equations is possible. This capability can be used most effectively in the design or simulation of chemical processes. Now we can consider as never before how designs of various units interact with one another, and we can very rapidly evaluate alternative suggestions for plant improvement.

To capitalize on these developments, we need to establish a strategy for solving such large and complex problems and to apply it to real processes. This book describes such a strategy and illustrates its application to the simulation of a sulfuric acid plant, using an executive information-handling system called PACER®. However, the principles can be applied to other processes and executive systems.

Many books have been written on related subjects but most have concentrated on the executive system [Motard, et al. (1968)], on simulating equipment or small sections of the process [Franks (1967) or Himmelblau and Bischoff (1968)], on simulating linear processes [Nagiev (1964)] or on simulating business problems in which technical questions are not asked [Meier, et al. (1969), Forrester (1961), or Tocher (1963)]. This book is unique in describing and illustrating the strategy of simulation; it answers

®A trademark of Digital Systems Corp., Hanover, N. H.

technical questions about real nonlinear processes, with realistic objectives and industrial data and constraints. Yet this is *not* just a case history of how we did a simulation. Rather, the acid plant case is used to illustrate the steps we recommend in simulating any process. To continually remind the reader that the principles are generally applicable, we have included problems at the end of most chapters.

For the past five years, we have worked extensively in the area of computer simulation and the design of real industrial processes. The systems studies include pulp and paper processing, a polymer reactor system, alkylation of hydrocarbons, boiling heavy water reactor systems, business systems, the Bayer process for the production of alumina, integrated steel plants, sulfuric acid production, hydrocarbon cracking and processing, petrochemical processing, ethylbenzene-styrene systems, and ammonia production. These projects were done as graduate research, as graduate and undergraduate projects, as mission-oriented research projects, as contract research, and as consulting projects with industry. From this experience we have selected the sulfuric acid plant to best illustrate the strategy, because the technology is not excessively complex.

Books of this kind are rare because combining the needed experience demands extensive cooperation between university and industry and among university faculty members who pool their knowledge to tackle ambitious but practical problems. The acknowledgements show the many people who contributed to this book.

The book is written both for engineers in industry who are interested in applying new computer techniques to the design and analysis of plants, and for senior and graduate chemical engineering students for use in design and simulation classes. A detailed knowledge of programming languages or of computer hardware is not necessary to apply the techniques discussed.

The reader will be shown a strategy for solving complex plant problems, an approach to simulation and design of any chemical process, and the principles for developing an executive system. Details of PACER and of the sulfuric acid plant will be given through the concrete examples that illustrate the ideas.

The techniques introduced here may indeed revolutionize the role of the engineer in process analysis and design.

The main theme is a strategy for solving large and complex problems through the use of computer-aided analysis. In Chapters 1 and 2 we introduce the strategy, simulation, and executive systems. Details of the example executive system (PACER) and the case process (the sulfuric acid plant) are given in Chapters 3 and 4. The strategy is outlined briefly in Sec. 4.2. It is then applied to the simulation and optimization of a contact sulfuric acid plant with the PACER executive system in Chapters 4 to 12. In Chapter 13 we summarize the strategy for the simulation of any plant by highlighting the findings of the acid plant study and illustrating them with examples of other processes. In Chapter 14 we discuss the new and future developments in computer-aided analysis. We encourage the reader to peruse the problems at the end of each chapter; these illustrate how the specific details given in each chapter can be generalized.

We have successfully used this material in three-credit graduate courses, and as a complement to a three-credit undergraduate course in process control and simulation and to a four-credit undergraduate course in cost estimation and process analysis. However, by far the greatest use has been in a four-credit senior project that has been on either design or simulation. In this project, which forms the basis for many of the examples given in the text, the students were introduced to the principles in about six hours of lectures and through working as team members on the project. The detailed mechanics and recommendations for such projects are given by Crowe, et al. (1969).

Hamilton, Ontario THE AUTHORS

Acknowledgements

It is with pleasure that we acknowledge the host of people who contributed to this book. Without their help and encouragement this book could not have been written.

Our industrial colleagues opened their doors to us and supplied technical details and assistance and gave us permission to share the experience gained.

From Canadian Industries Ltd., we thank Dr. N. E. Cooke, J. T. Woyzbun, C. D. Atherton, Dr. G. M. Cameron, C. A. Miller, H. M. Jones, J. A. Labash, R. H. Geist, and Dr. J. D. McIrvine.

From Shell Canada Ltd., we thank Dr. I. J. O. Korchinski, O. K. Miniato, R. C. Brawn, B. Kopp, J. B. Rico, G. G. Myers, and R. J. Leflar.

From the Aluminum Company of Canada Ltd., we thank Mr. H. S. Monahan, K. I. Verghese, R. P. Brown, P. F. Bagatto, and J. A. Fraser.

From the Polymer Corporation Ltd., we thank Mr. D. E. McLellan, Dr. P. M. Reilly, W. Komarnicky, N. J. Little, D. W. Montgomery, William Porter, W. N. Brown, and A. J. Newton.

At McMaster University many colleagues helped us: Professors G. L. Keech, D. J. Kenworthy, and W. H. Fleming of the Computation Center, and Professors R. B. Anderson, M. H. I. Baird, J. W. Hodgins, K. L. Murphy, J. D. Norman, K. G. Pollock, J. D. Raal, and J. Vlachopoulos of the Department of Chemical Engineering.

Mr. T. Toong of Canadian General Electric assisted in the documentation of the executive systems.

The extensive work of and suggestions by our students, both undergraduate and graduate, we gratefully acknowledge. These include M. G. Coome, P. A. DiSalvo, P. Friedman, J. McGregor, W. Pacey, G. K. Patel, G. Renaud, J. Yallop of the class of 1965, J. P. Allingham, G. R. Carpenter, T. E. Erme, L. W. Fitz, V. Leyerzapf, C. E. Love, T. H. Northcott, W. K. Taylor, H. A. Thibaudeau, R. G. Wilson, L. T. Wisnowski, H. W. Wong of the class of 1966, J. F. Chappell, R. H. Gilmour, A. F. Gleave, D. C. Haney, E. D. Kitcher, J. A. Powell, I. D. Shaw, T. L. Sutton of the class of 1967, A. G. Brown, H. I. Canning, J. E. Gates, P. B. Melnyk, A. J. Orlickas, D. E. Williams of the class of 1968, W. D. Akers, R. G. Earp, K. F. W. Kingdon, A. L. Nguyen, M. J. Rossignoli, R. F. Sivak, C. E. Slade, J. W. Suttie, D. J. Taylor, R. S. Wawrinty, G. E. Wolfe of the class of 1969, and our graduate students D. R. Prowse, M. Aizawa, W. F. Petryschuk, G. Chartrand, O. Orbach, J. D. Freel, A. W. T. Hui, B. P. LeClair, M. C. Li, C. R. Milburn, S. L. Pearce, A. Y. Yau, J. K. Khosla, J. P. Sheel, K. A. Burrill and P. Seto.

We received excellent suggestions about the material from the participants at the courses we have given on this and related material at Niagara Falls, New York; at Sarnia, Ontario; Quebec City; Montreal; at Dartmouth College and at McMaster University.

Our secretaries have patiently typed and retyped various versions of the text. We thank Mrs. Ora Orbach, Mrs. E. S. Gravestock, Miss Ruth Wier, Mrs. J. M. Double, and Miss J. Copeland. Mrs. Ellen Baker of the Computing center helped compile the indices.

We appreciate the assistance of Professor R. R. Hughes in polishing the text.

Financial support for our instructional and research activities has been supplied by the Chemical Engineering Division of the Chemical Institute of Canada, and the National Research Council of Canada. A grant from the Department of University Affairs of the Province of Ontario and the cooperation of the Canadian General Electric Company of Canada have made possible the documentation of the MACSIM system and the development of GEMCS.

Contents

1. Digital Simulation: A Boon to the Engineer 1

1.1 Satisfying an Engineering Need, 1. *1.2 What Is a Simulation?*, 6. *1.3 How To Create a Simulation*, 8. *1.4 The Plan of the Book*, 11.

2. From Plant to Computer: The Techniques of Digital Simulation 12

2.1 Developing a Description of Information Flow, 13. *2.2 From Process to Information Flow Diagram*, 18. *2.3 From Information Flow Diagram to Numerical Form*, 20. *2.4 Planning the Calculations—Finding Recycles*, 23. *2.5 Planning the Calculations of a Recycle Set*, 31. *2.6 Summary*, 37. *Problems*, 38.

3. PACER® (1966)—An Executive Program for Process Simulation 40

3.1 Specifications and Major Elements of the System, 41. *3.2 PACER Organization*, 42. *3.3 PACER Operation—Order of Calculation Specified*, 46. *3.4 How PACER Determines the Order of Calculation*, 58. *3.5 Summary of PACER Data Requirements*, 64. *3.6 Summary*, 65. *Problems*, 66.

4. The Acid Plant: An Example of a Plant Simulation 69

4.1 A Description of the Acid Plant, 70. *4.2 Strategy and Decisions for Solving Simulation Problems*, 76. *4.3 The Application*, 80. *4.4 Summary*, 93. *Problems*, 95.

5. Two Key Decisions About Modeling and Physical Properties 104

5.1 Collection of the Input Data, 104. *5.2 The Modeling Decision*, 110. *5.3 How To Handle Physical Properties*, 115. *5.4 Decisions Made for the Acid Plant Study*, 117. *5.5 Organization of Resources*, 123. *5.6 Summary*, 125. *Problems*, 126.

6. The Simulation of the Fluid Handling Equipment 133

6.1 Introduction to and Strategy for Modeling Equipment, 133. *6.2 Choice of Models and Physical Phenomena to be Included*, 134. *6.3 The Development of BLOW1 as an Example*, 137. *6.4 The Development of MIXER6 as an Example*, 143. *6.5 Summary*, 149. *Problems*, 149.

7. Simulation of Heat Exchangers 153

7.1 Statement of the Problem, 153. *7.2 The Approach to the Simulation Problem*, 153. *7.3 Formulation of the Model Equations*, 155. *7.4 Data Required for the Superheater and Economizer Model*, 162. *7.5 Programming the Model*, 168. *7.6 Matching the Model to the Plant*, 170. *7.7 Data Required for the Simulation of the Waste Heat Boilers*, 171. *7.8 Programming the Model BOILS2*, 173. *7.9 Matching the Model to the Plant*, 175. *7.10 Other Heat Exchangers in the Plant*, 175. *7.11 The Problem of the Oversophisticated Model*, 177. *7.12 Summary*, 177. *Problems*, 178.

8. Simulation of Single-Bed Reactors 179

8.1 Statement of the Problem, 179. *8.2 Choice of the Rate Processes to be Used in the Model*, 180. *8.3 Formulation of the Model Equations*, 183. *8.4 Data Required for the Model*, 188. *8.5 Programming the Model*, 191. *8.6 Matching the Model to the Plant*, 193. *8.7 Summary*, 195. *Problems*, 195.

9. Studies of a Multibed Reactor with Heat Exchange: Oxidation of Sulfur Dioxide 197

9.1 Introduction, 197. *9.2 Description of the Reactor*, 198. *9.3 Plant Data Available*, 199. *9.4 Studies of the Reactor using PACER*, 200. *9.5 Improve-*

ment of Convergence on Pressure—SETSP1, 208. *9.6 Summary*, 209. *Problems*, 209.

10. Simulation of Packed Gas Absorbers and an Absorber Plant 210

10.1 Statement of the Problem, 210. *10.2 Choice of the Transport Processes for the Model*, 212. *10.3 Formulation of the Rate Equations for the Comprehensive Absorber Model*, 218. 10.4 *Data Required for the Accuracy of the Model*, 220. *10.5 Programming the Model*, 220. *10.6 Matching the Model to the Plant*, 223. *10.7 Simulation of an Absorber Plant*, 223. *10.8 Summary*, 226. *Problems*, 226.

11. Studies of the Whole Plant and a Summary of the Acid Plant Case 228

11.1 Complete Plant Studies, 229. *11.2 Technical Summary and Conclusions of the Acid Plant Case*, 252. *11.3 Comments on PACER and the Modular Approach*, 253. *11.4 Summary*, 254. *Problems*, 254.

12. Optimization of the Plant Simulation 257

12.1 Introduction, 257. *12.2 Specification of the Optimization Problem*, 258. *12.3 Choice of the Optimization Technique*, 260. *12.4 Summary*, 269. *Problems*, 272.

13. The Strategy of Simulation—A Summary 274

13.1 Define the Problem and Learn the Process, 275. *13.2 Develop the Base Case and the Sensitivity Analysis*, 291. *13.3 Resource Analysis and Simulation Decisions Required to Answer the Questions*, 292. *13.4 Prepare Unit Computations* 294. *13.5 Answer Questions by Simulating All or Part of the Plant*, 295. *13.6 Summary*, 299. *Problems*, 299.

14. On the Horizon and Beyond 302

14.1 Variety of Applications of Computer-aided Executive Systems, 302. *14.2 Computer Hardware Developments*, 311. *14.3 Solution of Problems Shared in Many Applications*, 312. *14.4 Summary*, 316. *Problems*, 316.

Appendix A: Brief Descriptions of Processes 318

A.1 An Alkylation Plant, 318. *A.2 A Bayer Process Plant*, 322.

Appendix B: Coding Sheets for MACSIM and Sample Output 325

Appendix C: Symbols 334

Appendix D: Glossary 339

Appendix E: References 343

Author Index 355

Subject Index 359

Digital Simulation: A Boon to the Engineer

1

Innovation has always been an exciting and challenging part of engineering. The innovative engineer creates new processes and finds different processing conditions in order to increase profits. Often, evaluation of the ideas for improvement is straightforward and immediately rewarding. This is represented schematically in Fig. 1.1(a). But an evaluation can be anything but exciting when it involves the tedious task of long and repetitious calculations caused by the recycle of energy or material. A representation of recycle calculations is shown in Fig. 1.1(b).

1.1 Satisfying an Engineering Need

Recycle occurs frequently in chemical plants to conserve material and to improve the overall efficiency. Such recycle, however, introduces calculational difficulties. The engineer needs a technique to relieve him of the tedium of repetitive recycle calculations and to free him for creative work. Such a technique should combine the speed of digital computers with methods of handling information.

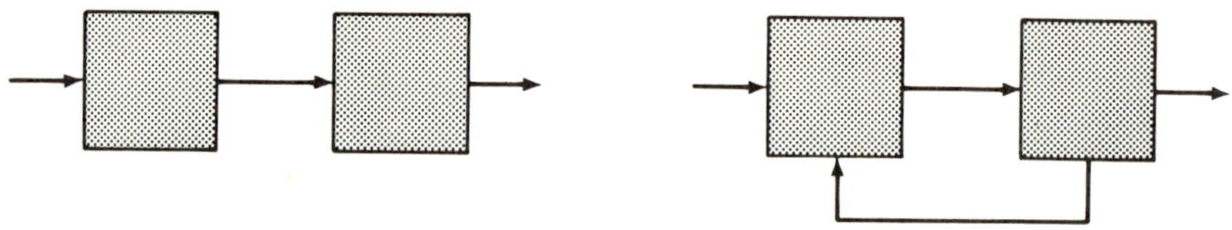

(a) Evaluation of new ideas is easy because no recycle is present

(b) Evaluation of new ideas is tedious because of recycle calculation

Figure 1.1. Representation of calculations.

To see how the need can be satisfied, consider how an engineer presently deals with calculations, whether they be done by hand or with the aid of computers. To analyze processes in the search for improvements, the engineer devises a mathematical description, that is to say, creates a simulation, of the proposed innovations. Traditionally, in simulating any process, the engineer starts with some input conditions and uses the principles of Unit Operations to calculate the outputs from the first equipment unit or stage. This is repeated for each succeeding unit with an output of one unit being an input to the next. For example, the engineering flow diagram shown in Fig. 1.2(b) might be used by an engineer to do material and energy balances for the process shown in Fig. 1.2(a). He uses Fig. 1.2(b) to keep track of the calculations and to summarize the results. While calculating the material and energy flows, the engineer creates a simulation of the process shown in Fig. 1.2(a). The simulation arrangement is shown in Fig. 1.2(c). For each set of input conditions in streams 1, 2, and 3 there is a corresponding set of calculated answers for streams 4, 5, and 6. The characteristics of this approach are:

1. The inputs to each unit are used to calculate its outputs from the principles of Unit Operations that describe the phenomena occurring in the unit.
2. The processing sequence and arrangements among units must be known.
3. The calculated outputs from one unit become the inputs to other units.
4. The engineer must recognize when he has a trial-and-error or recycle problem.
5. For these trial-and-error or recycle problems, the engineer tries to specify the sequence in which calculations should be done to minimize the work.

Of these characteristics, only the equipment (1), the arrangement of the equipment (2), and the input to the plant vary from problem to problem. The rest —the handling of numbers and the information transmission from calculation to calculation (3), the identification of recycle (4), and the specification of the sequence in which calculations are performed (5)—are common to all

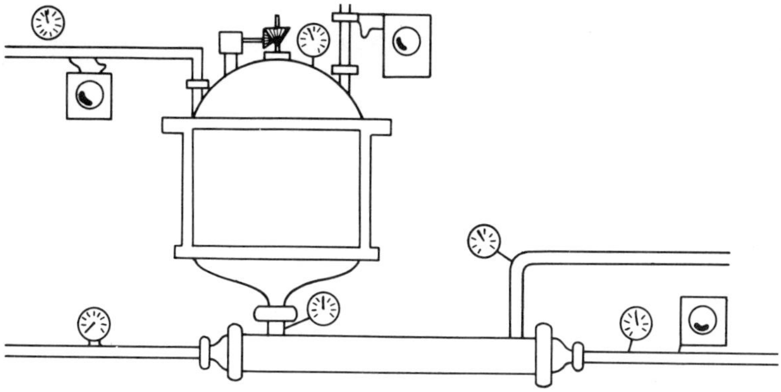

(a) Plant to be simulated

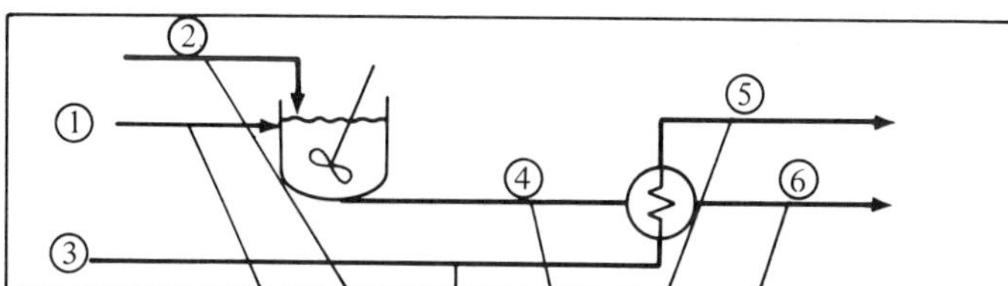

Stream no.	1	2	3	4	5	6	
Flow rate Temperatuıe Pressure	149 52 2	240 GIVEN 21	159 87 24				

(b) An engineer's process flow diagram

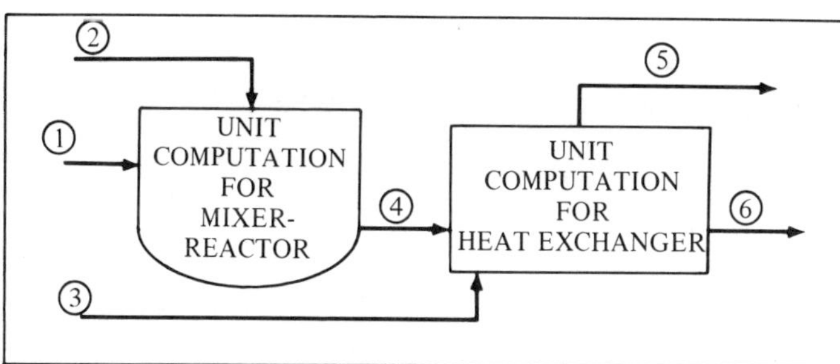

Stream no.	1	2	3		4	5	6	
Flow rate Temperature Pressure	145 52 3	245 130 24	159 57 24					

Input conditions

Blackboard for storing answers (supplied by executive program)

(c) Simulation arrangement

Figure 1.2. Devising a simulation.

problems and can profitably be handled by an executive computer program. Such a program satisfies an engineering need because it relieves the engineer of the bookkeeping and quickly handles the repetitive calculations required for the once-dreaded recycle problems. With such a system the engineer supplies the specified input conditions, indicates the equipment arrangement, and supplies a mathematical model which describes the Unit Operation for each equipment unit. This is shown in Fig. 1.3. The executive computer program is like a project director who obtains realistic results for the complete process by requiring each engineer to supply him with the calculated performance of one equipment unit.

WHAT THE ENGINEER MUST PROVIDE		WHAT THE EXECUTIVE PROGRAM DOES
Input conditions	→	Reads the input data into appropriate storage locations
Description of how the equipment is connected	→	Transmits information through streams connecting Unit Computations
Sequence of calculation (if desired)	→	Determines a sequence if it is not supplied
Stream and equipment information		
Mathematical models of all the equipment { Unit Computation, Unit Computation, Unit Computation }	→	Performs the Unit Computations in proper order
Control information to instruct executive program as to printout and other decisions	→	Prints and stores the results of the calculations ↓
Engineering interpretation of the results calculated	→	ANSWER

Figure 1.3. Relationship between the engineer and the executive program.

The combination of a digital computer and the information-handling executive programs presents a powerful technique to both the industrial engineer and the undergraduate student. The use of this tool makes possible the solving of problems involving tens of thousands of numbers. For example, the illustrative case presented later in this book required a solution of approximately 500 simultaneous equations, many nonlinear, involving about 1000 stream variables and 200 equipment parameters. A comprehensive executive-aided computer simulation can be used to predict the effect of changing conditions, physical layout, and capacity, to do mass and energy balances rapidly, to optimize the operation quickly and safely, to detect and eliminate production bottlenecks, to provide depth of knowledge about complete system behavior, to improve control and investigate the feasibility of on-line computer control, to facilitate cost calculations, budget control, and planning of operations, and to train operators and engineers.

Indeed, there are many purposes for developing a simulation. At one extreme is a simple simulation which can show the process designer how the equipment units interact and how sensitive the calculations are to changes in design parameters. An example of this might be patching together in 10 to 40 man-hours a process mass balance from an existing library of programs for various units. Although the questions that can be answered by this simulation are not sophisticated, the answers are nevertheless valuable. At the other extreme is a very accurate simulation for answering technically sophisticated problems. A simulation to supply such answers may take two to four man-years. This too is most valuable because without the computer such problems would simply not be attempted because they would be too complex. However, for such ambitious projects it is financially attractive to use an executive-aided computer simulation, mainly for problems complicated by recycle, for those involving many numbers, or for those requiring an analysis of a large number of alternatives. For example, Fig. 1.4(a) shows a large sectioned distillation column. Because of the recycle, even such a

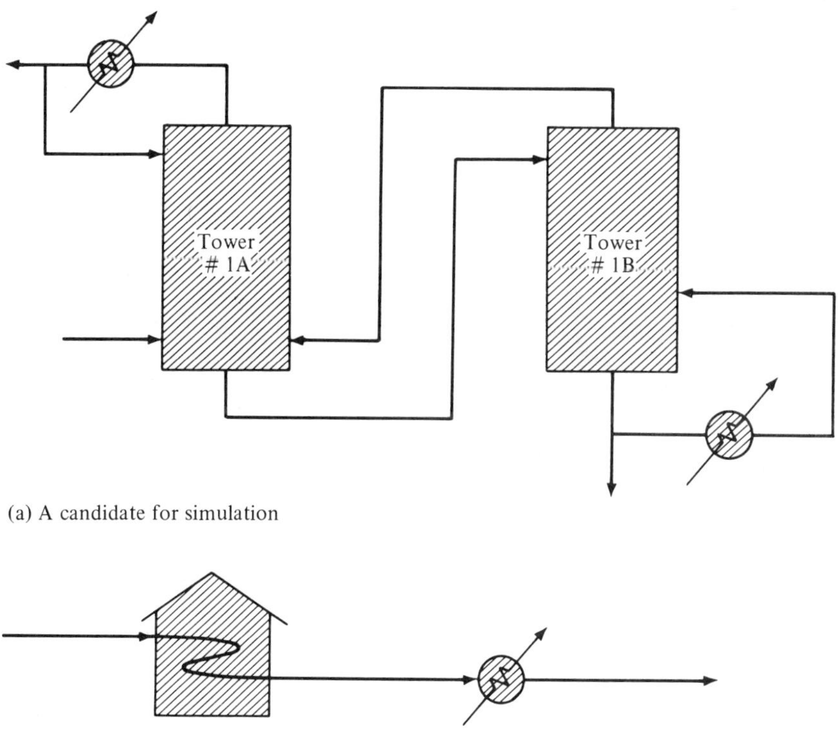

(a) A candidate for simulation

(b) Probably not a candidate for simulation

Figure 1.4. Typical problems for engineers.

small system is a good problem to tackle with an executive-aided simulation. However, the system shown in Fig. 1.4(b) is probably not a candidate for an executive-aided simulation because there is no recycle, there are relatively few numbers involved, and hence an engineer might solve the problem more cheaply without the executive-based system. Other factors which influence the decision of whether or not to create a very accurate executive-aided computer simulation include a need for corroborating data, ready access to a computer, the ability to describe equipment behavior, and the availability of people with computational and programming experience. An elaboration of these factors is presented in Chapter 13.

However, gaining the advantages offered by the executive for either simple or very accurate simulations requires strict discipline in supplying input information to the system. Even though an executive-aided computer simulation may demand much time, skill, and computation, it is a boon to the engineer which, when correctly employed, can increase profits. The applicability will also increase greatly as more use is made of digital computers for everyday engineering. This book describes the approach to creating an executive-aided digital computer simulation of the steady-state operation of a plant (hereafter referred to simply as a simulation), illustrates this approach through a case study, and provides guidelines for using the principles to tackle other steady-state problems.

1.2 What Is a Simulation?

Perhaps the easiest way to clarify the concept of a simulation is to relate it to the engineer's analysis of the process shown in Fig. 1.2(a). The engineer calculates the overall mass balance by calculating for each unit in turn what happens to the inputs to that unit. From the flow diagram he knows that the mixer-reactor is followed by the heat exchanger. He first calculates the answers for the units where all the inputs are known, in this case the reactor, and then proceeds to calculate the behavior of the heat exchanger. He uses Fig. 1.2(b) as a convenient place to store his answers, to relate answers to flows in pipes, and to remind himself of the relationship among the units.

In a simulation, calculated outputs are also produced from given inputs, but most of the decisions, bookkeeping, and calculations are done through computer programs rather than by hand. A simulation is a set of computer programs which calculates realistic values of the significant operating variables for the process being simulated. An analogy between the simulation and the real process is worth noting. Just as a plant is made up of equipment units connected by pipelines, so a simulation consists of sets of calculations, called Unit Computations, linked by streams that transmit information unchanged between the different sets of calculations. Material flows through the pipelines because of the laws dictated by nature; information flows through the stream network as prescribed through the executive program.

The simulation has three parts: the data set, the executive program, and the sets of mathematical models called Unit Computations. The relationship among these is indicated in Fig. 1.3. The data set includes the input conditions, the processing sequence, the order of calculation, and information pertaining to the calculations in the Unit Computations. The executive program, given the data set, calls upon different Unit Computations according to a calculating sequence (which it either devises or is given in the data set) and calculates the values of operating variables for all streams in the system. Consider now some characteristics of the executive program and the Unit Computations and in particular the demands they place on the engineer creating a simulation.

The executive program

The executive program transmits information through the streams and stores the calculated results. It can plan the sequence in which calculations are to be done if it is requested to do so. For example, in Fig. 1.2(c) the executive program would supply the blackboard for storing the answers, would know that the heat exchanger followed the reactor, and would know that the behavior of the reactor would have to be calculated before the operation of the heat exchanger could be predicted. Several executive systems exist: CHESS, CHEOPS, UWOPS, CHIPS, FLEXIFLOW, POWERFACTS, PACER, CHEVRON; Steward (1967), Evans, et al. (1968), and Rudd and Watson (1968) review some of these. Many programs are proprietary. Today the engineer's job is mainly one of effectively selecting and using an existing executive program. He needs to know what the executive program can and cannot do for him, the advantages and disadvantages of using a particular program, the required format for the input information, and the format used in printing out the answers.

The Unit Computations

A set of calculations that predicts what occurs within an equipment unit is called a Unit Computation. Normally there is a one-to-one correspondence between the Unit Computation and the plant equipment carrying out the Unit Operation. For example, in Fig. 1.2(c), the Unit Computation for the mixer-reactor calculates from the input information carried in streams 1 and 2 the exit conditions (stream variables) in stream 4. The values given for stream 4 should be the same as the hand calculation reported in Fig. 1.2(b) and the plant conditions of Fig. 1.2(a).

The chemist with his atomic model kit often constructs replicas of an infinite variety of molecules; the engineer can use his set of Unit Operations to devise a new process or he can patch together Unit Computations to calculate an existing process. This can be visualized as a building-block approach.

Each "block" or Unit Computation refers to only one Unit Operation, for example, heat exchanger Unit Computation, reactor Unit Computation, and so on. Each Unit Computation is self-contained, and once a collection of Unit Computations representing most Unit Operations is available, a variety of processes can be quickly assembled by connecting the appropriate Unit Computations together.

Consider briefly the concepts of creating a Unit Computation for the heat exchanger of Fig. 1.2(c).

Assume this Unit Computation should calculate the exit temperature of the hot process stream 6 of Fig. 1.2(c). The familiar equation from the Unit Operations of heat transfer

$$q = UA(\Delta T)_{lm} \tag{1.1}$$

could be used. But this would require a trial-and-error calculation which is unnecessary since the concept of effectiveness factor has been reported [Kays and London (1964)]. The exit temperature for this system can be calculated from

$$T_{h,\text{out}} = T_{h,\text{in}} - E(T_{h,\text{in}} - T_{c,\text{in}}) \tag{1.2}$$

where the effectiveness factor, E, is given by

$$E = \frac{1 - \exp[-NTU(1 - R)]}{1 - R\exp[-NTU(1 - R)]}$$

with $NTU = UA/(WC)_{\text{process fluid}}$
U = overall heat transfer coefficient
A = exchanger area
W = mass flow rate
C = specific heat
$R = (WC)_{\text{process fluid}}/(WC)_{\text{water}}$*

From this simple set of equations the exit temperature can be calculated. These equations, together with the necessary input-output formats imposed by the executive program, make up the Unit Computation.

An engineer can be very flexible in writing the Unit Computation. It can be general or specific, approximate or accurate, and based on fundamentals or on statistical correlations. Following chapters describe how to choose the model for a Unit Computation from the options available.

1.3 How To Create a Simulation

To create a simulation the engineer writes Unit Computations consistent with a selected executive program. The task is complicated by the overwhelm-

* The ratio is $(WC)_{\text{smaller}}/(WC)_{\text{larger}}$ but for this problem we assumed the process fluid had the lower thermal capacity. Details are given in Sec. 7.3.

ing amount of detail due to the very complexity which warranted the use of a computer simulation in the first place. Handling these details requires careful planning and allocation of time and people.

Figure 1.5 presents a suggested strategy for solving the problem; Fig. 1.6 shows a possible organization of the personnel. Figure 1.5 should be examined in some detail. While the problem is being defined, the questions to be answered are determined and the decision to adopt a simulation approach is made based on economics and the availability of information. The planning stage includes the organization of the personnel; training in the use of the computer-based executive system chosen; agreement on lines of communica-

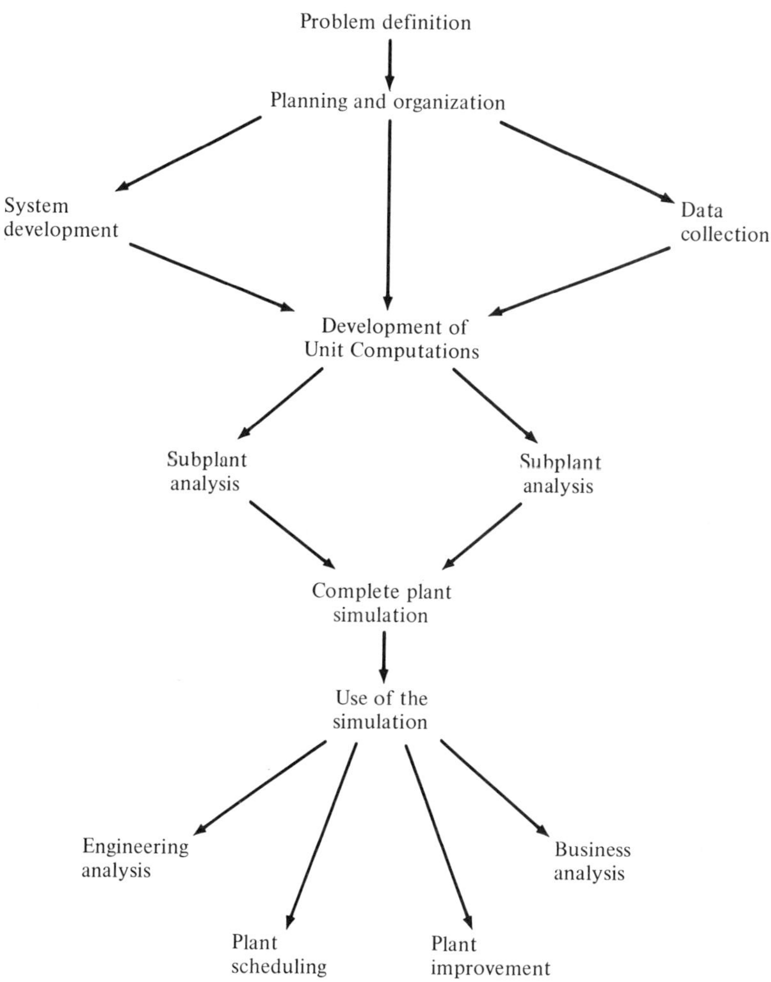

Figure 1.5. Suggested overall strategy.

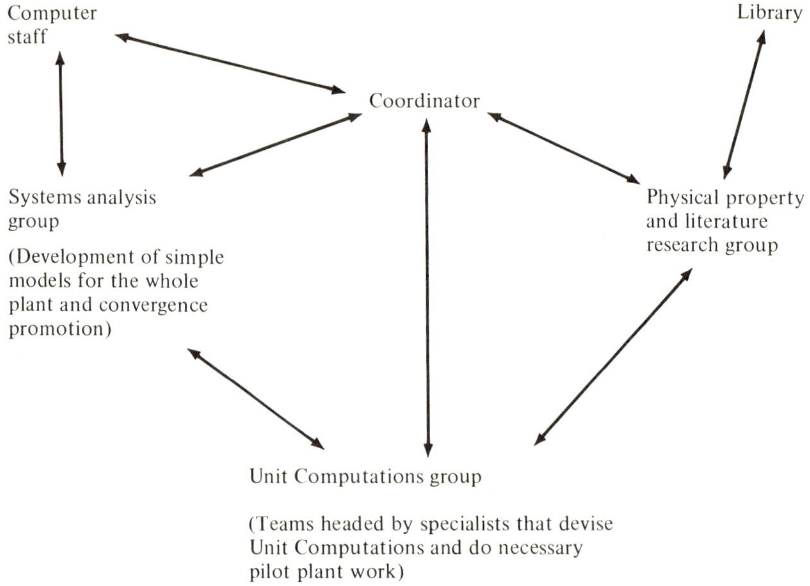

Figure 1.6. Organization of the main lines of communication.

tion, the form of reports, nomenclature, and job responsibility; construction of a critical-path schedule; and the collection of additional background information, physical property data, and plant operating data used in building and checking the models. The next stage is the development of the individual Unit Computations. Two support activities—devising reasonable methods for manipulating the physical properties and creating a simple simulation of the plant to indicate areas where complications develop—should be done concurrently since these necessary jobs are common to all the Unit Computations.

It is perhaps at this stage that the organization of personnel is most important, provided, of course, that a simulation team is available. Figure 1.6 illustrates a reasonable organization for a team effort. One group maintains close liaison with the computation center, writing extremely simple models for all Unit Computations and simulating the complete plant. From this, the group gains experience in how the system behaves and in the importance of its interactions, and gains a feeling for the sensitivity of the system to the accuracy of individual Unit Computations. Another group can develop a physical property handling system that incorporates reliable tested data. It is useful to create teams which devise individual Unit Computations in their own fields of expertise. Models are prepared and tested for a variety of conditions. It is essential that these teams be well informed about the accuracy and variety of model needed, and about the activities of the other project workers. All these activities need to be supervised by a team coordinator.

The individual models are combined to simulate portions of the plant and finally to simulate the complete plant. This activity is accompanied by the corroboration of assemblies of plants by comparing calculated with measured results, by use of techniques to reduce the total computation time, by promoting the convergence of recycle calculation, and by any required elaboration or simplification of the individual Unit Computations.

1.4 The Plan of the Book

This book shows how to simulate a complex plant and how to use the result to optimize plant operation. The general principles and guidelines are presented through a case study, with the PACER executive program utilized to develop a sophisticated simulation of a contact sulfuric acid plant. The approach, the typical decisions, and the problems encountered are adequately illustrated by the case provided. These principles are then summarized and illustrations of how they can be extended to other problems are given.

Chapters 2 and 3 discuss what simulation entails, describe features common to most executive programs that can be used, and give some details of one particular executive program, PACER.

Chapters 4 to 11 outline the approach to the problem in detail. The concrete example of simulation given with the PACER executive program is the 315 ton per day contact sulfuric acid plant owned by Canadian Industries Ltd. in Hamilton, Ontario, Canada.

Chapters 12, 13, and 14 discuss the application of the simulation work for plant improvement and suggest ways the engineer can extend the techniques of simulation. The problems at the end of each chapter also illustrate the extension of the simulation techniques.

The emphasis throughout the book is on the strategy and approach to the problem; the case study of the acid plant with the PACER executive is the medium through which the points are discussed.

From Plant To Computer: The Techniques of Digital Simulation 2

There are three distinct requirements for the simulation of a chemical plant. The first is an executive program, which handles the numbers and stores them, passes data amoung the Unit Computations, and determines the sequence of calculation if it is not specified by the user. The second is a set of Unit Computations which describes the individual types of equipment found in the plant. The third is the set of data describing the initial stream variables and the equipment parameters to be used in the Unit Computations.

In Chapter 1 the idea of an executive program was introduced and its usefulness was indicated. This chapter examines factors common to many executive programs and their Unit Computations, preparing the way in Chapter 3 for a detailed discussion of PACER, the particular executive program used by the authors. Later chapters describe the detailed preparation of Unit Computations and the gathering of data for the sulfuric acid plant as an example and consider many aspects of achieving and exploiting a digital computer simulation of such a plant.

An executive program as it is used in this book is a collection of subprograms which has some or all of the following features:

1. *Modularity*, which allows changes to be made in how units are connected, independently of what the units are—provided information is passed between units in a standard format;
2. Ability to determine a *sequence of calculation* of the Unit Computations;
3. *Easy alteration* of the parameters in Unit Computations, of the types of Unit Computation, and of the connections between them;
4. *Efficient solution* of the case to give a meaningful answer;
5. *Easy interpretation* of the results of the simulation.

If the executive program is supplied with the feed conditions, with the types of Unit Computation and how they are connected, and with the subroutines needed, it must be able to produce the desired results.

Every executive program will have its own rules about how the data are prepared, how the connections between units are recorded, and how the Unit Computations must be written. Most executive programs either allow the user to specify the sequence in which the calculations are to be done or can automatically find a sequence for him. The hardest work in achieving a simulation is to prepare the Unit Computations and to ensure that they obey the rules of the particular executive program. Since a workable, if not optimum, sequence of calculation can normally be specified by the user or found by the executive, Secs. 2.4 and 2.5, which discuss planning of calculations, may be omitted on first reading.

Whatever executive program is used and whatever Unit Computations are written, there are several common problems in a simulation, namely, the existence of recycle, the analysis of information flow, and the finding of a way to do the calculations. These three topics are discussed in turn in the rest of this chapter.

2.1 Developing a Description of Information Flow

Information is a set of values for the variables of a process needed to describe fully the state of a stream at any point in the process. Examples are flow rate, temperature, pressure and composition of a process fluid, or temperature and flow rate of saturated steam. Information can flow from place to place not only through actual streams but through control lines, across heat exchanger surfaces, and by any other mechanism which alters or fixes values of stream variables. The study of information flow in a process is necessary when recycle exists. Without recycle, calculations could be done one unit after another with no repetition. Therefore recycle is discussed before information flow is examined.

The problem of recycle in chemical plants

Most chemical processes have recycle of either matter or heat. Recycle means that a stream leaving a process unit affects a stream entering that unit. For example, in Fig. 2.1 the output stream from the mixer affects its input stream coming from the overhead of tower #2 because there is a connection through three pieces of equipment. The output from a unit can only be calculated if its input is known, but for a process with recycle its input is only known once its output has been calculated. The classical chemical engineering approach has been to assume values for as many streams as are required to compute a unit and then proceed until the calculated values of stream variables agree with the assumed values. Although steps can be taken to accelerate the solution of a recycle problem, unaided convergence to solution is still very widely used. Examples of typical recycles are:

1. The preheating of the feed of a distillation column by the bottoms stream (recycle of heat);
2. The use of sulfuric acid product to dry the air fed to the sulfuric acid plant (recycle of material);
3. The generation of steam from the heat of combustion of sulfur which is then used to drive the turbine supplying the air to support that combustion (recycle of matter and heat).

As bases for further discussion, consider the process flow diagrams with recycle shown in Figs. 2.1 and 2.2 and called Plants A and B, respectively. The first, Plant A, includes several different Unit Operations and has recycle of material. The second, Plant B, consists of simple mixers and surge tanks and is used in Sec. 2.4 to discuss planning the calculation.

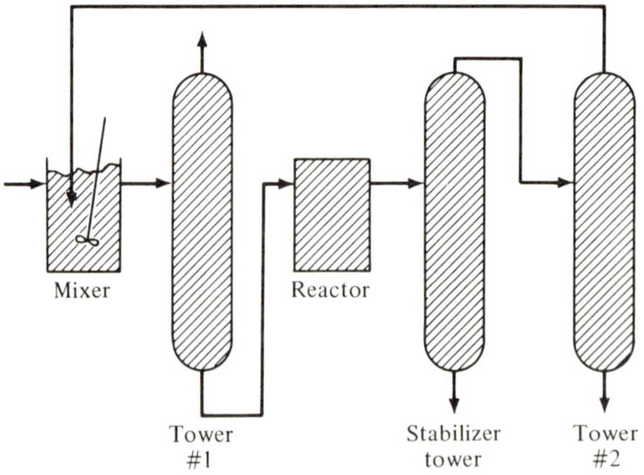

Figure 2.1. Process flow diagram—Plant A.

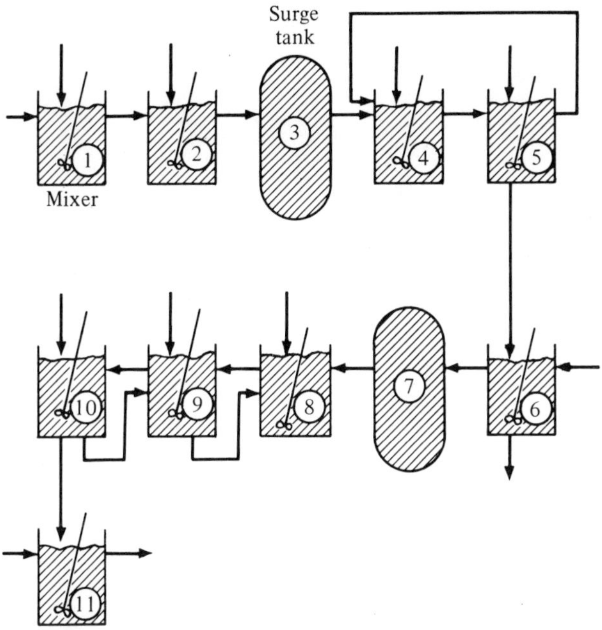

Figure 2.2. Mixing process flow diagram—Plant B.

Mixing—a simple Unit Operation

As a typical element of a recycle process in Fig. 2.2, consider the mixer, shown in Fig. 2.3. In each such mixer the agitation is assumed to be sufficiently intense so that the output streams are at exactly the same concentration as the contents of the tank. In addition to combining all the input streams to a uniform mixture, the mixer splits the outflow into two streams. Two control valves are shown on the output streams to indicate that the relative amounts of fluid in the outputs can be varied. The mixer is assumed to operate in the steady state so that there is no accumulation in the tank, or alternatively, the capacity of the mixer is assumed to be very small. Temperature changes are ignored in this simple case.

If the input stream flows and compositions are known and if the proportion of the total flow rate leaving in each output stream is specified, then the flows in the output streams can be calculated by a simple material balance. Note that just as the physical tank "solves" simultaneously the material balances for every component, so the model of the tank solves the equations for material balance simultaneously.

The Unit Computation

The physical mixer can be modeled by a Unit Computation called MIXER. Thus the streams into the mixer of Fig. 2.3 carry information about flow rate and composition. The mixer modifies this information and puts it

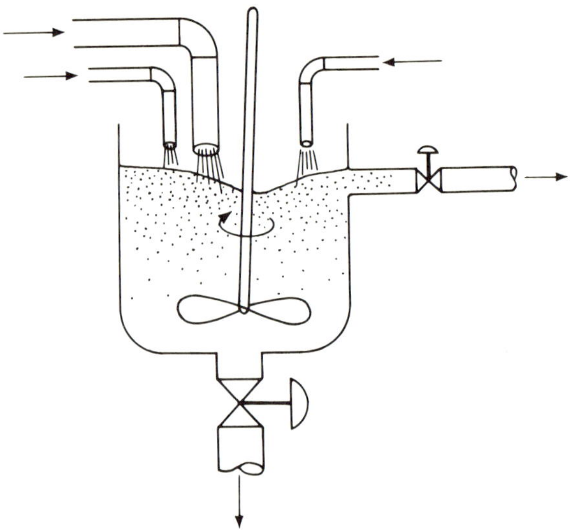

Figure 2.3. A simple mixer.

in the output streams. A stream can then be seen as a carrier of information and an equipment unit as a modifier of information. They can be represented schematically with lines for streams and a diamond for the mixer, as shown in Fig. 2.4. The arrows show the direction of material flow. Associated with

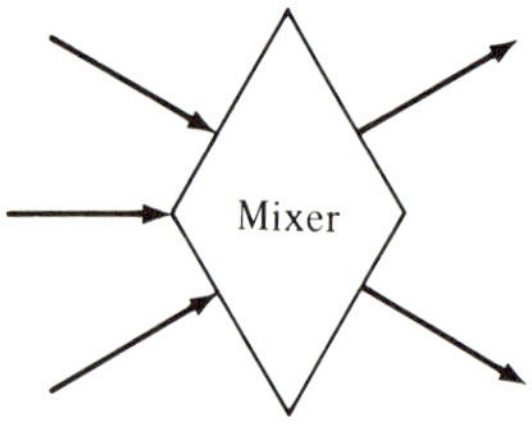

Figure 2.4. The mixer as a Unit Computation.

each equipment unit or Unit Computation is a set of equipment parameters, such as the proportion of the total flow in each output stream of the mixer. Other examples of equipment parameters would be the heat transfer coefficient of a heat exchanger or the bed depth of a catalytic reactor. A one-sentence abstract for the Unit Computation MIXER is as follows: Given the flows in any number of input streams and the equipment parameters specifying the proportion of the flow in each output stream, find the flows in the several output streams by simple material balance.

The computer programming necessary to implement this simple Unit Computation is straightforward. Note that MIXER can be used for mixing or splitting streams in such a way that all the output streams have identical composition and temperature.

TABLE 2.1
PROCESS MATRIX FOR PLANT A—FIG. 2.6

Unit number	Unit Computation name	Associated stream numbers		
1	MIXER	1	7	−2
2	DISTIL	2	−8	−3
3	REACT	3	−4	
4	DISTIL	4	−5	−9
5	DISTIL	5	−7	−6

The order of the input and output stream numbers in a row of the process matrix is important for the transfer of information into the Unit Computation. For example, in a heat exchanger the first input and output may be the process fluid and the second input and output the heat exchanger fluid. Or in the distillation calculation of Fig. 2.6, the first output is the overhead and the second the bottoms. Thus the process matrix encodes the entire structure of the information flow diagram:

1. What stream links what units;
2. The name of the Unit Computations representing each unit;
3. The order of the input and output streams of a unit.

In fact, given only the process matrix, the information flow diagram can be completely reconstructed. Another real advantage of the process matrix is that the engineer can readily understand and read it.

The stream connection matrix*

The stream connection matrix is an array with three entries per row. The first entry is the stream number and the second and third are the numbers

TABLE 2.2
STREAM CONNECTION MATRIX FOR PLANT A—FIG. 2.6

Stream number	From unit number	To unit number
1	0	1
2	1	2
3	2	3
4	3	4
5	4	5
6	5	0
7	5	1
8	2	0
9	4	0

* May be omitted on first reading.

of the equipment units from which that stream comes and to which it goes, respectively. Thus the stream connection matrix for Plant A as seen in Table 2.2 can be constructed either from the information flow diagram of Fig. 2.6 or the process matrix of Table 2.1. Note that feed streams are shown as coming from, and product streams as going to, unit number zero.

It can be seen that of the three items of information in the process matrix, only the first is retained in the stream connection matrix. Thus there is neither indication of the type of Unit Computation nor of the order of input and output streams of a unit.

The incidence matrix*

Another way of encoding the information flow diagram is the incidence matrix, shown in Table 2.3 for Plant A of Fig. 2.6. The left column contains the equipment number and the remaining columns correspond to stream numbers. A symbol "+1" shows that the stream number given by that column enters the equipment number given by the row. A symbol "−1" shows conversely that the stream leaves the equipment. A blank or zero shows that the stream does not connect to that equipment. Thus since stream number 4 leaves unit number 3 and enters unit number 4, the column for stream 4 has "−1" in row 3 and "+1" in row 4.

If the sum of a column is zero, the stream connects two units. If the sum is +1, the stream is a feed, and if it is −1, the stream is a product. Recycle in the process can be detected immediately in the incidence matrix if no rearrangement of rows would put the −1 above the +1 in each column

TABLE 2.3
INCIDENCE MATRIX FOR PLANT A—FIG. 2.6

Unit number	Stream number 1	2	3	4	5	6	7	8	9
1	1	−1					1		
2		1	−1					−1	
3			1	−1					
4				1	−1				−1
5					1	−1	−1		
Sum	1	0	0	0	0	−1	0	−1	−1

having a zero sum. Thus stream 7 in Table 2.3 indicates that there is recycle although which stream is responsible will depend partly on the numbering of the units and streams.

The incidence matrix contains the same information as the stream connection matrix and thus has less information than the process matrix.

* May be omitted on first reading.

The adjacency matrix*

The adjacency matrix for Plant A of Fig. 2.6 is shown in Table 2.4. It is a square matrix in which a given row and column number corresponds to a specific unit. A "1" shows a connection from the unit given by the row number to the unit given by the column number. A zero shows that there is no connection in that direction. Thus unit number 3 is connected to unit number 4 in Fig. 2.6 and so a "1" appears in row 3, column 4 of Table 2.4.

TABLE 2.4
ADJACENCY MATRIX FOR PLANT A—FIG. 2.6

		To unit number				
		1	2	3	4	5
From	1		1			
	2			1		
unit	3				1	
	4					1
number	5	1				

But unit 4 is not connected to unit 3 so that a zero appears in row 4, column 3. Since no unit is connected to itself diagonal elements are zero. The adjacency matrix is most useful for mathematical manipulation to find recycle, as will be shown in the next sections. However it has the least content of all the matrices because no feed or product streams and no stream numbers appear. Although the adjacency matrix and the incidence matrix contain many zeros and could waste computer memory, binary bit notation can be used to reduce drastically the memory requirements.

2.4 Planning the Calculations—Finding Recycles*

This section and the next show how to find a sequence of calculation from the numerical form of the information flow diagram. These sections may be omitted on first reading if the reader assumes that a sequence of calculation is specified externally by the user. In Plant B of Fig. 2.7, it is clear that unit 1 can be calculated directly since the inputs are known and then unit 2 can be calculated. It can also be seen that units 4 and 5 form a recycle loop. Unit 6 could be done if units 4 and 5 were calculated. Units 8, 9, and 10 could be done together if unit 6 were calculated and when they were completed, unit 11 could be calculated. This example shows that some units can be

* May be omitted on first reading.

directly calculated and some must be calculated together as part of a recycle loop.

The computer unfortunately cannot look at the information flow diagram for Plant B of Fig. 2.7 but must work on its numerical equivalent. How can the computer be instructed to carry out the calculations if it has only some numerical encoding such as the process matrix?

The planning of the calculations falls naturally in two parts:

1. Separating the units into those which form part of a recycle loop and those which are (or will be) directly calculable;
2. Planning the calculation of each of the separate recycle loops.

Consider first what is meant by serial and recycle processes.

No recycle—serial processes*

If there is no recycle in an information flow diagram, the units can be calculated one after another, starting from a unit which has only (known) feed streams entering it. At least one such unit must exist; otherwise there is recycle. Once a unit has been calculated, its output streams become known inputs to subsequent units. At least one of these units has all its inputs known, and can be calculated. All units can be sequentially calculated if and only if there is no recycle.

A *serial process* is one with no recycle. A part of a process which has no recycle can also be called serial and its units can be called a *serial set.*

Recycle processes*

A recycle process is one in which an output stream of a unit affects at least one of its input streams. Even if there is recycle, there may be a serial set of units which can be directly calculated because the first unit has only feed streams entering it. An example is units 1 and 2 in Plant B of Fig. 2.7. Such a set can be calculated until a unit is reached which has an unknown input stream. If all possible units have been directly calculated and there are still some units uncalculated, then the process has at least one recycle loop and is a *recycle process.*

There may be a serial set of units which is downstream from the recycle loop. An example is unit 11 of Plant B of Fig. 2.7. Once the recycle loop has been solved, this downstream set will be directly calculated. Therefore we place all such units in a list apart from the recycle problem, to be calculated directly, after the recycle problem has been solved.

The various recycle loops of a plant may not all be mutually connected. A set of units which are connected such that there is some path following the arrows from every unit to every other unit is called a *recycle set.* There may be several recycle loops within a recycle set and the basic problem is how

* May be omitted on first reading.

best to do the calculation of such a set. Examples of two recycle sets are units (4, 5) and units (8, 9, 10) in Fig. 2.7. There are actually two recycle loops in the set (8, 9, 10).

The units of any information flow diagram may then be divided into two groups: those in serial sets and those in recycle sets. No pair of recycle sets can be connected in both directions or they would together form a single larger recycle set. Thus if the recycle sets are considered as pseudo-units, the information flow diagram is a serial set of units and pseudo-units which can be calculated sequentially, provided each recycle set can itself be calculated. For example in Plant B of Fig. 2.7, the serial set is 1, 2, (4, 5), 6, (8, 9, 10), 11, where the parentheses indicate pseudo-units.

Thus the plan of calculation of a process consists of:

1. Identifying serial sets and recycle sets and placing them in a feasible calculation sequence;
2. Finding how to calculate each recycle set.

There are direct ways of achieving 1, which will be described next. There are two basically different ways to do 2, namely to calculate the entire recycle set simultaneously or to find a sequence of calculation within the recycle set. Both simultaneous and sequential methods will usually require repetitive calculations (iterations) until the stream variables have reached values which satisfy a pre-specified criterion of error or which change by no more than a preset amount from one iteration to the next.

The methods available for separating a flow diagram into serial and recycle sets are first discussed. Then the problem of solving a recycle set itself is examined in Sec. 2.5.

Separating the serial and recycle sets*

A number of ways have been proposed for separating the units in serial sets from those in recycle sets. The process matrix can be used for a systematic search as described below. The list-processing languages of computation [Bobrow and Raphael (1964)] have been applied by Sargent and Westerberg (1964) to determine the recycle structure. The adjacency matrix has been the basis for several articles on this problem [Cavett (1964), Norman (1965), Himmelblau (1966)] and for discussion in recent books [Himmelblau and Bischoff (1968), Rudd and Watson (1968)]. The separation of serial and recycle sets of units is discussed using both the process matrix and the adjacency matrix.

The process matrix

The separation of serial and recycle sets can be illustrated using Plant B of Fig. 2.7, whose corresponding process matrix is shown in Table 2.5. The

* May be omitted on first reading.

method consists in repeatedly scanning the rows of the process matrix to find units which can be calculated.

Examine each row of Table 2.5 in turn. A unit can be calculated if all its input streams are known, either because they are feeds or because they

TABLE 2.5
PROCESS MATRIX FOR PLANT B—FIG. 2.7

Unit number	Unit Computation name	Associated stream numbers				
1	MIXER	1	2	−3		
2	MIXER	3	4	−5		
4	MIXER	5	6	7	−8	
5	MIXER	8	9	−7	−10	
6	MIXER	10	11	−12	−13	
8	MIXER	12	14	16	−15	
9	MIXER	15	17	19	−16	−18
10	MIXER	18	20	−19	−21	
11	MIXER	21	22	−23		

come from other units which can already be calculated. Unit 1 has only feed streams into it and thus can be calculated. This makes stream 3 known. Then unit 2 can be calculated because stream 3 is now known and stream 4 is a feed. But unit 4 cannot yet be calculated because stream 7 is not known, even though stream 5 is now known and stream 6 is a feed. There are no more units which can be calculated among those remaining. Since some units remain uncalculated, there is at least one recycle set. This is true by the definition of recycle; an input stream (7) depends on the values of variables of an output stream (8) of the same unit. Thus a serial set composed only of units 1 and 2 has been identified.

Serial sets which follow all recycle sets (and will be directly calculable) can be discovered by scanning the remaining units to find one whose output streams are either not connected to any unit or are connected to a unit in such a serial set. This is the converse of finding a serial set at the inlet. In Table 2.5 it is clear that unit 4 does not qualify because stream 8 is an input to unit 5. Neither does unit 5 because stream 7 is an input to unit 4. Only unit 11 is in a serial set because stream 23 is not the input to any other remaining unit. With unit 11 removed, no more units can be found in the serial set with unit 11. For example, unit 10 has output stream 19 connected to unit 9 even though stream 21 is connected to unit 11 just found.

In general, those units which remain after the serial sets at the beginning and end of the process have been removed may be part of a recycle set, or of a serial set joining two recycle sets. For example, in Fig. 2.7, units 4, 5, 6, 8, 9, and 10 would remain after serial sets (1, 2) and 11 were removed. The process matrix can be used to find which units are in which recycle sets by trying to find a sequence of calculation for those units. This is discussed in Sec. 2.5.

The adjacency matrix

The adjacency matrix is another tool for separating serial and recycle sets of units. The information flow diagram for Plant B of Fig. 2.7 again serves as the example and its adjacency matrix $\mathbf{A}'$ is shown in Table 2.6. By definition, there is a unity in row i, column j of $\mathbf{A}'$ (i.e., element $a'_{ij} = 1$) if and only if a stream goes from the unit corresponding to row i to the unit corresponding to column j and otherwise $a_{ij} = 0$.

TABLE 2.6
ADJACENCY MATRIX $\mathbf{A}'$ FOR PLANT B—FIG. 2.7

		To unit								
		1	2	4	5	6	8	9	10	11
	1		1							
	2			1						
	4				1					
From unit	5			1		1				
	6						1			
	8							1		
	9						1		1	
	10							1		1
	11									

Suppose the column corresponding to a unit in $\mathbf{A}'$ has only zeros, as for unit 1 of Table 2.6. This means that no unit is connected to unit 1, or in other words, only feed streams enter unit 1. Then unit 1 can be recorded on a list for later calculation and removed from $\mathbf{A}'$ by striking out its column and its row. This is seen in Table 2.7(b) where unit 1 is removed from $\mathbf{A}'$. Once unit 1 is removed, unit 2 has a zero column and can thus be removed as well. There are no more zero columns.

Suppose the row corresponding to a unit has only zeros, as does unit 11 of Table 2.6. This means that unit 11 is not connected to any other unit and will be directly calculable once all other units have been calculated. Then unit 11 may be recorded on a list for later calculation in front of the previously found unit with a zero row. The unit 11 is removed from $\mathbf{A}'$ by striking out its row and column as Table 2.7(d). There are no more zero rows and so matrix $\mathbf{A}'$ has been reduced to matrix $\mathbf{A}$ where all the remaining units 4, 5, 6, 8, 9, and 10 are either in recycle sets or in a serial set between a pair of recycle sets.

The same separation of initial and final serial sets has been achieved both by the process matrix and by the adjacency matrix. The further unraveling of the remaining units into distinct recycle sets can be done directly by use

TABLE 2.7
SEQUENTIAL REMOVAL OF ZERO COLUMNS AND ZERO ROWS

(a) Matrix $\mathbf{A}'$

		To unit								
		1	2	4	5	6	8	9	10	11
	1		1							
	2			1						
	4				1					
	5			1		1				
From	6						1			
unit	8							1		
	9						1		1	
	10							1		1
	11									

Remove node 1 (zero column)

(b) Matrix $\mathbf{A}''$

		To unit							
		2	4	5	6	8	9	10	11
	2		1						
	4			1					
	5		1		1				
From	6					1			
unit	8						1		
	9					1		1	
	10						1		1
	11								

Remove node 2 (zero column)

(c) Matrix $\mathbf{A}'''$

		To unit						
		4	5	6	8	9	10	11
	4		1					
	5	1		1				
	6				1			
From	8					1		
unit	9				1		1	
	10					1		1
	11							

Remove node 11 (zero row)

(d) Matrix $\mathbf{A}$, the reduced adjacency matrix

		To unit					
		4	5	6	8	9	10
	4		1				
	5	1		1			
From	6				1		
unit	8					1	
	9				1		1
	10					1	

(e) The resulting list of calculations is:
1,2,?,?,?,?,?,?,11

of the adjacency matrix but only by exhaustive search using the process matrix. The identifying of different recycle sets with the adjacency matrix is now examined.

Identifying recycle sets—powers of the adjacency matrix*

The reduced adjacency matrix **A**, formed by removing zero rows and columns, can be used to identify the different recycle sets and, if desired, the recycle loops within each set. If the recycle set is to be solved by treating all the units of the set simultaneously, then it may not be necessary to identify recycle loops within the set. However, if the set is to be solved by sequential calculation of the units, then the recycle loops may need to be identified.

The method of finding recycle sets and their loops requires finding the powers of the adjacency matrix by matrix multiplication. The usual rule applies, namely that the (i, j) element of the product of two matrices **A** and **B** is $\sum_{k=1}^{m} a_{ik} b_{kj}$ with a_{ik}, b_{ik} as the (i, k) elements of **A** and **B**, respectively. The arithmetic, however, is Boolean; that is:

Boolean addition:	$0 + 0 = 0$
	$0 + 1 = 1$
	$1 + 1 = 1$
Boolean multiplication:	$0 \cdot 0 = 0$
	$0 \cdot 1 = 0$
	$1 \cdot 1 = 1$

This is equivalent to using ordinary arithmetic and replacing all positive integers by unity.

When the nth power of **A** has been found by $(n - 1)$ successive multiplications of **A** with itself, it can be proved [Norman (1965)] that $\mathbf{A}^n$ shows the paths which exist from any unit to any other unit through n streams. As with matrix **A**, a unity in the (i, j) element of $\mathbf{A}^n$ implies that at least one path via n streams exists from the unit corresponding to row i to the unit corresponding to column j. Conversely, a zero implies no such path.

If $\mathbf{A}^n$ $(n = 2, 3, \ldots)$ is generated for a recycle process, some diagonal elements of $\mathbf{A}^n$ will eventually become unity. But any unit i corresponding to a diagonal unity has a path via n streams from unit i back to itself—that is, a recycle loop. All the recycle loops can thus be found. The powers of **A** for Plant B are shown in Table 2.8(a).

Identifying recycle sets—the reachability matrix

If the Boolean sum of the powers of **A** is formed, an element of unity results wherever unity exists in any power.

* May be omitted on first reading.

TABLE 2.8
POWERS OF THE ADJACENCY MATRIX

(a) The powers of matrix **A**

A

	4	5	6	8	9	10
4		1				
5	1		1			
6				1		
8					1	
9				1		1
10					1	

$\mathbf{A}^2$

	4	5	6	8	9	10
4	1		1			
5		1		1		
6					1	
8				1		1
9					1	
10				1		1

$\mathbf{A}^3$

	4	5	6	8	9	10
4		1		1		
5	1		1		1	
6				1		1
8					1	
9				1		1
10					1	

(b) The n-reachable matrix, $\mathbf{R}_n$

$\mathbf{R}_1$

4		1				
5	1		1			
6				1		
8					1	
9				1		1
10					1	

$\mathbf{R}_2$

1	1	1			
1	1	1	1		
			1	1	
			1	1	1
			1	1	1
			1	1	1

$\mathbf{R}_3$

1	1	1	1		
1	1	1	1	1	
			1	1	1
			1	1	1
			1	1	1
			1	1	1

$\mathbf{R}_\infty$

	4	5	6	8	9	10
4	1	1	1	1	1	1
5	1	1	1	1	1	1
6				1	1	1
8				1	1	1
9				1	1	1
10				1	1	1

Recycle loops: (4, 5), (8, 9), (9, 10)

The resulting matrix,

$$\mathbf{R}_n = \sum_{j=1}^{n} \mathbf{A}^j \tag{2.1}$$

records the connections from unit i to unit j via some number of streams $\leqslant n$, as shown in Table 2.8(b). Then element r_{ij} in matrix $\mathbf{R}_n$ equals 1 if and only if at least one such connection exists. In the limit as $n \longrightarrow \infty$, $\mathbf{R}_n \longrightarrow \mathbf{R}_\infty$, the reachability matrix records whether *any* connection exists from unit i to unit j via *any* (finite) number of streams. Thus $r_{\infty, ij} = 1$ if and only if *some* connection $i \longrightarrow j$ exists.

Now the transpose of either **A** or $\mathbf{R}_n$, that is, $\mathbf{A}^T$ or $\mathbf{R}_n^T$, has the effect of reversing the directions of all the $i \longrightarrow j$ connections, since rows become columns and vice versa. In a recycle set there is by definition not only some connection $i \longrightarrow j$ but also some connection $j \longrightarrow i$. The reachability matrix $\mathbf{R}_\infty$ and its transpose $\mathbf{R}_\infty^T$ are superimposed and a unity kept only where unity is in both $\mathbf{R}_\infty$ and $\mathbf{R}_\infty^T$. The result, called the intersection, is written

$$\mathbf{W} \equiv \mathbf{R}_\infty \cap \mathbf{R}_\infty^T \tag{2.2}$$

and is shown for the example in Table 2.9. Now $r_{\infty, ij} = 1$ only if $i \longrightarrow j$ in the flow diagram but $r^{\mathrm{T}}_{\infty,\,ij} = r_{\infty,\,ji} = 1$ only if $j \rightarrow i$ in the flow diagram. Then $w_{ij} = 1$ if and only if both $r_{\infty,\,ij}$ and $r^{\mathrm{T}}_{\infty,\,ij} = 1$, that is $i \rightleftharpoons j$. This will

TABLE 2.9
INTERSECTION MATRIX **W**

$\mathbf{R}_\infty$

	4	5	6	8	9	10
4	1	1	1	1	1	1
5	1	1	1	1	1	1
6				1	1	1
8				1	1	1
9				1	1	1
10				1	1	1

$\mathbf{R}^{\mathrm{T}}_\infty$

	4	5	6	8	9	10
4	1	1				
5	1	1				
6	1	1				
8	1	1	1	1	1	1
9	1	1	1	1	1	1
10	1	1	1	1	1	1

$\mathbf{W} = \mathbf{R}_\infty \cap \mathbf{R}^{\mathrm{T}}_\infty$

	4	5	6	8	9	10
4	1	1				
5	1	1				
6						
8				1	1	1
9				1	1	1
10				1	1	1

Recycle sets: (4, 5), (8, 9, 10)
Serial set: 6

eliminate any unit which is not in some recycle set because then there will not be a connection from that unit to any other in both directions. But by the definition of a recycle set, all members are mutually connected. Thus any nonzero row of **W** will list all members of one recycle set. Any unit not yet included in a recycle set and having a nonzero row of **W** will reveal a further recycle set. Any unit not in some recycle set and not an inlet or outlet unit is then in a serial set between a pair of recycle sets and will be directly calculable when the preceding recycle set has been calculated.

2.5 Planning the Calculations of a Recycle Set*

Once the recycle sets of units have been identified, the single question remains: how should an arbitrary recycle set be calculated? Remember that every unit in a recycle set is connected to every other unit by some path.

There are two basically different ways to answer this question. The calculations could be done sequentially, that is, one unit after the other, continuing until the change in any given stream variable from one iteration to the next is less than a preset tolerance.

Alternatively, the calculations could be done simultaneously, that is,

* May be omitted on first reading.

all the units in the set could be solved at once by some numerical method. Since processes are usually nonlinear, this method would also require iteration. These two distinct ways, the sequential and the simultaneous, are discussed in turn and then compared.

Sequential calculation*

Suppose that the calculation of a recycle set is to be accomplished in some sequence, one unit at a time, until all units have been done. The sequence is repeated until the change in every stream variable from one iteration to the next is less than some preset tolerance. This is *sequential calculation* of a recycle set. For each sequence of calculation, there is a unique set of streams, the *cut streams*, whose variables have not been calculated before they are needed on the first iteration. These variables must be given starting values.

What is the "best" sequence? How can it be found? These are key questions and the subject of much recent research. The best sequence is the one which minimizes the time of computation for a given accuracy, but unfortunately, to know the *actual* minimum time, one must already have done the same calculations several different ways. This obviously defeats the purpose of finding the best sequence beforehand.

Minimum computation time, the real criterion, must be translated into some equivalent criterion which is applicable initially. Several workers have chosen to minimize either the number of cut streams [Shannon, et al. (1966)] or more generally, the total number of stream variables in the cut streams [Rubin (1962), Sargent and Westerberg (1964), and Lee and Rudd (1966)]. Of course the latter criterion includes the former as a special case when all streams have the same number of variables. It has not been proved in general that such criteria imply minimum computation time. Moreover, it seems unlikely to be provable since the disturbance on the calculations caused by assuming a value of a stream variable must depend on how far the assumed value is from its ultimate converged value. Much research needs to be done on the convergence of calculations and on defining criteria for the best sequence.

Whatever criterion is chosen for the best sequence, that sequence can be found either by systematic search [Sargent and Westerberg (1964), Shannon, et al. (1966)], or by numerical manipulation of matrices [Rubin (1962), Lee and Rudd (1966)]. The algorithm of Sargent and Westerberg and the more general one by Christensen (1967) use a list-processing language to find the minimum number of recycle parameters to assume known. Rubin (1962) produced his own counterexample and Lee and Rudd (1966) have to resort to a search technique for some cases. Christensen's algorithm could find

* May be omitted on first reading.

a best sequence where these other methods failed. To illustrate the ideas in the numerical approach, the process matrix and the cycle matrix of Lee and Rudd (1966) are applied in turn to Fig. 2.8 from Lee and Rudd, which is a recycle set.

The process matrix can be used to find a workable sequence of calculation for a recycle set by exhaustively searching among the streams of the recycle set. Each stream is examined in turn to see, if that stream were known, whether a scan of the process matrix would show that all units in the set could be calculated. Thus in the process matrix for Fig. 2.8, Table 2.10, even if stream 1 were known, no further calculations could be done. In fact no single stream, assumed known, can enable the set to be calculated. All pairs of streams are then examined to see, if they were known, whether the set could be calculated.

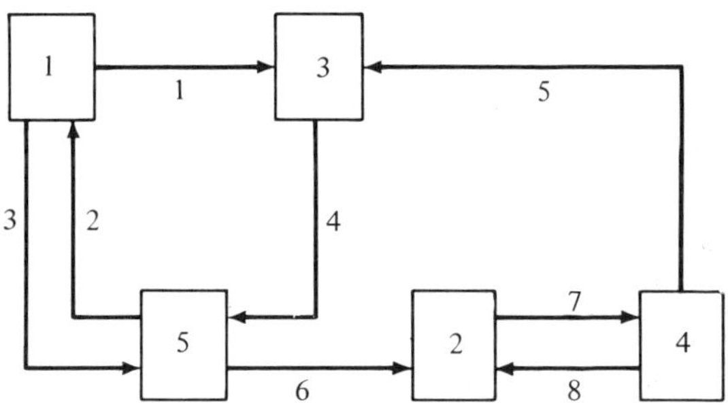

Figure 2.8. Recycle set [Lee and Rudd (1966)].

In Table 2.10, if streams 1 and 5 were known, only unit 3 could be calculated. But if streams 2 and 7 were known, the entire set could be calculated in the sequence (1, 4, 3, 5, 2). Notice that streams 2 and 7 are recalculated when units 5 and 2 are done so that the assumed values can be replaced

TABLE 2.10
PROCESS MATRIX FOR LEE AND RUDD'S EXAMPLE—FIG. 2.8

Unit number	Unit Computation name	Associated streams			
1	UNIT1	2	−1	−3	0
2	UNIT1	6	8	−7	0
3	UNIT1	1	5	−4	0
4	UNIT1	7	−5	−8	0
5	UNIT1	3	4	−2	−6

by newly calculated ones and the sequence repeated. If no pair of streams could lead to a complete calculation, then all combinations of three would be examined. For large recycle sets the number of combinations of streams four or more at a time would be too large to warrant an exhaustive search for an arbitrary number of streams required to be assumed known.

For comparison, the cycle matrix of Lee and Rudd (1966) is applied to the same recycle set in Fig. 2.8. Suppose that in Fig. 2.8 the recycle loops have been found by using the adjacency matrix as in Sec. 2.4. There are four loops which are, by the streams comprising them: (2, 3), (7, 8), (1, 4, 2), and (4, 6, 7, 5). These loops can be represented in the cycle matrix given in Table 2.11, where the rank of a loop is the number of streams in it and the stream frequency is the number of times a stream is in a loop.

TABLE 2.11
CYCLE MATRIX FOR LEE AND RUDD'S EXAMPLE—FIG. 2.8

Loop number	Stream number 1	2	3	4	5	6	7	8	Loop rank
1		1	1						2
2							1	1	2
3	1	1		1					3
4				1	1	1	1		4
Stream frequency	1	2	1	2	1	1	2	1	

The minimum number of streams to be cut (assumed known) in order to eliminate all recycle can be found in the following manner. A stream i is said to be contained in another stream j if each loop in which stream i is found also involves stream j. Thus streams 1 and 3 are contained in stream 2, streams 5, 6, and 8 are contained in stream 7, and streams 1, 5, and 6 are contained in stream 4. Since no more recycle loops could be cut by any stream than by the stream which contains it, streams 1, 3, 5, 6, and 8 can be eliminated. Then the remaining cycle matrix is shown in Table 2.12. Since a loop of rank 1 can only be cut by cutting the one remaining stream, streams 2 and 7 must be cut. Fortunately this breaks loops 3 and 4 at the

TABLE 2.12
REDUCED CYCLE MATRIX FOR LEE AND RUDD'S EXAMPLE—FIG. 2.8

Loop number	Stream number 2	4	7	Loop rank
1	1			1
2			1	1
3	1	1		2
4		1	1	2
Stream frequency	2	2	2	

same time. Therefore, assuming values for the variables of streams 2 and 7 will allow a direct calculation of all the units in Fig. 2.8 in the sequence (1, 4, 3, 5, 2). If all the loops cannot be broken by cutting loops of rank 1, Rudd and Watson (1968) and Christensen (1967) should be consulted for the algorithms needed.

Simultaneous calculation*

The calculation of a recycle set by simultaneously solving its equations has been discussed by Nagiev (1957), Rosen (1962), Naphtali (1964), and Ravicz and Norman (1964). It can be shown that if the equations involved are linear, then calculating the recycle set requires only a direct solution by available matrix methods to find the answer. However, for nonlinear equations, the answer must be obtained by successive trials. Some insight into the simultaneous approach can be gained by considering the method of Nagiev (1957, 1964).

Nagiev has been a leading proponent of the application of the theory of linear systems to chemical plants. His method may be summarized as follows. For any complex system involving recycle, sets of algebraic equations are formulated by taking a balance for each component or for the total flow of mass or energy at each unit. Referring to a general stagewise operation shown in Fig. 2.9, the following are defined:

α_{ij} is the fraction of the stream leaving unit i, returned to unit j;
F_i is the flow leaving unit i;
F_{oi} is the flow entering unit i as feed, and assumed known;
α_{io} is the fraction of flow leaving unit i as a product.

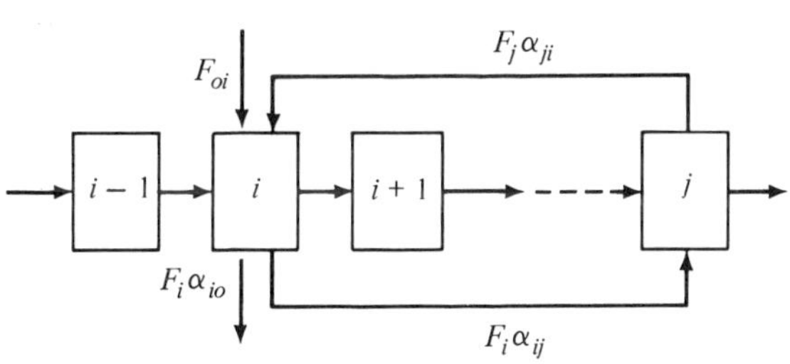

Figure 2.9. General stagewise process with recycle.

If there are m units in the calculation, the following set of algebraic equations is readily produced to describe the steady state of the process:

$$\begin{aligned}
(1-\alpha_{10}-\alpha_{11})F_1 \quad & -\alpha_{21}F_2 - \quad \cdots \quad -\alpha_{m1}F_m = F_{01}\\
-\alpha_{12}F_1 + (1-\alpha_{20}-\alpha_{22})F_2 & - \quad \cdots \quad -\alpha_{m2}F_m = F_{02}\\
\vdots \qquad\qquad & \qquad \vdots \qquad\qquad\qquad\qquad \vdots\\
-\alpha_{1m}F_1 \quad & -\alpha_{2m}F_2 - \cdots + (1-\alpha_{m0}-\alpha_{mm})F_m = F_{0m}
\end{aligned}$$

* May be omitted on first reading.

This set of m linear equations can be solved by standard matrix methods to find the flows F_i. For a linear system the simultaneous approach is certainly the better one.

The main difficulty is that few chemical plants are truly linear in their behavior. For a unit with nonlinear equations, the values of the splits α_{ij} must be obtained from a linear approximation of the model equations of the unit. The simultaneous calculation of a plant would then involve alternately the solution of the linearized equations for each set of α_{ij} and the repeated adjustment of the values of all the α_{ij}.

Comparison of sequential and simultaneous calculation of a recycle set

The sequential approach—iteration involving successive substitution

Fig. 2.10 represents a mixer followed by a splitter returning part of the output to the mixer. In the splitter the material entering the unit is divided according to the fraction α. Note that both the mixer and the splitter can be described by the Unit Computation MIXER defined in Sec. 2.1.

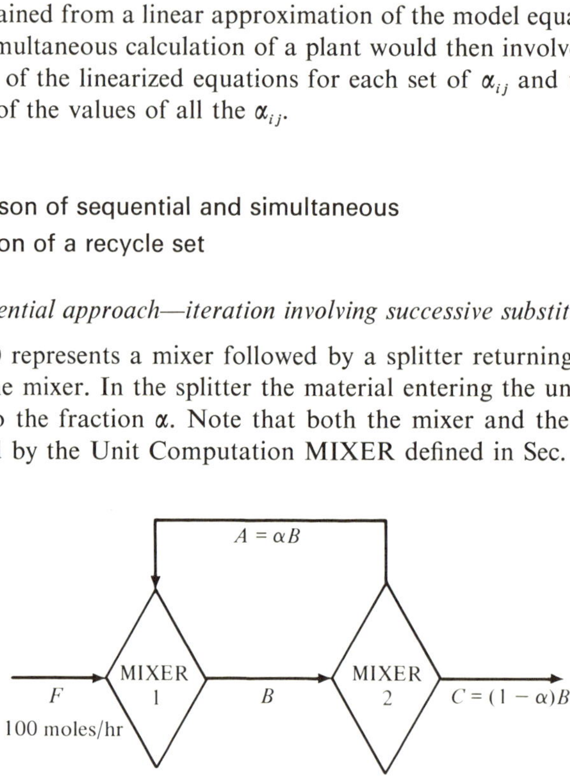

Figure 2.10. Application of the simultaneous approach to a simple recycle system.

In the method of successive substitution, the flow in stream A is initially zero, since it is unknown. Thus flow B is 100 moles per hr as a first approximation, yielding a first guess for the flow A of 100α.

In the second iteration around the recycle section, the next approximation of the flow in B is $100\,(1 + \alpha)$ moles per hr and the second approximation of flow A is $100\alpha\,(1 + \alpha)$. This calculation can be continued giving the results in Table 2.13 for two values of α, 0.33 and 0.90.

If α is the fraction returned in line A, the flow in A in iteration N is $100\alpha\,(1 + \alpha + \alpha^2 + \cdots + \alpha^{N-1})$ or, with the sum of the geometric series, $A_N = 100\alpha(1 - \alpha^N)/(1 - \alpha)$. The limit A_N as $N \longrightarrow \infty$ is $A_\infty = 100\alpha/(1 - \alpha)$. It is seen that with $\alpha = 0.33$, after three or four iterations, the change in A

TABLE 2.13
EFFECT OF RECYCLE ON CONVERGENCE

$\alpha = 0.33$ Iteration	Flow A	$\alpha = 0.90$ Iteration	Flow A
0	0	0	0
1	33.3	1	90
2	44.3	2	171
3	48.1	3	244
4	49.4	4	310
⋮		5	369
∞	50.0	6	422
		7	470
Converges rapidly		⋮	
		∞	900
		Converges slowly	

from one loop to the next is very small. This recycle process is said to converge rapidly. If α is increased to 0.90, the recycle flow is increased and it is seen in Table 2.13 that the process converges very slowly. Processes with relatively large recycle flows may be expected to converge slowly in such iterative calculations.

The simultaneous approach—direct solution

It is apparent that the problem represented by Fig. 2.10 can be solved directly by inspecting the two balance equations $100 + A = B$ for unit 1 and $A = \alpha B$ for unit 2, whence $A_\infty = 100\alpha/(1 - \alpha)$. Thus for $\alpha = 0.33$, $A = 50$ moles per hr and for $\alpha = 0.90$, $A = 900$ moles per hr.

For real nonlinear systems, no direct comparison on the same case can be made of the sequential and simultaneous methods of calculating a recycle set. The ultimate criteria, speed of computation and accuracy, will most likely depend on the particular case studied, and each method will probably always be superior to the other for some types of problems.

2.6 Summary

This chapter has introduced several concepts and definitions common to the simulation of complex chemical plants with any executive program. These were the problem of recycle, the flow and modification of information, and the planning of the calculations.

The translation of the process flow diagram into an information flow diagram was described. Alternative methods were described for representing the information flow diagram numerically and methods for locating recycle loops from such numerical representations were examined. The next chapter will consider the method used by PACER in greater detail.

Using a simple example, the sequential iterative approach to solution of recycle problems has been compared to a simultaneous approach. The iterative approach used by PACER allows nonlinear modeling of any desired sophistication.

PROBLEMS

2.1. Take a process flow diagram having recycle from *Hydrocarbon Processing and Petroleum Refiner* **48**, 11 (for example, the ammonia process, p. 150, or ethylene/propylene process, p. 180). Construct an information flow diagram and number the units and streams.

(a) Specify from this diagram alone the sequence in which you would ask the computer to perform the calculations, grouping together those units in the same recycle loop.

(b) Construct the process matrix, incidence matrix, and adjacency matrix for this process. Give the adjacency matrix to a colleague and have him draw the information flow diagram from it. Then give him the incidence and process matrices in turn and have him add to the information flow diagram what he can from each matrix.

(c) From the adjacency matrix for this process, find the serial sets of units at the inlet and outlet of the process. By using powers of the reduced adjacency matrix, find the recycle sets in the process. Find the minimum number of streams to be cut in order that each recycle set can be calculated in sequence. Write the resulting overall sequence of calculations and compare with that of part (a).

2.2. Write an algorithm for finding the serial sets and recycle sets of a given adjacency matrix. How many words of computer memory would be required for the adjacency matrix of a process with 70 units, if each element occupied one word? How many words would be needed if each element occupied one binary bit of a 35-bit word?

2.3. (a) When two recycle loops containing n and m lines, respectively, share one or more (say k) units, they form together a recycle loop of up to $(n + m)$ lines. How could such composite loops be ignored in finding all the

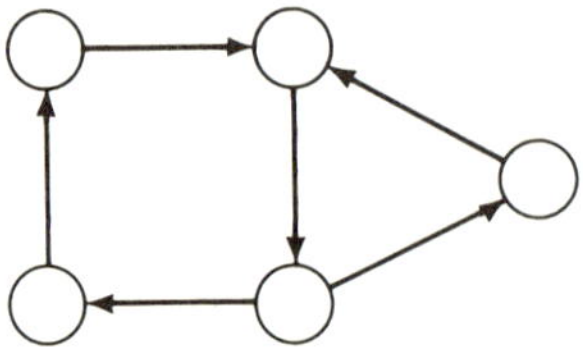

Figure 2.11. A recycle system.

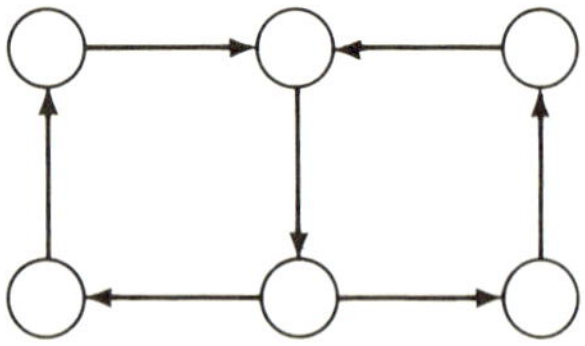

Figure 2.12. A recycle system.

recycle loops from diagonal elements of the powers of **A** equal to 1? For example, consider the network in Fig. 2.11.

(b) In (a), when $n = m$, there will be fewer than $2n$ diagonal elements of $\mathbf{A}^n$ which equal 1. How would one find out, using $\mathbf{A}^n$ and $\mathbf{R}_{n-1}$, which units are shared? Use as an example the system in Fig. 2.12.

2.4. The order of a recycle system is defined as the minimum number of streams to be cut so that the calculations can be performed in a direct sequence. What is the order of recycle for the systems shown in Fig. 2.7 and 2.8? Can the process matrix, the adjacency matrix, the stream connection matrix, or the incidence matrix be used to identify the order of recycle of a given information flow diagram?

PACER® (1966)— An Executive Program for Process Simulation 3

Chapter 2 described the characteristics of executive programs. Particular consideration was given to the concept of the Unit Computation and its interaction with the executive program, and the method of converting the process flow diagram into numerical form. For the interested reader, methods of planning the calculations for recycle sets were described.

In this chapter we describe in detail a modular executive system, PACER, using the process matrix to encode the information flow diagram. The original version of PACER was written by P. T. Shannon and H. Mosler [Mosler (1964)] at Purdue University. The 1966 version of PACER described in this book was used to evaluate techniques and concepts of simulation and modeling of chemical processes. The currently available commercial PACER system, while inspired by the concepts of the original PACER, is a major extension and development of those concepts and differs significantly in structure and scope from PACER (1966).

® A registered trademark of Digital Systems Corporation, Hanover, New Hampshire.

3.1 Specifications and Major Elements of the System

The name PACER—Process Assembly Case Evaluator Routine—indicates the intended use: to simulate complex chemical processes by assembling appropriate mathematical models of the equipment units in a building-block fashion and then running case studies on the process assemblies.

In general, executive systems evolve with the needs of the problems to be solved, the preferences of the user, and the capability, size, and speed of the digital computers available. Despite many advances in machine and programming, the original specifications summarized in the list below might still be applied to the development of similar systems today.

Consider how PACER satisfies the specifications in the list below. The process sequence, feed conditions, and equipment characteristics are described respectively in the process matrix, the stream variable list (called SN), and a set of Unit Computations and their associated equipment parameter lists (EN and ENC) one for each equipment unit. These are input data that can be easily changed by the user.

The PACER System Design Specifications (1961)

1. Process sequence, feed conditions, and set of equipment are to be part of input data.
2. Process sequence, operating conditions, and equipment used are to be easily changeable by user.
3. Executive is to determine a calculation sequence, execute this sequence, and print requested results. (User can specify sequence calculation.)
4. System must be open-ended (able to be expanded and modified as required).
5. System must be usable during development.
6. System must be able to incorporate past work.
7. System must be easy to use and understand.
8. System must ultimately have a very large information retrieval capacity.
9. Calculations must be easily traceable in detail.

PACER can locate recycle streams and the sequence of calculation as described in Chapter 2 and elaborated on in Sec. 3.4. The PACER system also allows the user to specify the sequence of calculations.

Because PACER was given a modular structure, it is open-ended, usable during development, and is capable of incorporating past work. PACER is open-ended because the matrices for storing stream and equipment parameters can be redimensioned, because the process flow diagram can be easily changed, and because each Unit Computation can be connected to any other. The selection of the process matrix and the Unit Computation as the main concepts of the program makes it easy for the engineer to understand because the format and terminology are familiar to him. The engineer can control the amount and variety of output printing by setting constants in the input data and can easily trace the calculations if the computation goes astray. Thus PACER satisfies the design specifications which were established for an executive program.

3.2 PACER Organization

To introduce the structure and operation of PACER we start with the single Unit Computation MIXER. It is important to distinguish between the Unit Computation and the equipment unit to which it is applied. The Unit Computation MIXER is a computer subroutine. There may be several different equipment units in the information flow diagram which call on MIXER for their calculation.

The specific MIXER problem is shown in Fig. 3.1. The output flows from the MIXER are to be calculated from the input stream variables and the equipment parameters. All stream information is stored in the respective SN list. The equipment parameters are available from the EN list. Auxiliary lists* AEN and ENC can be used at the programmer's discretion for extra equipment information.

The EN list associated with each Unit Computation specifies the parameters to be used by the general relationships in that Unit Computation. These relationships are made specific by the actual information stored in the EN list for a particular equipment unit. Thus if we consider Fig. 3.1, the

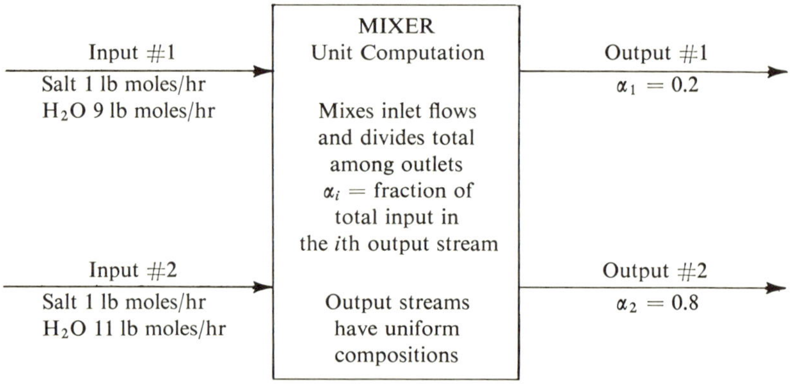

Figure 3.1. Schematic diagram of a simple mixer problem.

equations in the box could be written in the form

$$\left.\begin{aligned} F_1 &= 0.2F \\ F_2 &= 0.8F \end{aligned}\right\} \tag{3.1}$$

* Early versions of PACER included an auxiliary stream list called SNC. We have not found this flexibility particularly useful. Hence, no reference is made to this feature in this textbook.

or in the form

$$F_i = \alpha_i F \tag{3.2}$$

with α_i being supplied from the EN list that contains the values

$$\alpha_1 = 0.2$$
$$\alpha_2 = 0.8$$

and where

F_i = moles/hr of output stream i
F = total moles/hr flowing through the MIXER.

Equation 3.1 is specific; such equations would mean the MIXER Unit Computation could only be used when 20% goes out the first stream and 80% leaves in the second stream. Equation 3.2 is general; such a MIXER can be applied anywhere in the process provided values for α_i can be obtained. The EN list is the source of such information. Particular values for the equipment parameter list for MIXER are shown in Table 3.1.

Each Unit Computation in PACER usually requires a list of equipment parameters. In general the format of the EN list differs with each type of Unit Computation. PACER does require that the first element of the EN list be the equipment unit number but the rest of the format may be defined by the programmer. PACER stores the equipment parameters list for equipment unit number NE in the row numbered NE of the EN matrix (composed of all EN lists).

Consider now the SN list for the stream information. PACER requires that the first and second elements be the stream number and stream flag respectively. The stream flag designates the type of stream: feed streams to the process assembly are automatically flagged 1; product streams from the process are flagged 2; streams connecting two units are flagged 0. In PACER the stream number is the same as the row number of the SN matrix (composed of all SN lists) in which the corresponding SN list is stored. Table 3.2 gives a sample SN list for the MIXER problem under discussion.

TABLE 3.1
MIXER EN LIST FORMAT

Element	Meaning	Values for the unit in Fig. 3.1
1	Equipment number from process information flow diagram	1
2	Fraction of total input in first output stream	0.2
3	Fraction of total input in second output stream	0.8
4	Fraction of total input in third output stream	0
5	Fraction of total input in fourth output stream	0

TABLE 3.2
STREAM VARIABLE LIST FORMAT FOR MIXER STUDY

Element	Meaning	Values for the unit in Fig. 3.1
1	Stream number	1
2	Stream flag	1
3	Total flow, lb moles/hr	10
4	Flow of water, lb moles/hr	9
5	Flow of salt, lb moles/hr	1

All SN lists generally have the same format so that any stream can connect any Unit Computation to any other in the information flow diagram. Besides the SN matrix, PACER uses separate matrices designated STRMI and STRMO to store the input and output stream lists, respectively, needed by the Unit Computation currently being calculated. The advantage of using STRMI and STRMO rather than the SN matrix itself is in preventing inadvertent destruction of numbers in SN by the Unit Computation; in allowing the programmer to refer easily to input streams independently of the information flow diagram or the SN matrix; and in distinguishing conveniently between new and old values of stream variables when testing for convergence. By convention, the first input stream in the process matrix is the first row of the STRMI matrix, the second input stream is the second row, and so on. There are NIN input streams. The same convention applies for the NOUT output stream information transferred to or from the STRMO matrix. Thus there are four main matrices: SN and EN for relatively permanent data storage and STRMI and STRMO for temporary storage of stream lists.

In MIXER the programmer secures his input information from the NIN input streams in STRMI, uses the equipment parameters stored in the NE row of the EN matrix and stores the calculated results in the NOUT streams of STRMO. The equipment number NE is the row number of that unit in the process matrix. This organization of the data is summarized in Fig. 3.2. PACER transfers all the stream information needed by the Unit Computation work sheets shown in Fig. 3.2, and removes the answers to SN. Just how this is done is described in Sec. 3.3.

The matrices used in PACER are summarized in Fig. 3.3. The variable names shown at the upper right and lower left corners of the arrays denote the maximum lengths of the columns and rows of the arrays. These dimension variables are part of the input control data for a given problem. Thus the engineer can dimension his problem, up to the largest dimensions permitted by his executive, at execution time so that small problems are run efficiently with respect to handling the SN and EN lists.

Thus far, three of the five sets of information with which PACER works

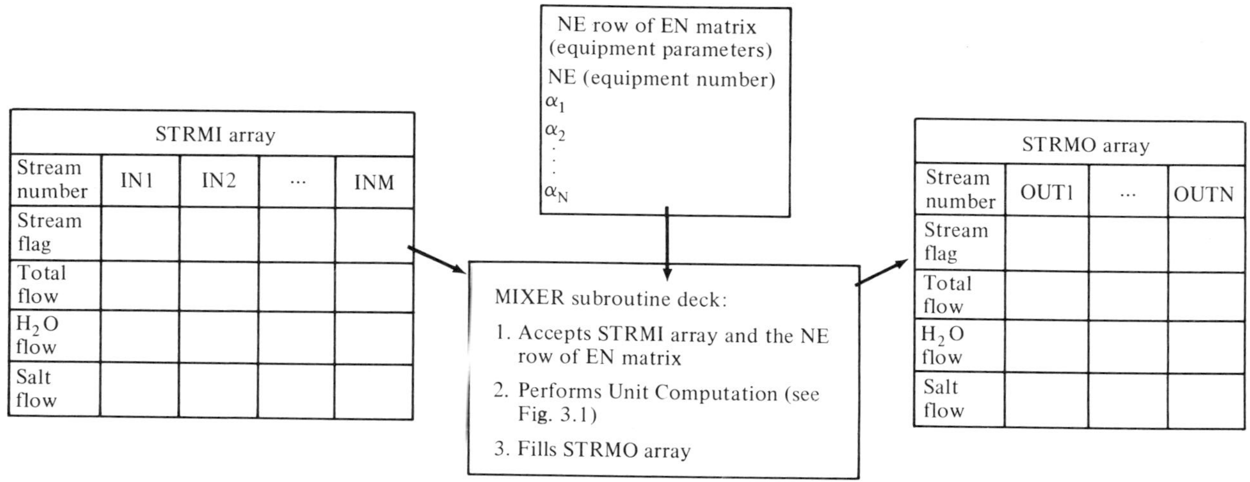

Figure 3.2. Data organization for a Unit Computation (MIXER) called by process matrix row NE, MIXER, +IN1, +IN2, . . . , +INM, −OUT1, −OUT2.

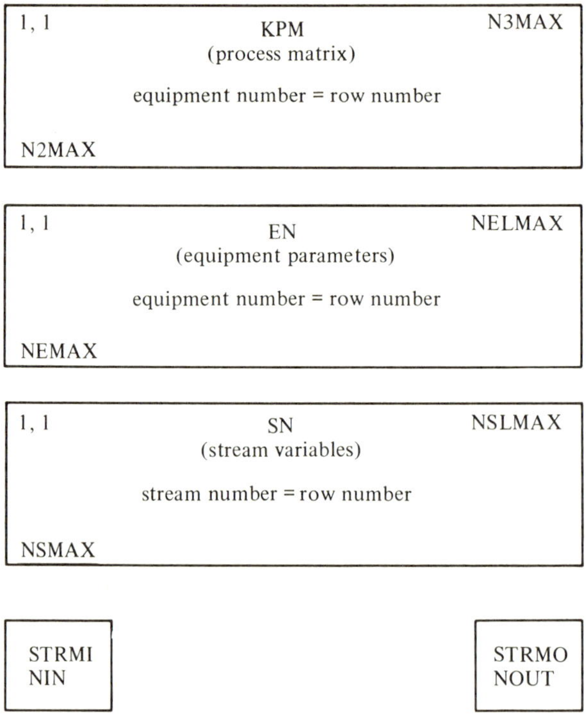

Figure 3.3. Summary of PACER arrays.

have been discussed; the engineer must understand and supply these in each simulation. The three are the process matrix (described in Chapter 2), the SN list, and the EN list. The remaining two sets are the PACER control information and the Unit Computations. To appreciate the control information, it is necessary to understand how PACER operates.

3.3 PACER Operation—Order of Calculation Specified

How the executive program moves the numbers to and from the work area can be understood by considering the setup and execution of a single MIXER Unit Computation. Then a simple recycle problem is studied. Finally, the functions of the subroutines comprising the executive program itself are described, along with how the whole system is organized on a computer and the type of control that the programmer can specify. It is assumed that the order of equipment calculation is known. Section 3.4 shows in detail how PACER finds a sequence of calculation if the user does not specify the sequence as part of the input control data.

Setup and execution of a single unit computation

Suppose PACER has been instructed to execute the MIXER Unit Computation for unit #5. PACER sets NE equal to 5. Suppose the corresponding fifth row in the process matrix is 5 MIXER 10, 8, —7. The first input stream is 10, and the second input stream is 8. The number of input streams, NIN, is 2.

The first output stream is 7, and the number of output streams, NOUT, is 1. The stream numbers correspond to the row numbers in the SN matrix. Equipment parameters for unit #5 are located in the fifth row of the EN matrix. This organization of the data is shown in Fig. 3.4.

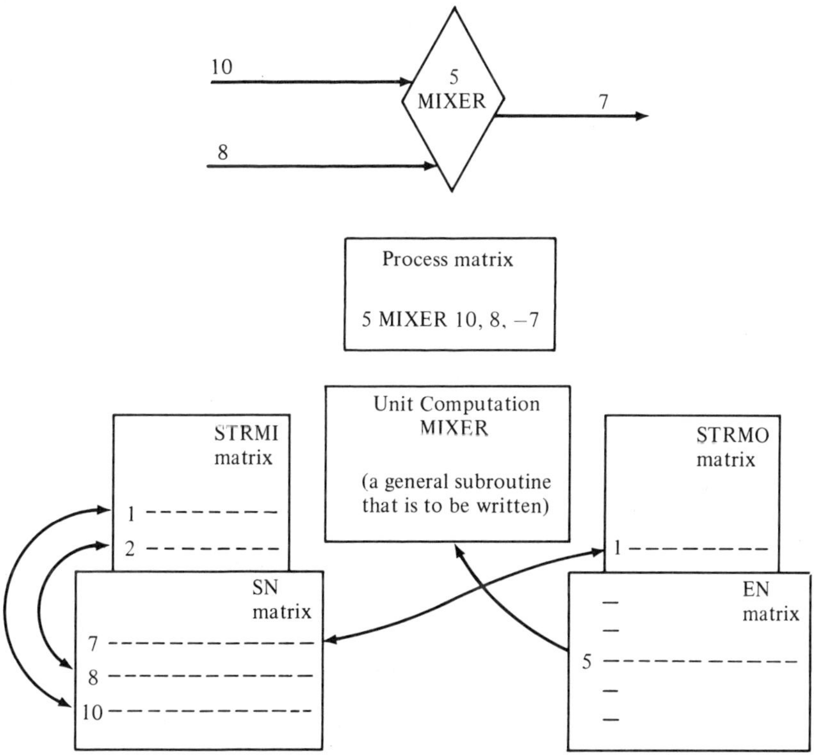

Figure 3.4. A mixer and the related information blocks within PACER.

PACER uses the following procedure to execute every Unit Computation:

1. Using row number NE of the process matrix, PACER determines the ordered set of input stream numbers and NIN. For each input stream, PACER transfers the corresponding row from the SN master matrix to the STRMI work matrix. For this example, NE = 5, rows

10 and 8 of the SN matrix are placed in the first and second rows of STRMI, and NIN is set equal to 2.

2. Using row number NE of the process matrix, PACER determines that there is only one output stream, so NOUT is set equal to 1 after row 7 of SN has been stored in the first row of STRMO.
3. PACER then calls the Unit Computation MIXER and turns program control over to the subroutine of the Unit Computation. The Unit Computation has been written so that it finds equipment parameters for equipment number NE in row number NE of the EN matrix. Within the Unit Computation the stream variables and equipment parameters are transferred to the subroutine's local variables as required. The subroutine is executed. The results of the calculations produce new numbers in the STRMI and STRMO work matrices (and possibly in row NE of EN and ENC). Control is then returned to the PACER executive for more bookkeeping.
4. If the Unit Computation is being executed as a serial calculation, the NIN rows of the STRMI and the NOUT rows of STRMO are returned directly to the rows of SN, as determined by the stream numbers.
5. If the Unit Computation is being executed in a recycle calculation, then before returning the stream lists from the work matrices to the master matrices, all new input and output stream variables in the STRM (that is, STRMI or STRMO) work matrices are compared to their corresponding values in the master SN matrix. The SN matrix contains the values from the last iteration. To test for convergence after the execution of each Unit Computation, the PACER executive computes the fractional change in each newly calculated variable in an entering or leaving stream as

$$\frac{|\text{New stream value} - \text{Previous value}|}{\text{New stream value}}$$

This is compared with the tolerance for that variable which is provided in the input data as the list DELS. When all the stream variables in a recycle set have passed this test during a single iteration, then that set is said to have been converged. Note that it is necessary only to test for convergence of variables in the cut streams, but then some other variables may not converge to the same tolerance.

The Unit Computation may call the physical properties necessary to perform its calculations. Once a Unit Computation is set up for execution (steps 1 and 2), control is transferred to the Unit Computation. Most often, Unit Computations for equipment units use only the single row of the EN matrix and the stream values contained in the STRM work matrices. However, when necessary, the Unit Computation can use, change, or collect information contained in any or all rows of SN and EN matrices. Examples are a cost calculation of the process and a Unit Computation for optimization. In general the modular Unit Computations can perform any kind of information modification.

Formulation of the actual Unit Computations is demonstrated in Chapters 6 to 10. At this stage, however, it can be seen that to devise Unit Computations within PACER, access must be had to stream information in matrices called STRMI and STRMO and equipment parameters stored in EN (NE, __). Which information is in which STRMI list is dictated by the order of the streams in the process matrix.

One additional feature is COMMON; this is computer storage in core memory that is available to both the PACER executive and the subroutines within the Unit Computations. This is the source and destination of information to be communicated between subroutines by PACER. A listing of PACER COMMON used in the sulfuric acid plant case is given in Table 3.3.

TABLE 3.3
PACER COMMON AND DIMENSION DECK FOR THE SULFURIC ACID PLANT CASE

```
COMMON STRMI (5, 20), STRMO (5, 20)
COMMON SN (80, 20)
COMMON EN (60, 20), ENC (60, 10), KPM (60, 7)
COMMON AEN (4, 40), KUS (80), KPS (3), DELS (20)
COMMON KVOID
COMMON KSEM (80, 3), NELIST (60, 2), NAME (60)
COMMON KEFLAG (60), KSFLAG (80), NECALL (60)
COMMON TITLE (24), KE1 (60), KE2 (60), KE3 (60), KSPRNT (10)
COMMON NIN, NOUT, NSLMAX
COMMON NELMAX, NECLMX, NEMAX, NSMAX, NEQUIP
COMMON N2MAX, N3CLMX
COMMON NOAEN, NOEN, LIMIT, LIMIT2, LIMIT3
COMMON LOOP, KBACK, KBACK2, KARDS, KPRINT, KONV
COMMON KRUN, KSETS, KCLEAN, LOOPS, MCHAIN
COMMON NCHAIN, KOUNT2, NE, NLIST, LIST
COMMON NUS, NUSMAX, NPS, NPSMAX, NUS1, NUS2
COMMON NUS3
COMMON NE2MAX, NE3MAX, KES, KES2, KPLOOP
COMMON NEUSED, NSUSED
COMMON T
```

The meanings for some of the terms are familiar; others are defined in the section immediately following.

The following is a list of the functions of PACER other than its bookkeeping responsibility:

1. Reads the data and sets up the matrices;
2. Prints the data;
3. Determines the sequence of calculation if it is not supplied by the user;
4. Uses the determined order of calculation to perform the bookkeeping necessary for doing serial and direct calculations;

5. Prints the results or traces the calculations and goes on to the next case to be evaluated;
6. Performs all computer organization and information transfer, thus freeing the user from the thousands of detailed statements required to do this apparently simple task.

PACER has been devised so that the extent to which it performs some of these functions can be controlled by the engineer. To this end, the engineer needs to supply a set of control instructions. Examples of the type of control instructions that might be incorporated into an executive system are given in the section immediately following. This section summarizes control instructions for PACER (1966). The values shown are those used for the acid plant studies described beginning in Chapter 4; they will vary with the size and type of plant being simulated.

Sample control instructions

The following is a list of sample control instructions used in the sulfuric acid plant study discussed beginning in Chapter 4. Each control instruction is given in FORTRAN and is accompanied by its definition.

KRUN Run number to identify problem for the user.

KSETS Controls intermediate printings; normal value = 0.
= 0; prints initial conditions, stream matrix every KPRINT loops in a recycle calculation, and answers.
= 1; in addition to above, prints warnings and error messages from within Unit Computations.
= 2; in addition to above, prints the input and output stream lists, STRMI and STRMO, respectively, as each equipment unit is calculated; the number of deviations from test list values is also printed.
= 3; in addition to above, prints tag lists for streams and equipment units; for large problems, KSETS should not be greater than zero except for debugging, since a large amount of printing otherwise results.

KCLEAN = 0; all matrices and lists are set equal to zero before reading the data following; KCLEAN would not be set equal to zero when executing a second case, using the converged values from a case just calculated; normal value = 0.

LOOPS Maximum number of iterative loops to be allowed in a recycle calculation; normal value = 20.

KCARDS If positive, causes the stream variables matrix to be punched on cards at the conclusion of a run; normal value = 0.

KPRINT Number of loops in a recycle between printout of the entire stream variables matrix; it is automatically made equal to 5 unless otherwise specified.

KONV A number used to control the initiation of convergence promotion during calculation of a recycle set; normally it is preferable to wait for some number of loops KONV before applying a technique for convergence promotion; the number KONV is supplied in the input data but could be altered during the calculations; normal value = 0.

Printout control

Control instructions were built into the MACSIM executive program to reduce the printing output and to aid in developing programs. Printout is suppressed by inserting unity rather than zero in elements.

Element number	Information suppressed
1	Process matrix
2	Preferred streams and streams in each iteration
3	Stream connection matrix
4	Control variables
5	Test list DELS
6	Initial EN
7	Initial ENC
8	Initial SN
9	
10	Calculation order
11	Final EN
12	Final ENC
13	Final SN
14	
15	Calculation order

Dimension variables

The size of the problem is controlled by the following ten variables. Their maximum values are limited by the COMMON statements and may be changed by redimensioning the COMMON decks, up to the limit of the available computer storage.

N2MAX Number of rows used in the process matrix; it must be greater than or equal to the highest equipment number, NEMAX, used ($\leqslant 60$).

N3MAX	Number of numerical columns in process matrix ($3 \leqslant$ N3MAX $\leqslant 10$).
NEMAX	Highest equipment unit number; not necessarily the total number of equipment units ($\leqslant 60$).
NSMAX	Highest stream number; not necessarily total number of streams ($\leqslant 80$).
NELMAX	Maximum length of equipment parameter list ($\leqslant 20$).
NECLMX	Maximum length of auxiliary equipment list ($\leqslant 10$).
NSLMAX	Length of stream variables list ($\leqslant 20$).
NPSMAX	Maximum number of preferred stream numbers ($\leqslant 3$).
NOCOMP	Number of flow components ($\leqslant 10$).

Process matrix, KPM

NOKPM	Number of rows in the process matrix to be read ($\leqslant$N2MAX).

Equipment information, EN and ENC

NOEN	Number of equipment parameter lists to be read ($\leqslant$NEMAX).
NOENC	Number of auxiliary equipment lists of length NECLMX to be read ($\leqslant$NEMAX).

Stream information, SN

NOSN	Number of stream variable lists to be read, usually the input streams ($\leqslant$NSMAX).

Stream variables test lists, DELS

NDELS	$= 0$ if no test list; > 0 if stream variable test list is to be read in to test for absolute fractional convergence of stream variables list in recycle calculation; maximum number of elements = NSLMAX.

Preferred streams

NPS	Number of preferred stream numbers ($\leqslant 3$).

Additional equipment lists, AEN

NOAEN	Number of additional equipment lists to be read ($\leqslant 4$); maximum length is 40.

Calculation order list, NELIST

KK	$= 0$ if no list; > 0 if calculation order list is to be read in.

KSPRNT Number of stream variables lists to be printed ($\leqslant 9$) after each loop in a recycle calculation.

Setup and execution of a simple recycle problem*

A simple recycle problem is shown in Fig. 3.5. Consider the flow of only two components in each stream and an equal split among the two output streams of unit #2. Thus the stream variables list has the format

Stream number	Stream flag	Flow of water, lb moles/hr	Flow of salt, lb moles/hr

which become the column headings for the SN and STRM matrices. PACER can be instructed to do the recycle calculation in the sequence (1, 2) until the stream variables change less than absolute fraction specified in DELS. The fractions of the total input in the first and second output streams, α_1 and α_2, are located in rows 2 and 3 of the EN matrix shown in Table 3.1.

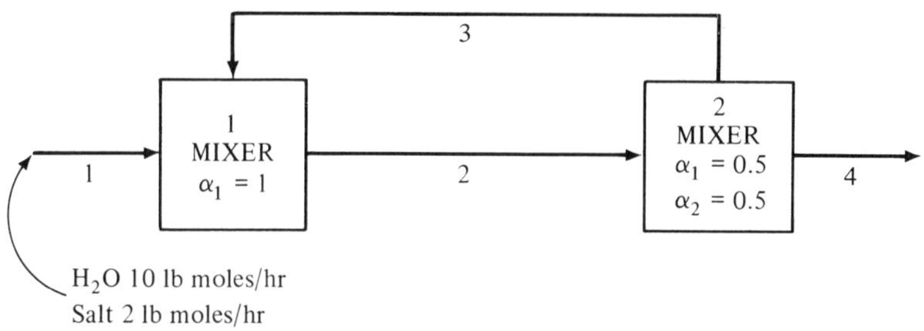

Figure 3.5. A simple recycle process.

The three matrices SN, STRMI, and STRMO are given in Table 3.4 to present the successive answers for the iterative solution to this recycle problem. This is shown after each of the following steps in the execution of each Unit Computation:

1. Set up STRMI and STRMO;
2. Execute the Unit Computation and test for convergence using DELS;
3. Store the answers in the SN matrix.

The first two iterations of the process are discussed to demonstrate the information flow. Note that as the stream information moves from STRMO to SN and then to STRMI for stream 2, the output information of unit #1 becomes the input information to unit #2, and the stream information flows through the process assembly. Each Unit Computation performs a local material balance for both components. However, the material balance for

* May be omitted on first reading.

Table 3.4
Stream Matrices for Simple Recycle Problems

	SN				STRMI				STRMO			
	No.	Flag	H_2O	Salt	No.	Flag	H_2O	Salt	No.	Flag	H_2O	Salt
Set up initial condition	1	1	10	2	0	0	0	0	0	0	0	0
	2	0	0	0	0	0	0	0	0	0	0	0
	3	0	0	0	0	0	0	0	0	0	0	0
	4	2	0	0	0	0	0	0	0	0	0	0

Feed stream 1 is flagged 1.
Product stream 4 is flagged 2.

	SN				STRMI				STRMO			
1. Set up unit #1	1	1	10	2	1	1	10	2	2	0	0	0
	2	0	0	0	3	0	0	0				
	3	0	0	0								
	4	2	0	0								
2. Execute unit #1 and test	1	1	10	2	1	1	10	2	2	0	10	2
	2	0	0	0	3	0	0	0				
	3	0	0	0								
	4	2	0	0								

Note that two stream values exceeded DELS.

	SN				STRMI				STRMO			
3. Store results unit #1	1	1	10	2	1	1	10	2	2	0	10	2
	2	0	10	2	3	0	0	0				
	3	0	0	0								
	4	2	0	0								
4. Set up unit #2	1	1	10	2	2	0	10	2	3	0	0	0
	2	0	10	2					4	2	0	0
	3	0	0	0								
	4	2	0	0								
5. Execute unit #2 and test	1	1	10	2	2	0	10	2	3	0	5	1
	2	0	10	2					4	2	5	1
	3	0	0	0								
	4	2	0	0								

Note that four stream values exceeded DELS.

	SN				STRMI				STRMO			
6. Store results unit #2	1	1	10	2	2	0	10	2	3	0	5	1
	2	0	10	2					4	2	5	1
	3	0	5	1								
	4	2	5	1								

Process iteration complete; did any stream values exceed DELS?
Yes. Then do another process iteration.

	SN				STRMI				STRMO			
7. Set up unit #1	1	1	10	2	1	1	10	2	2	0	10	2
	2	0	10	2	3	0	5	1				
	3	0	5	1								
	4	2	5	1								

TABLE 3.4 (CONT.)

8. Execute unit #1 and test	1	1	10	2	1	1	10	2	2	0	15	3
	2	0	10	2	3	0	5	1				
	3	0	5	1								
	4	2	5	1								
9. Store results unit #1	1	1	10	2	1	1	10	2	2	0	15	3
	2	0	15	3	3	0	5	1				
	3	0	5	1								
	4	2	5	1								
10. Set up unit #2	1	1	10	2	2	0	15	3	3	0	5	1
	2	0	15	3					4	2	5	1
	3	0	5	1								
	4	2	5	1								
11. Execute unit #2 and test	1	1	10	2	2	0	15	3	3	0	7.5	1.5
	2	0	15	3					4	2	7.5	1.5
	3	0	5	1								
	4	2	5	1								
	Note four stream variables exceeded DELS.											
12. Store results unit #2	1	1	10	2	2	0	15	3	3	0	7.5	1.5
	2	0	15	3					4	2	7.5	1.5
	3	0	7.5	1.5								
	4	2	7.5	1.5								

the process is said to be achieved only after the component flows in streams 2 and 3 have an absolute fractional change less than that specified in DELS. The calculations are then said to have converged. However, the error between the converged solution and the "correct" solution depends not only on DELS but on how fast the converged solution was achieved. If the convergence were slow, the solution reported for a given set of DELS may be appreciably in error compared to the true, steady-state solution. Therefore, the final answers should always be checked independently.

The executive program subroutines*

PACER has been written in FORTRAN II†. A complete understanding of how PACER works demands a detailed study of the FORTRAN listing. However, briefly, the PACER (1966) subroutines under the control of a main calling program are:

DREAD reads input data for a case. The stream connection matrix is produced and an equipment-calling directory is established.

* May be omitted on first reading.

† Available from Dr. P. T. Shannon, Digital Systems Corp., P.O. Box 986, Hanover, N.H. 03755.

EXREAD is called after DREAD to read any extra equipment lists and the specified sequence of calculations. The extra equipment lists allow selected Unit Computations to have extra-long parameter lists.

SEFLAG searches the stream connection matrix to set the stream flags and records a list of unused equipment numbers.

DPRINT prints all of the input data.

EXPRNT prints information read by EXREAD.

MAIN3 controls the calculation of the process if the calculation sequence has been specified by data read in through EXREAD.

MAJOR determines sequence of equipment calculation, if the user does not specify it, by searching the process matrix. If the input streams to an equipment unit are known, the output streams can be calculated. If the equipment unit can be calculated directly, MAJOR then calls subroutine EQUIP to do the requisite bookkeeping prior to calling the Unit Computation for execution. For recycle calculations the ordered list of units is determined by GUESS1 and GUESS2 and the calculations are done in SETUP. When MAJOR is completed, the entire process has been calculated.

GUESS1 systematically takes up to three streams and temporarily assumes they are *known* in order to find a feasible sequence of calculation for recycle problems. GUESS1 calls GUESS2.

GUESS2 sees if it is possible to find a list of equipment units which has as output streams the assumed known input streams. If possible, it calls for a sequential iterative calculation to be done by SETUP, involving all Unit Computations in a recycle loop, after establishing the shortest list of Unit Computations involved.

EQUIP does the bookkeeping necessary for *direct* calculations when transferring stream information from the SN matrix to temporary STRM lists. After a Unit Computation has been executed, it transmits the computed information back to the SN matrix.

SETUP does the bookkeeping for *sequential iterative* calculations. It calls Unit Computations in a proper sequence, using EQCAL, and tests for convergence using TEST after each Unit Computation is executed.

EQCAL calls the proper Unit Computation corresponding to the equipment numbers in the calculation order list. These Unit Computations are stored on disc or tape.

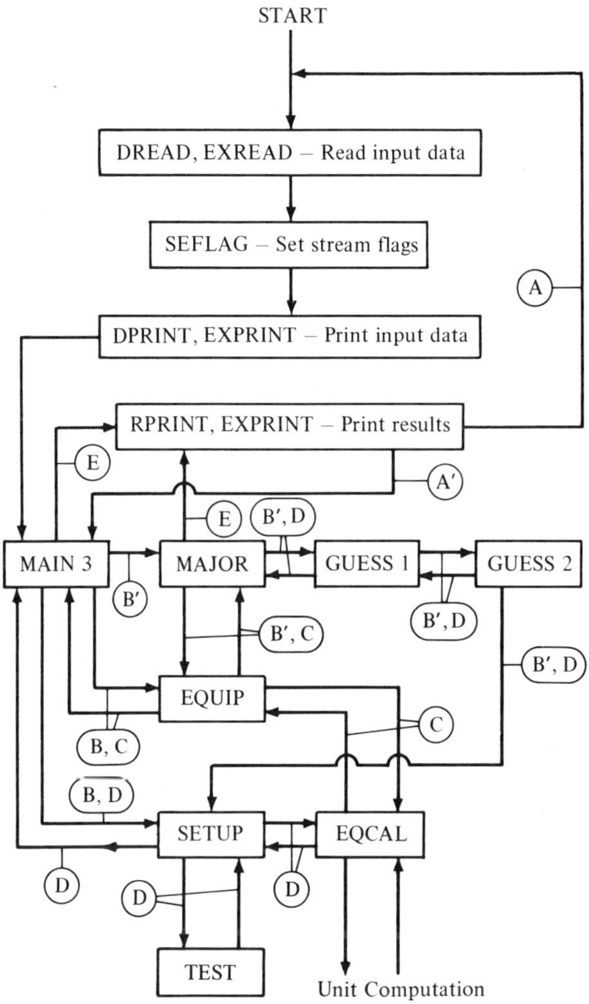

A – Only one case evaluation with the given input data
A′ – More than one case evaluation with the given input data
B – Calculation sequence specified
B′ – Calculation sequence *not* specified
C – Direct calculation of a unit can be done
D – Iterative calculation of a recycle set must be done
E – Calculation of current case is complete

Figure 3.6. Information flow diagram for the PACER (1966) group of subroutines.

TEST tests for convergence by calculating the absolute fractional change in all stream variable values from one iteration to the next in recycle problems. When all stream variables in a recycle loop have passed a tolerance DELS, the recycle calculation under consideration has stabilized and the executive can move on to other calculations in the problem.

RPRINT prints final values of stream variables and equipment parameters.

Fig. 3.6 is an information flow diagram of PACER (1966). The total size of the PACER executive program and its Unit Computations may exceed the core memory of some computers. In such a case, the simulation may be split into separate blocks called LINKS. The LINKS are chained together so that the required LINK will automatically be called into the computer's main memory and executed as required. The chain LINKS are stored on magnetic tape or disc. The LINK organization of PACER and the Unit Computations should be organized to minimize the number of changes of links, to make chaining as fast as possible, and to provide a standard exit for all Unit Computation subroutines.

The variables in COMMON remain in core memory at all times and hence are accessible to all subroutines in the system. This is how PACER transfers information between subroutines.

3.4 How PACER Determines the Order of Calculation*

If all input streams to an equipment unit are known, the output streams can be calculated using the Unit Computation. Thus information flows into and out of an equipment unit. In general, information flows forward although it can be transmitted backwards through a process.

PACER operates entirely from the process matrix. The current status of calculation of each stream and equipment is denoted by two lists of tags (KSFLAG for streams and KEFLAG for equipment). The scanning operation used to search out recycle loops is shown in Fig. 3.7(a). Initially, all elements of KEFLAG and KSFLAG are set equal to zero. KSFLAG for a stream is set greater than zero if it is a feed or a product, if it has been already calculated, or if it is not used at all. The process matrix is then scanned row by row to find any unit with a positive integer in KSFLAG for each input stream, since that unit can be calculated. When such a unit is found, the proper Unit Computation is called to calculate the output streams of the unit. KSFLAG for each output stream is set greater than zero since

* May be omitted on first reading.

they are now all known. KEFLAG for this unit is also set to a positive integer. Scanning the process matrix continues until one complete scan has performed no new calculation of a unit. If all units have been calculated, the case is solved. If, however, one or more units are still uncalculated, then there is a recycle problem and a sequence of calculation for it must be found.

For a recycle problem an iterative procedure shown in Fig. 3.7(b) is employed. To determine the sequence of recycle calculations, PACER tentatively assumes that a stream is *known* and again scans the process matrix to find any more units that could now be calculated. If the stream which has been assumed known is an output of one of the units which could now be calculated, then a sequence of units has been found which contains a recycle set. (Some units in this sequence may follow the recycle set but are not part of it.) If PACER has not yet found a recycle set, it returns to the list of unknown streams to assume another stream *known*. Unknown streams are assumed *known* first singly, then in pairs, and finally in threes, in an attempt to find a sequence of units which would recalculate the assumed streams as outputs. This sequence contains a recycle set, and the assumed streams provide a starting point for iterative calculations. If no such sequence of units can be found, PACER must be given a specified sequence to follow.

If a sequence of units has been found, PACER removes from this list all units which are not in the recycle set and so are unnecessary to the iterative solution. The initial values of the variables for the assumed known streams are used in performing the first iterative loop. Upon completion of the loop, new values for the assumed *known* streams have been calculated. These values are used in the next loop of the iterative solution. The values in the outputs streams of each unit are tested for convergence by comparing them with the values from the previous loop. When the absolute fraction change is less than DELS for each stream variable, convergence has been obtained. Upon convergence of an iterative solution, the process matrix is again scanned to determine whether any new units can be calculated directly. With this technique, PACER identifies recycle and determines a sequence of calculations to solve the problem.

These principles are now applied to both serial and recycle processes.

TABLE 3.5
PROCESS MATRIX FOR FIG. 3.8

Unit Number	Unit Computation name	Associated streams			
1	MIXER	1	−2		
2	MIXER	2	3	−4	−7
3	MIXER	4	−5	−6	
4	MIXER	7	−8	−9	

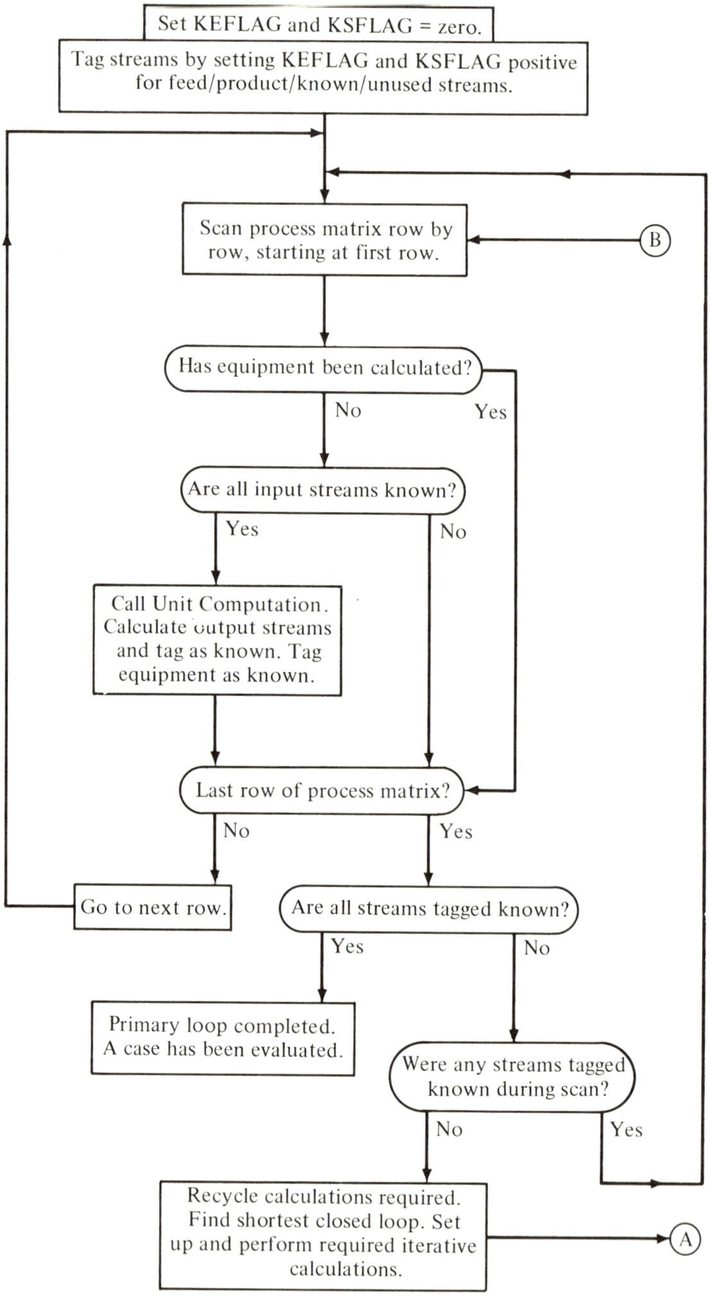

Figure 3.7(a). Program logic for primary loop of PACER.

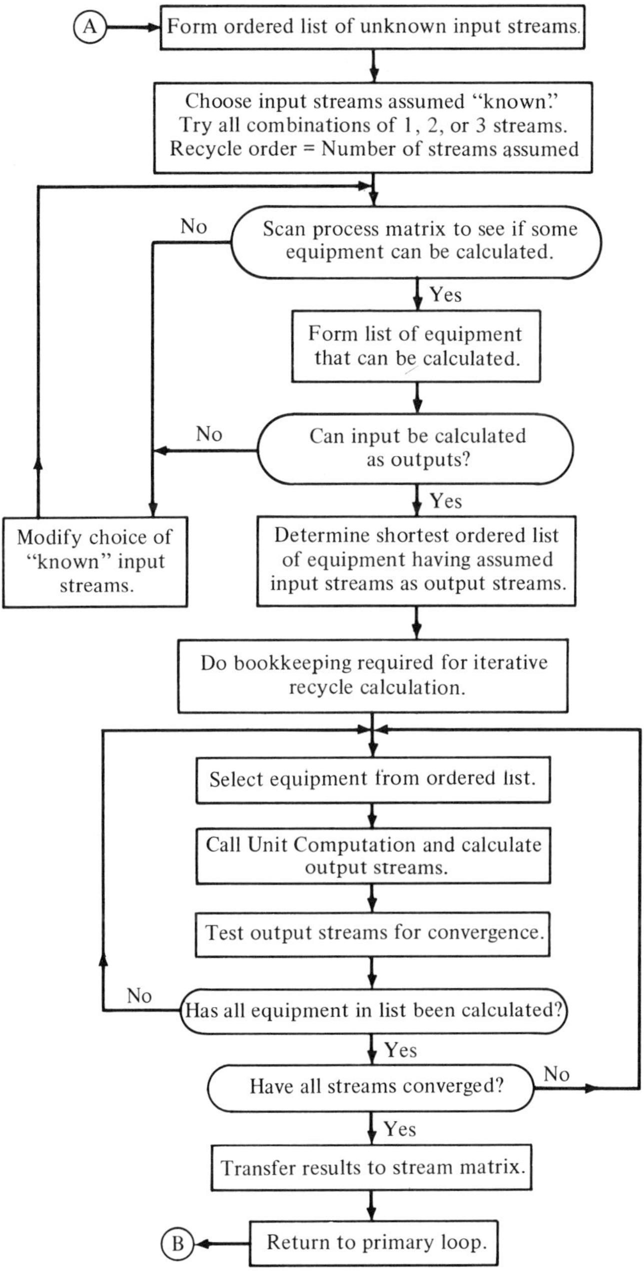

Figure 3.7(b). PACER program logic for recycle calculations.

Figure 3.8 illustrates a simple serial process of four MIXER Unit Computations. Table 3.5 is the corresponding process matrix. PACER first produces a stream connection matrix. Streams coming from the "outside," notably streams 1 and 3, are recognized as feeds and are flagged by a "1." Streams going to the outside are product streams and are flagged "2"; these are streams 5, 6, 8, and 9. This is carried out in DREAD. Connecting streams are flagged "0." The stream connection matrix and the flags are summarized in Table 3.6.

PACER now examines the process matrix repeatedly row by row in MAJOR to identify Unit Computations for which the input streams are known and hence the output streams can be calculated. In this example, unit #1 can be calculated, and it is tagged *known*, and all of the streams leaving it become *known*.

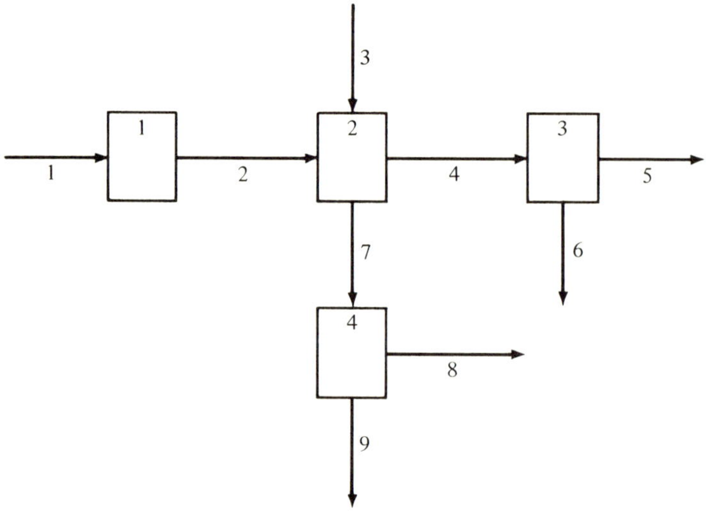

Figure 3.8. A serial process.

TABLE 3.6
STREAM CONNECTION MATRIX AND SAMPLE FLAGGING

Stream	From	To	Flag
1	0	1	1
2	1	2	0
3	0	2	1
4	2	3	0
5	3	0	2
6	3	0	2
7	2	4	0
8	4	0	2
9	4	0	2

It is apparent that the next row in the process matrix now contains a unit with a feed stream and a *known* stream and is itself *known*. As indicated in Fig. 3.5, when each unit is found to be *known*, it is also calculated at that time. In the example of Fig. 3.8, all units will have been calculated when the last row of the process matrix has been examined.

Figure 3.9 illustrates a simple recycle process made by connecting the output from equipment #3 to equipment #2. Again, a stream connection matrix is produced, and stream 1 is flagged as a feed stream, and streams

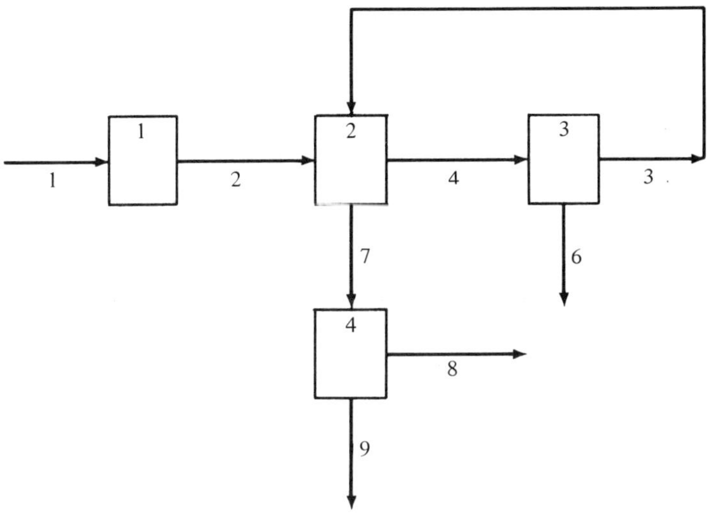

Figure 3.9. A recycle process.

TABLE 3.7
PROCESS MATRIX FOR FIGURE 3.9

Unit Number	Unit Computation name	Associated streams			
1	MIXER	1	−2		
2	MIXER	2	3	−4	−7
3	MIXER	4	−3	−6	
4	MIXER	7	−8	−9	

6, 8, and 9 are flagged as product streams. Other streams are flagged "0" as connecting streams that are not *known*. MAJOR recognizes that unit #1 can be calculated. However, units #2, #3, and #4 cannot be calculated directly. Within GUESS1, an ordered list is formed of the inputs to these units, namely, streams 3, 4, and 7. Streams marked "preferred" by the programmer are moved to the head of this list. Each stream in this list is assumed *known*, and GUESS2 is called to scan the process matrix to identify a list of units that can be calculated and whose outputs include the assumed streams. PACER allows up to three streams in the list to be assumed *known* simultaneously. After GUESS2 has found such a list of units, it finds the shortest ordered list having assumed input streams as output streams of other units in the list. Although units #2, #3, and #4 are *unknown*, the shortest list having *unknown* inputs which are outputs of other units contains only units #2 and #3. Hence a recycle loop containing these units has been found. Sequential iterative calculations involving units #2 and #3 are carried out so that after convergence to the satisfaction of TEST, streams 3, 4, 6, and 7 become *known*. Finally, in a reexamination of the process matrix it is found that unit #4 can now be calculated. All units have thus been calculated.

3.5 Summary of PACER Data Requirements

The engineer supplies five sets of information to PACER to simulate a process: process matrix, SN list, EN list, control specifications, and a set of Unit Computations.

The *process matrix* is a numerical representation of the information flow diagram. This has been discussed in Sec. 2.3 and examples are shown in Table 3.5 for Fig. 3.8 and Table 3.7 for Fig. 3.9. PACER requires that input streams be first and positive, followed by output streams which are negative. The order of listing the streams indicates the order of transfer into the STRM matrices.

The meaning of the elements of the *SN list* is specified by the engineer. This is not fed into the program but rather is constantly referred to by the engineer in writing Unit Computations and in reading the printed results. PACER requires that element 1 represent the stream number, and element 2 the stream flag. The SN information is used in the execution of the Unit Computations. The *EN list* is similar to the SN list.

The *control specifications* need to be specified. Reasonable values for these specifications and their meaning are given on pp. 50–53 and an example layout is given in Appendix B.

The working models of the *Unit Computations* listed in the process matrix need to be supplied to PACER. In developing these Unit Computations, the programmer realizes that in the PACER system he has access to four sources of information.

1. *Stream information.* Stream information can be obtained from any location but normally the information in the STRM matrices is used. Input stream lists are transmitted from the SN matrix into the stream transfer matrix, STRMI, in the order in which they appear in the process matrix. PACER also transmits the lists for the output streams from SN to STRMO, also in the order given in the process matrix. Within each Unit Computation the programmer must transfer infor-

mation across from STRMI to STRMO even if no change occurs in the information.

2. *Equipment parameter information.* Unit Computations can be generally applied to various equipment units because constants and quantities specific to each unique unit are supplied through the EN list corresponding to the equipment number NE in the process matrix. There is an auxiliary equipment parameter list called the ENC that is also available. Unit Computations may modify data in both the EN and ENC matrices although this is done in design more often than in simulation problems.
3. *The auxiliary equipment parameters.* Because some parameter lists may be very long, PACER allows a limited number of Unit Computations to use a maximum of 38 additional parameters. In the 1966 version, this was allowed in up to four Unit Computations, but this number can be increased if space is available.
4. *Information transfer by means of COMMON.* All information within COMMON is available to each Unit Computation. In general, in devising the Unit Computations it is best that transfers be made from the STRM matrices into and from new variables meaningful to the programmer such as:

   FLOW = STRMI (1, 3)
   UA = EN (NE, 4)

   Not only are the transferred numbers defined, but a conventional program can thus be adapted to the PACER format.

This summarizes the four sets of data required by PACER.

3.6 Summary

PACER is a modular executive system designed to be open-ended, flexible, and easy to understand and to use. For input data, it uses the process matrix to encode the flow diagram, some integers to control printout and dimension the matrices, stream lists for feed conditions, and Unit Computations with their associated lists of equipment parameters. Use of the equipment lists means that the Unit Computations can be made general enough to be used for several equipment units of the same type.

The SN matrix stores stream information. In the matrix there is one SN list for each stream. The STRM matrices, STRMI and STRMO, provide temporary storage locations for that stream information while it is required for the execution of a Unit Computation. These working matrices are provided to prevent the inadvertent destruction of numbers in SN by the Unit Computation.

Details are given of how PACER determines the sequence of calculations and executes recycle calculation. A list is given of the group of subroutines that make up PACER.

The programmer needs to assign meaning to the elements in the SN and EN lists so that he can devise the Unit Computations and interpret the results. Before evaluating a case he also needs to supply PACER with the input data and a set of Unit Computations. Details of how to devise the Unit Computations are given in the succeeding chapters.

PROBLEMS

3.1. The details given in this chapter describe how PACER works. To actually use an executive program we need to learn the formats required for the input data. In the modified version of PACER used at McMaster, called MACSIM, the overall arrangement of the deck is given in Table 3.8. The MACSIM control data include, in the order given on pp. 50–53:
Titles—first two cards; start at column 2, end at 72; must be present even if blank.

TABLE 3.8
ARRANGEMENT OF CARDS FOR THE MACSIM EXECUTIVE VERSION OF PACER

$JOB	003214 MACSIM	(user's name in cols. 34-45)	
$IBJOB	PLDRA		
$CHAIN MAIN			
↕ Object deck of PACER3	(11/28/66 version)		
↕ Object deck of SPACEP	(2/20/68 version)	subroutine needed for the MAP version of PACER and MACSIM	
$ENTRY			
$LINK LINK1			
↕ Object decks of Unit Computations			
↕ Source decks of Unit Computations			
$ENTRY			
$ENDCH			
↕ Source deck of MACSIM COMMON deck	(2/20/68)		
ENDCO			
↕ MACSIM control data			
↕ Any other data, such as physical property data or convergence data.			
$IBSYS			

Control variables—one card, format 8I5, 15I1 starting with KRUN and ending with KONV followed by, in columns 41 to 55, printing suppression instructions starting with process matrix and ending with calculation order.

Dimension variables—one card, format 10I5.

Process matrix—one card for NOKPM on format I5; the process matrix follows on the next NOKPM cards, in format I5, 1X, 2A3, 1X, 2A3, 1X, 9I5.

Equipment information—one card for NOEN on I5; next, on "list format" (5 F 15.0), come the EN lists; each new list begins on a separate card (hereafter called list format); one card for NOENC format I5; ENC lists follow with list format.

Stream information—one card for NOSN in format I5 followed by NOSN stream lists on list format.

Stream variables test lists—one card for NDELS in format I5; the test list follows with NSLMAX elements; the first two elements on the card cannot be used with the list format.

Preferred streams—NPS on one card format I5 followed on the next card by the list of up to three streams on format 3I5.

Additional equipment lists—one card for NOAEN on format I5 followed on the next cards by AEN in list format.

Calculation order list—one card in format 2I5 gives KK and LISTT, the number of units in the list; the NELIST is then given on the next cards in format 10I5.

KSPRNT—one card with format 10I5.

Given these format specifications, prepare a set of coding sheets that will make it easy to prepare a case for keypunching.

3.2. (a) Prepare an algorithm for a Unit Computation SETST that adjusts the temperature in a stream but otherwise transmits flows unaltered. The program should work with PACER.

(b) Write the program in FORTRAN IV so that it will work within PACER or the executive program of your choice.

(c) Prepare a simple case study that will adequately test your SETST program.

3.3. Once one executive system is understood, it is interesting, and indeed helps to improve an existing executive system, to study the characteristics of other executive systems. Some PACER-like executive systems include:

—CHESS by Motard, Lee, Barkley, and Ingels (1968);

—Executive systems for water and waste-water systems by Smith, Eilers, and Hall (1968);

—GEMCS by Johnson and associates (1970).

To aid in a study of these systems consider the following questions:

(a) Compare the system with PACER for similarities and differences.

(b) Does it use stream and equipment lists? If so, what are the equivalent names for SN, EN, ENC, AEN? If not, how is the information handled?

(c) Does this system use the STRM working matrix concept? If so, what are the names for these? If not, is the function equivalent to the STRM working matrices satisfied? How?

(d) What control do we have over the printout?

(e) How is convergence tested?

(f) Prepare a set of coding sheets that are applicable to this executive program.

3.4. How difficult is it to convert a Unit Computation written to work within PACER so that it will work within CHESS? Smith, et al.'s executive system? GEMCS? MACSIM? Any other executive system of interest?

The Acid Plant: An Example of a Plant Simulation 4

Simulation requires an executive program and a problem. An executive program, PACER, has been described in detail in Chapter 3. Now a practical problem is considered: improvement of the operation of a contact sulfuric acid plant. This chapter and the following seven chapters describe how a simulation using the PACER executive program was made of the contact sulfuric acid plant owned and operated by Canadian Industries Limited in Hamilton, Ontario, Canada. The plant is described in Sec. 4.1. The general strategy used in tackling a simulation problem is outlined in Sec. 4.2 and is applied in Sec. 4.3 through Chapter 11. In Chapter 12 some optimization of portions of the plant is introduced to illustrate how the simulation can be used for improving plant operation. The approach described in this case study assumes that no previous computer simulation has been done and that a physical property package and a library of Unit Computations are not available.

4.1 A Description of the Acid Plant

A sufficient steady-state description of a process includes a process description and flow diagram, information on the economic conditions, and data about the environmental conditions and specifications.

Process description and flow diagram

In this plant sulfuric acid is made from molten sulfur, air, and water in a three-step process: feed preparation, reaction, and absorption. In the feed preparation section, air is compressed, dried, and burned with molten sulfur to produce a water-free gas feed of sulfur dioxide and oxygen for the reactor section. The combustion produces heat. In the reactor section the gas feed is cooled to about 400°C and sent through a series of fixed catalyst beds. The reaction generates heat that must be removed. In the absorber section the cooled product gas is absorbed in sulfuric acid to produce either more concentrated acid or oleum. Heat is also generated in the absorption step. The equations for the reactions in the three sections are:

$S + O_2 \longrightarrow SO_2 + \text{Heat}$	Feed preparation
$SO_2 + \frac{1}{2}O_2 \rightleftharpoons SO_3 + \text{Heat}$	Reactor
$SO_3 + H_2O \longrightarrow H_2SO_4 + \text{Heat}$	Absorber; this is usually done through the use of sulfuric acid instead of water.

The heat produced at several stages generates steam, part of which drives the blower in the feed preparation section and part of which is sent as service steam to other parts of the plant. The raw materials are molten sulfur, ambient air, and water; the products include 99% sulfuric acid, oleum, and steam. The Hamilton Works plant is a Monsanto Type S 200 ton per day plant constructed in 1948 by Leonard Construction Company. Since then, the plant has been modified and presently produces about 315 tons per day. The normal products are 93% acid (66°Bé), 77% acid (60° Bé), 25% oleum and 110 psig steam. Fig. 4.1 is a flow diagram which illustrates the three major sections of the plant. Fig. 4.2 is a photograph of the plant. The equipment units are enumerated below.

The feed preparation section includes:

1. Blower 1*: Air is supplied by a single-stage centrifugal blower driven directly by a steam turbine. The air is not cleaned. Superheated steam which is generated in the reactor at 325 psig and superheated to about 300°C drives the turbine and is discharged at 113 psig and 230°C. When the plant capacity was increased, more nozzles were added to the turbine to increase the output of the blower. The excess steam not

* The numbers refer to equipment units shown in Fig. 4.1.

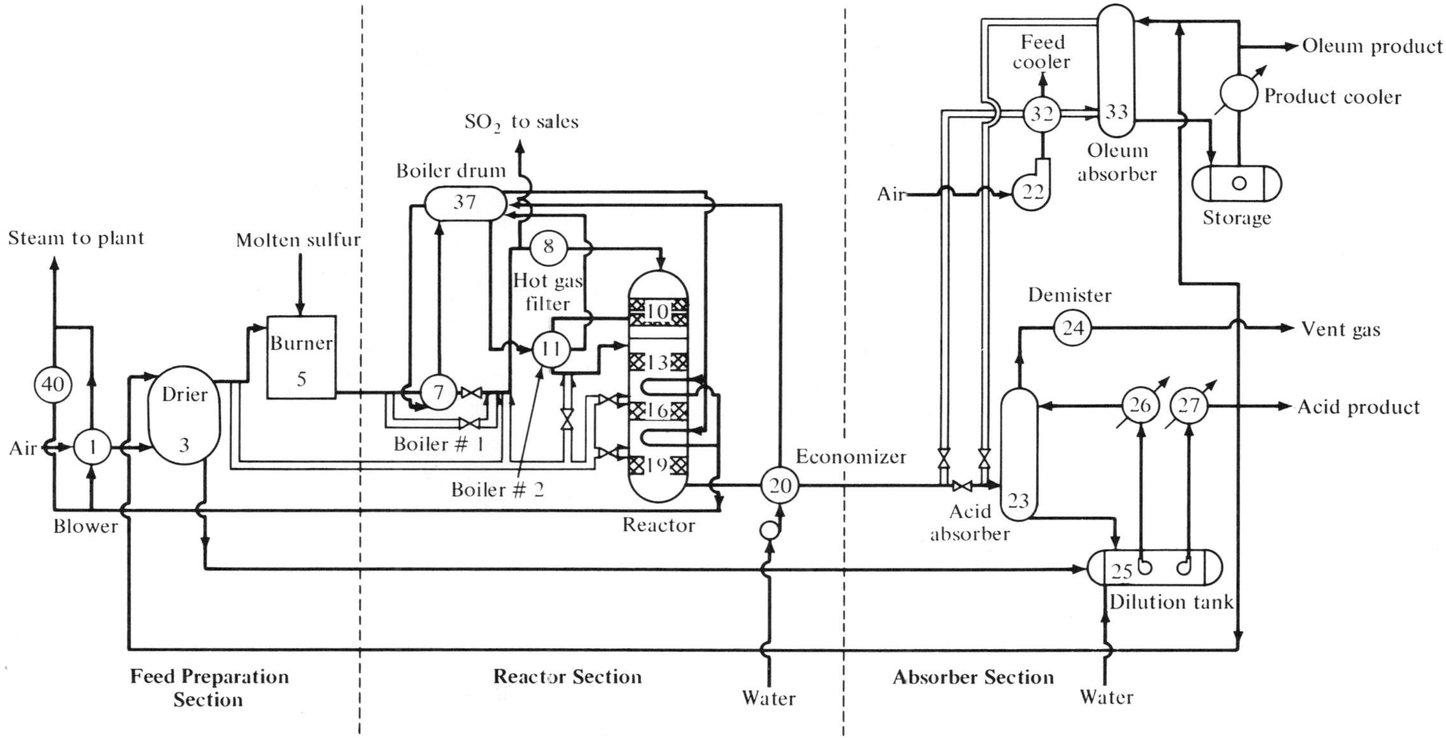

Figure 4.1. Process flow diagram of the Canadian Industries Limited contact sulfuric acid plant.

Figure 4.2. The Monsanto-designed contact sulfuric acid plant of the Hamilton works.

needed to drive the turbine is bypassed and is reduced (40) to the service steam pressure of 110 psig. The combined steam is at 110 psig and 255°C.

2. Drying tower 3: Air is dried by contacting it with sulfuric acid in a packed tower of inside diameter 14 ft-7 in., randomly packed with 12 ft of 3 in. stoneware partition rings. A 3 ft-6 in. layer of 1 to $2\frac{1}{2}$ in. quartz pebbles above the packing serves to demist the dry air, and there is also a knockout drum before the burner. The air is dried by a countercurrent flow of 99% acid entering at 55° to 60°C with a flow of about 400 US gals per min. Moisture content of the leaving air is low.
3. Sulfur burner 5: The burner is a so-called film type, designed by Monsanto. It is a steel cylinder lined with insulation and fire brick and containing a brick checkerwork over which the burning sulfur flows. The molten sulfur is not filtered since it is free of organic material, and no difficulty is encountered in the operation of the burner. The exit temperature of 971°C is carefully controlled.

The reactor section includes:

1. Boiler #1 and boiler #2: These are shell and tube boilers: boiler #1 cools the gas from the sulfur burner from 971° to about 406°C; boiler #2 cools the gas from the first reactor from about 613° to 452°C. Both boilers are supplied with water to the tube side by natural circulation from a common steam boiler drum (37). Both boilers can be partially bypassed to provide temperature control of the gas feed to the reactor. Manually operated butterfly valves are in the bypass lines.

   Presently, cooling of the process gas by boiler #2 limits the capacity which could be increased by improved water circulation rate.

   After passing through boiler #1, the hot gas is filtered by being passed down through a 19 ft diameter bed of crushed firebrick housed in a low cylindrical vessel. This hot gas filter (8) protects the catalyst bed from solid impurities such as ash, but since molten sulfur has been used in the plant, the rate of fouling of this filter has been much decreased.

   Immediately before the hot gas filter, three streams are combined: gas from boiler #1, gas bypassing this boiler, and cool air from the blower. These streams are manipulated by dampers so that the concentration of SO_2 in the gas leaving the burner is 10.4% while the concentration and temperature of the gas entering the reactor are 9.5% SO_2 and 424°C.
2. Reactors 10 through 19: The reactor has a 17 ft internal diameter and contains five layers of catalyst; layers 1 and 2 are considered as the first bed.

   Gases from the first bed are removed from the reactor and cooled in boiler #2; cool air is also added to these gases as they leave the boiler and enter the second bed.

Gas cooling after the second and third beds is accomplished by finned tube exchangers in the reactor; the heat released superheats the steam flowing inside the tubes from the boiler drum. Cool air is added after the first superheater and before the third bed; it is also possible to add cool air before the fourth bed.

3. Economizer 20: Gases leaving the reactor are further cooled from 427° to 250°C by this unit which heats the boiler feed water (35,500 lbs per hr including blowdown) from 105° to 200°C.

The absorber section includes:

1. Acid absorption tower 23: The absorption tower is 14 ft-7 in. in diameter and has 12 ft of 3 in. partition rings (dumped). This tower has a wire mesh demister (24) mounted in the stack above the tower.
2. Dilution tank 25: Acid leaving both the drying and absorption towers flows to a common dilution tank. Water is added to adjust the concentration to the product specification. Vertical submerged pumps (25) distribute the acid to the top of the absorber (23) and the drier (3).
3. Acid coolers 26 and 27: These are banks of horizontal cast iron pipes through which the acid flows while cooling water flows over the outside of the pipes. One such unit is visible on the left side of Fig. 4.2.
4. Oleum system: Although the oleum system was shut down during the plant tests which were part of this study, this part of the plant normally operates four days per week to produce up to 100 tons per day of 25% oleum. The oleum absorption tower (33) is 8 ft in diameter, is unlined, and is packed with 18 ft of $1\frac{1}{2}$ in. quartz pebbles. (This tower is shown on the right side of Fig. 4.2.) Before entering this tower, the gas is cooled to 90° to 100°C in a shell and tube exchanger (32); the coolant on the shell side is air, delivered by a motor-driven blower (22). Inside the tower the gas is absorbed in a countercurrent stream of concentrated acid. The absorbing acid is a combination of cooled tower bottoms and acid from the dilution tank (25). Undissolved gas from the oleum tower returns to the acid absorption tower where the remaining SO_3 is removed.

Details of the operation of sulfuric acid plants are given by Duecker and West (1959), Farkas and Dukes (1968), Connor (1968), Burkhardt (1968), and Industrial and Engineering Chemistry (1950).

Economic* and environmental conditions

It is essential to appreciate fully the economic conditions under which the process is operating. For a 315 ton per day plant operating in 1967 with a

* These figures are from the literature and any similarity between them and those of Canadian Industries Limited is purely coincidental.

TABLE 4.1
EXAMPLE TONNAGE COST OF SULFURIC ACID FOR A 315 TON PER DAY PLANT (1967), BASED ON DATA REPORTED BY IEC (1950) AND BY FAITH, ET AL. (1966)

Item	Unit cost	Usage per ton	Acid cost per ton
Raw sulfur	\$40/ton	668 lb	\$13.36
Utilities:			
Electricity	\$0.54/100 kwhr	5 kwhr	0.03
Steam (110 psig)	\$0.50/1000 lb	200 lb	0.10
Water			
For boiler	\$0.20/1000 gal	350 US gal	
For acid	\$0.20/1000 gal	34 US gal**	0.15
For cooling	\$0.02/1000 gal	3,600 US gal	
Labor (L)	\$3.50/hr	0.64 man-hours	2.24
			15.88
Credit from net sale of 120 psig steam	\$0.50/1000 lb	1,700 lb	—.85
			15.03
Maintenance 0.04 I_F^*, annually			1.48
Supervision 0.1 (L)			0.23
Depreciation 0.08 I_F^*, annually			2.95
Indirectly attributable costs 0.75 (1.1 L + 0.04 I_F^*)			2.95
Total manufacturing cost (M)			22.64
General expense 0.1 M			2.26
Total cost per ton 100% sulfuric acid			\$24.90

* The fixed capital investment, I_F, is \$3.3 million [Dryden and Winfield (1963)] for a 300 ton per day plant in 1967. Assume plant operates 300 days of the year.

** The stoichiometric amount is 44 US gal but water absorbed from the humid air by acid in the drier accounts for the difference.

steam turbine used to drive the blower, the contributions to the unit cost per short ton of acid are summarized in Table 4.1.

There are specifications on the product: 66°Bé commercial grade acid with a minimum of 50 ppm of iron. Some constraints on the streams are:

1. An air pollution constraint so that the maximum concentration of sulfur dioxide is 0.3 volume percent in the vent gas from the acid absorber;
2. A fixed concentration of 10.4 volume percent SO_2 for the gas leaving the burner so that a portion of this stream could be sold for another manufacturing process;
3. Maintenance of the stream pressure at 110 psig.

Environmental conditions include an unlimited amount of cooling water from the bay, city water, and ambient air. The temperature variations for these are given in Table 4.2.

TABLE 4.2
AVERAGE TEMPERATURES FOR THE ENVIRONMENT, °C

	Feb.	May	July	Nov.
Bay water	12.0	15.0	25.0	13.5
City water	2.2	5.6	8.9	6.7
Atmospheric air	5.0	6.1	25.6	7.8

4.2 Strategy and Decisions for Solving Simulation Problems

Since a simulation is undertaken for the purpose of plant improvement, the strategy of plant improvement is discussed first to illustrate what role simulation plays. Then the plan of attack for any simulation problem is outlined. Related to the plan of attack are two key decisions that have to be made. These are discussed in turn.

A strategy for plant improvement and for simulation

The strategy for solving any plant improvement problem is similar to that used in solving any problem including a design problem. Polya (1956) suggests that the steps used in solving any problem are:

1. Define the problem.
2. Devise a plan for solving the problem.
3. Carry out the plan.
4. Look back at the answer, its implications, and its validity.

The details for these four steps when applied to design are given in Fig. 4.3.

Overall strategy	Details of process design strategy
1. Define	1. Identify need or questions to be answered. 2. Specify criteria. 3. Identify system parameters (inputs, outputs, transforming agents, constraints). 4. Collect information.
2. Plan	5. Gain conception of alternatives.
3. Carry out	6. Make prediction. 7. Make evaluation (comparison and selection) Physical basis Economic basis Financial basis.
4. Look back	8. Iterate: Repeat portions of or the whole strategy at different accuracy levels or for optimization of equipment size or conditions. 9. Communicate and implement results.

Figure 4.3. Strategy for process design.

This is a composite of the design strategy proposed by Modrey (1960), Nenniger (1966), Peters (1958), Rase and Barrow (1957), Rosenstein, et al. (1960), Shannon (1965), Sherwood (1963), Smalter (1964), Taylor (1961), Woodson (1964), and Vilbrandt and Dryden (1949). It is not only important to realize the similarity in strategy between problem solving and design, but also to realize that this can be used as a strategy for plant improvement. Figure 4.3 summarizes the approach for process design. Figure 4.4 suggests the details of these steps for plant improvement. In the present context, plant improvement means:

1. Giving engineers a better understanding of plant operation and in particular of the interaction of equipment; the interaction affects technical feasibility, economics and product purity. Although individual processing units can be simulated separately, the great advantage of the simulation techniques described in this book is that interaction among pieces of equipment can be studied.

Overall strategy	Details of plant improvement strategy
1. Define	1. Identify need.
	2. Specify criteria.
	3. Identify system parameters.
	4. Collect information to describe present condition of plant: flow diagram, mass, energy and time balance, temperatures, pressure, concentrations; describe the present operation through functional analysis plus mathematical models for each piece of equipment. Devise a complete description of present operation.
2. Plan	5. Gain conception of alternative operating conditions, causes of loss, or actions to improve operations.
	6. Make prediction.
	7. Evaluate actions.
3. Carry out	8. Implement most attractive proposals.
4. Look back	9. Communicate to engineers, operators, and management.
	10. Iterate: repeat strategy when cost structure or process network changes.

Figure 4.4. Strategy for plant improvement.

2. Giving the operators a feeling for the sensitivity of the variables and indications of the potential trouble spots under a wide range of operating conditions other than those now employed.
3. Improving the yield and reducing operating costs for any set of economic and environmental conditions.

A completed simulation facilitates steps 6 and 7, prediction and evaluation. The problem at hand is how to devise a working simulation such that it fits into the overall strategy of plant improvement outlined in Fig. 4.4. In this context a working simulation, by itself, is an elaboration of steps 1 to 4 in Fig. 4.4. Thus the plan of attack will be:

1. Identify the need or define the questions to be answered for plant improvement (Sec. 4.3).

2. Specify the criteria or the basis for judging improvement (Sec. 4.3).
3. Identify the significant variables and parameters (inputs, outputs, transforming agents, constraints, and the parts of the simulation) which are important for the criteria. An easy way to identify the variables and parameters is to study equipment interactions in a simple simulation. This identifies the areas where efforts need to be concentrated to achieve the needs. The information needed to prepare an approximate simulation using simple Unit Computations includes:
   (a) The process flow diagram (which needs to be converted into an information flow diagram and then a process matrix);
   (b) Operating data;
   (c) A set of simple models to achieve a starting simulation. Simple models include mixers, separators or splitters, and converters or exchangers. These are used much as a designer uses specifications of separations, conversion, or exchange to establish a preliminary mass balance as he gains a feel for the process flow diagram. Table 4.3 presents a very brief description of these simple models.

TABLE 4.3
SIMPLE UNIT COMPUTATIONS OF UNIVERSAL APPLICATION FOR SIMPLE SIMULATIONS

Name	Operation	Typical physical equipment
MIXER1 	(As described in Sec. 3.2) Specify the total flow into various exit streams from a uniform mixture of all feed streams.	Tees, splits, mixers, dividers, distributors
SEPA0 1, 2, 3 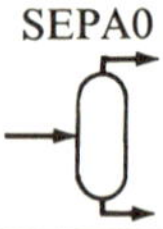	Specify the split or concentration of each component in a single feed for overhead or underflow streams.	Distillation columns, filters, settlers, decanters, flash evaporators
REACT1 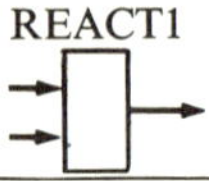	React one or more feeds to produce a new stream according to a specified conversion.	Reactors, crystallizers, digesters, converters
EXCH01 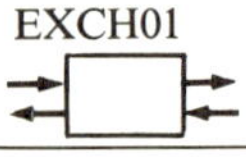	Transfer heat or material between phases flowing through the unit.	Absorbers, extractors, leachers

These models are then used to simulate a complete plant and to identify important equipment and control areas through a sensitivity analysis. In this analysis, different values of separation, conversion, and exchange are substituted into the different models to see where minor variations create major consequences. For the system in Fig. 3.5, the fractional splits, α, in the MIXER model would be the variable changed.

More details of this technique are given in Sec. 4.3 and in Chapter 13.

4. Collect information required for the simulation to satisfy the needs. Put the information into a form that is useful for the simulation. This is where the extensive work begins.
   (a) Describe the process flow diagram, convert the problem into an information flow diagram, and devise the process matrix (Sec. 4.3).
   (b) Collect physical properties, constants, and correlating coefficients (Chapter 5).
   (c) Collect the additional plant data that the sensitivity analysis reveals as important. Data are also needed at a variety of conditions away from present operation. This is essential in plant improvement so that confidence can be placed in the predictions (Sec. 4.3).
   (d) Devise and test single Unit Computations under a wide range of input information (Chapters 6 to 10).
   (e) Simulate small groups of Unit Computations (Chapters 9 and 10).
   (f) Simulate the Complete Plant (Chapter 11).

This completes the four steps needed to devise a simulation of a complete plant.

Two key decisions

Decisions that must be made before the modeling of individual units is attempted are (for phase 4):

1. The accuracy, generality, and the ability for extrapolation required of the models;
2. The method of handling physical property information.

These two decisions are related because gaining generality in the model implies that the physical properties cannot be stored within the model program. The details and implications of these decisions are discussed in Chapter 5.

The necessary organization of resources

Once the problem has been defined, the strategy selected, and the key decisions made, the available manpower and resources should be carefully allocated to ensure an efficient group assault on the project. Definitions of terms, formats for subprograms, long-range planning for the collection of corroborating data, the compiling of reliable physical properties, arrangements within the company or between branches of the company to facilitate smooth operations—these are some of the conditions that should be set up. Above all, good communication should be maintained among the computation center, the project engineers, and the plant operators.

Not only should the people and facilities be carefully organized, but a time schedule should be arranged. Such systems as Critical Path Method and PERT are useful.

4.3 The Application

The application of the strategy to a simulation problem is now given. To introduce what decisions have to be made, when they arise, and how they are made, the case study of the sulfuric acid plant described in Sec. 4.1 will be used. This plant is well suited as an illustration because the process is simple enough that the technological know-how does not overwhelm the simulation principles. Yet most of the decisions, characteristic of simulation, and important unit operations are introduced. In Chapter 13 and in the problems at the end of each chapter, topics that did not receive enough attention through the acid plant case are discussed. The strategy followed is that outlined in Sec. 4.2.

Identifying the need

Simulation for the sake of simulation is not useful. Simulations are made to answer questions. The questions selected for the acid plant illustration are specific to the plant yet are of general applicability. The questions chosen for this model simulation are:

1. Can the operation of the sulfuric acid plant be improved?
2. Would it be possible to determine the reasons for some of the observed behavior? For example, why do the tubes of Boiler #2 fail mechanically?

In addition, general questions concerning the feasibility of carrying out simulations include:

3. Could the sulfuric acid plant, with its recycle loops and complex interaction of transport phenomena (especially in the reactor section) be simulated on an IBM 7040 computer with the PACER executive program?
4. What are the advantages and disadvantages of modular executive programs?

In general, this case attempts to answer questions related to the technology of a given process and to illustrate principles and strategy.

Specifying the criteria

Criteria must now be chosen for judging whether the needs thus identified have been met. The criteria which correspond to the particular needs listed above are now discussed.

1. The cost per ton of product is a useful criterion for evaluating improvement. A study of the cost analysis in Table 4.1 suggests that the yield of sulfur trioxide from sulfur dioxide is significant because the raw material cost is about 50% of the total cost.
2. Simulation with relatively simple models allows the interaction of models to be studied and new feed and operating conditions to be specified. On the other hand, the simulation of a particular piece of equipment, say Boiler #2, is used to indicate whether operation under these conditions is possible. The success of the interaction between the whole system simulation and the individual model simulation depends on how well the operating conditions agree between the two simulation problems. Hence, a selection of the physical parameters and stream variables that complement both simulation purposes is required.
3. The simulation will be considered successful if the major pieces of equipment can be described in such detail that the temperatures, pressures, flow rates, and concentrations can be estimated in every major stream. The calculated data should agree with a consistent set of measured data within 10%. In general, however, the selected percentage depends on the use of the simulation. The problem is simplified by excluding the oleum section, neglecting the cooling water loop on the acid coolers, and omitting the complications of selling a portion of the SO_2 stream.

Identifying the variables and other system parameters

The choice of stream variables depends on the need and the criteria. Variables that could be used in the simulation are flow rate, temperature, pressure, concentrations of sulfur, sulfuric acid, sulfur dioxide, sulfur trioxide, water, nitrogen, oxygen, inert gases, iron dissolved in the acid and mist, and the particle size distributions. Of this list, the chosen stream variables and the corresponding units are:

total flow rate	lb moles/hr
temperature	degrees centigrade
pressure	psia
flow rate sulfur	lb moles/hr
flow rate water	lb moles/hr
flow rate oxygen	lb moles/hr
flow rate nitrogen	lb moles/hr
flow rate sulfur dioxide	lb moles/hr
flow rate sulfur trioxide	lb moles/hr
mole % liquid in gas-liquid mixture	dimensionless

The units should be selected with care. Because there were many reactions in this plant, lb moles per hr rather than lb per hr was selected as the unit for flow rate. In addition, this choice allowed utilization of the equivalence between volume percent, mole percent, and partial pressure percent to simplify gas flow calculations. Degrees centigrade was selected for the temperature unit because many data were available in this system. The heat unit, therefore, was the PCU or Pound Centigrade Unit. The pressure was expressed as lb per sq in. absolute. An additional problem was how to handle the concentrations for both gases and liquids conveniently and yet with the minimum of storage space. Since the concentration of acid can be expressed in terms of water and sulfur trioxide, the components chosen were sulfur, water, oxygen, nitrogen, sulfur dioxide, and sulfur trioxide. That is, the concentration of liquid sulfuric acid was not used in itself. Concentration can be expressed either as weight percent, mole percent, or as total molar flow rates. The molar flow rate method was chosen. One complication arises in the steam lines because both water liquid and vapor can exist. Hence, the quality of the steam expressed as mole percent liquid was a final variable.

The choice of units is not always apparent. Process technology, terminology, and company practice have often introduced combinations of units that have local appeal only, for example, mg per liter for water and waste water treatment, causticity for the Bayer alumina process, barrels per day for petroleum processing, degrees Baumé for acid production, or the freeness of pulp in the pulp and paper industry. Our experience indicates that it is better to standardize on lb or lb moles per hr, degrees centigrade, and a convenient pressure within the Unit Computations and within the complete simulation. This standardization means that general purpose Unit Computations and physical property data can be used. Then, if—for easy interpretation by the process engineers, operators, or manager—the local terminology is desired, the standard results of the simulation can be converted to local units by a special subroutine just before they are printed out.

After the variables have been selected, the stream list (SN list) can be prepared as illustrated in Table 4.4. The blank location 6 was sometimes used as a special flag to keep track of streams that enter and leave a unit. For example, both a gas stream and a liquid stream enter and leave an absorber. The gas information should follow the gas flow. A special flag, say number 25, could be used in location 6 to indicate a gas stream. However, the need for the special flag can be eliminated if care is taken in ordering the streams in the process matrix such that the process gas is the first stream to enter and the first stream to leave a Unit Computation. In this work the stream list length was constant at 13 locations, but it could be increased to, say, 30 components or be made of variable length if in different parts of the plant different information is transmitted.

Now consider the inputs, outputs, constraints, and most sensitive variables. The *inputs* to a simulation are the information flow diagram, the operat-

TABLE 4.4
ALLOCATION OF VARIABLES TO THE ELEMENTS IN THE SN LIST

Element	Meaning
1	Stream number
2	Stream flag
3	Total flow, lb moles/hr
4	Temperature, °C
5	Pressure, psia
6	—
7	Flow of sulfur, lb moles/hr
8	Flow of water, lb moles/hr
9	Flow of oxygen, lb moles/hr
10	Flow of nitrogen, lb moles/hr
11	Flow of sulfur dioxide, lb moles/hr
12	Flow of sulfur trioxide, lb moles/hr
13	Steam quality

ing data, equipment parameters, physical properties, and the cost information. The main concern is how they are supplied to the simulations, but no overall rule applies. Some data are fed into the PACER executive in a prescribed format; for example, the information flow diagram is fed in as the process matrix. The format of some inputs can be chosen in various ways. For example, the equipment data can be fed in through the equipment lists or can be programmed into the subroutine of the Unit Computation. The usual methods for the input data transmission are summarized in Table 4.5. Detailed descriptions of the mechanics of this transmission are given in Chapters 4 and 5.

TABLE 4.5
INPUTS TO A SIMULATION

Information input	Usually fed into programs	Topic discussed in
Flow diagram	Process matrix	Sec. 4.3
Operating data	SN list	Sec. 4.3
Equipment data	EN list	Chapter 5
Physical properties	—	Chapter 5
Cost data	ENC list	Chapter 5

The *outputs* from the simulation depend on the needs and on the criteria established to evaluate the needs. Normally, the most significant outputs are the calculated costs and calculated operating data. The costs usually are reported through a costing unit computation or a cost list whereas the stream list records all the stream variables. Occasionally, specific calculated information about the behavior of one unit can be reported in the equipment list.

The *constraints* range from the deceptively simple, such as water cannot exist as a liquid if the temperature and pressure are such that only a vapor should exist; or mass, mole, volume, weight fractions are between 0 and 1, to the realistic environmental constraints summarized in Sec. 4.1. Normally, these are handled within individual Unit Computations.

The last part is to identify the *significant variables* from a sensitivity analysis of a simple simulation. Normally, a simple mass balance can provide the necessary information. However, this selection depends on the process. The important principle is to start simply, obtain a feel for the system, and gradually increase the accuracy of the models as necessary.

To perform the approximate simulation, the information flow diagram must first be devised and then the process matrix. To start, an information flow diagram is drawn that exactly parallels the process flow diagram shown in Fig. 4.1. Note that such physically insignificant-looking equipment as a pipe tee or cross becomes a MIXER Unit Computation. Then the information flow diagram is changed to match the simulation purpose. For the acid plant, only mass balances are considered first so that the function of each equipment unit is to change mass flows and compositions by simple models such as those shown in Table 4.3. Each Unit Computation is given a mnemonic name related to the function the calculations perform. For example, a reactor bed Unit Computation could be called REACT. Successively more complex models of any one type of Unit Computation are distinguished by a numerical suffix to the mnemonic name, for example, REACT1, REACT2, REACT3, etc.

All equipment that does not alter the mass flows and any superfluous recycle lines are removed at this simple simulation stage. The resultant information flow diagram is shown in Fig. 4.5. It is worth noting that if the purpose of the simulation is to study the steam production, the simple simulation of the mass balance is trivial.

Each Unit Computation and each stream on the information flow diagram is arbitrarily assigned a number. After the streams and simple Unit Computations have been identified, the process matrix can be set up. For example, as shown in Fig. 4.5, the information in streams 116 and 115 enters the REACT1 101 with stream 102 leaving. Therefore the process matrix should include the entry 101 REACT1 + 116, +115, −102. The complete process mairix is shown in Table 4.6.

The next problem is to obtain reasonable estimates from existing plant data for all the conversions, splits, exchanges, and separations that occur in the plant. These data are supplied to the appropriate locations in the EN list for the simple Unit Computations listed in Table 4.3. The operating data available for two production capacities are given in Table 4.7.

The simple simulation suggested that it was essential to know the flows in streams 108 and 105. Furthermore it indicated that for this particular

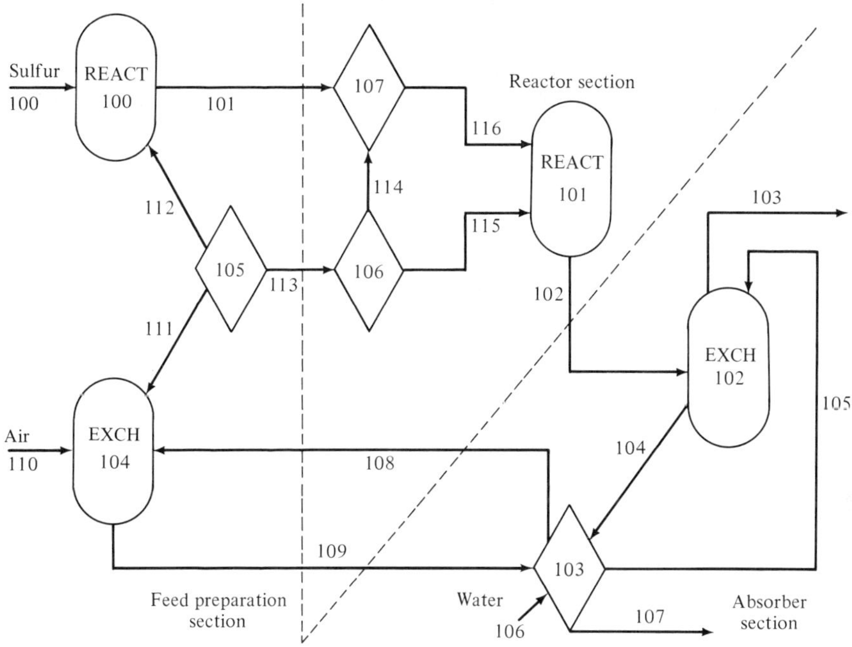

Figure 4.5. Information flow diagram for approximate mass balance simulation.

process an energy balance was very important. Hence the simple first simulation was useful to start the problem, and it indicated some flows that had to be measured.

TABLE 4.6
PROCESS MATRIX FOR THE PRELIMINARY SIMPLE SIMULATION

Unit number	Unit Computation name	Streams					
100	REACT1	100	112	−101			
101	REACT1	116	115	−102			
102	EXCH01	102	105	−103	−104		
103	MIXER1	104	109	106	−105	−108	−107
104	EXCH01	110	108	−111	−109		
105	MIXER1	111	−112	−113			
106	MIXER1	113	−114	−115			
107	MIXER1	114	101	−116			

TABLE 4.7
OPERATING DATA FOR THE SULFURIC ACID PLANT OWNED BY CANADIAN INDUSTRIES LIMITED

Location in acid plant	Pressure, in. H_2O		Temperature, °C	
	316 tons/day March 16, '65	150 tons/day Oct. 20, '65	316 tons/day March 16, '65	150 tons/day Oct. 20, '65
Enter blower	$-5\frac{3}{4}$	-1	0	14
Enter dry tower	$95\frac{1}{2}$	$28\frac{1}{2}$	29.2	
Leave dry tower		—	60	—
Enter sulfur burner	$89\frac{1}{2}$	$27\frac{1}{4}$		
Leave sulfur burner	$84\frac{1}{4}$	$25\frac{3}{4}$	972	948
Leave boiler #1	78	$24\frac{3}{4}$	424	293
Enter filter	76	$24\frac{1}{4}$	434	427
Leave filter	$67\frac{3}{4}$	$22\frac{1}{4}$	—	—
Enter pass #1	$66\frac{3}{4}$	22	426	405
	$60\frac{3}{4}$	$19\frac{3}{4}$	476	476
Leave pass #1	$60\frac{1}{4}$	$17\frac{3}{4}$	—	—
	$56\frac{1}{4}$		612	603
Air enter pass #2	$46\frac{1}{2}$	$15\frac{3}{4}$	450	440
Leave pass #2	$37\frac{1}{4}$	$12\frac{1}{4}$	510	491
Air enter pass #3	$26\frac{1}{4}$	12	439	434
Leave pass #3	29	$9\frac{3}{4}$	447	439
Enter pass #4	$28\frac{1}{4}$	$9\frac{1}{2}$	422	401
Leave pass #4	$19\frac{3}{4}$	4	426	400
Enter economizer	$16\frac{3}{4}$	$3\frac{1}{4}$	—	—
Leave economizer	7	$1\frac{1}{4}$	259	198
Enter 98 absorber	$6\frac{1}{2}$	—	240	—
Leave 98 absorber	$4\frac{3}{4}$	$\frac{1}{2}$	—	—
Pressure drop (knock out)	1	—	—	—
Pressure drop demister	$3\frac{3}{8}$	—	—	—
Leave burner			971	948
Leave boiler			423	294
Enter hot gas filter			434	430
Enter "A" layer #1			425	405
Leave "A" layer #1			476	476
Leave "B" layer #1			612	603
Enter layer #2			438	440
Leave layer #2			500	491
Enter layer #3			439	434
Leave layer #3			448	439
Enter layer #4			423	401
Leave layer #4			427	400
Leave economizer gas			259	198

In summary, the selection of the variables depends on the need and the criteria. Once the variables have been chosen, reasonable units and methods of reporting the information are selected. The variables are then assigned to elements in the SN list. A feel for other important system parameters and

TABLE 4.7 (CONT.)

Parameter	316 tons/day March 16, '65	150 tons/day Oct. 20, '65
Blower rpm	4225	2400
% SO_2 after boiler #1	10.3	10.3
% SO_2 into reactor	9.5	8.5
% SO_2 out stack	0.25	0.19
% Conversion	97.55	98.0
Stack appearance	good	good
Boiler #1 bypass valve	28° open	32° open
Boiler #2 bypass valve	closed	$\frac{3}{4}$ open
Superheater #1 valve	full open	full open
Superheater #2 valve	$\frac{1}{2}$ open	$\frac{1}{2}$ open
Saturated steam bypass	closed	closed
Sulfur consumption	212,550 lb/day	
Production rate @ 100% H_2SO_4	632,250 lb/day	
Temperature of sulfur in pit, °C	140	140
Temperature of sulfur entering burner, °C	140	133
Saturated steam, psig	338	325
Superheated steam, psig	322	319
Superheated steam, °C	307	315
110 psig steam to plant, °C	254	285
Rate of flow to plant, lbs/hr	31,800	15,400
Feed water entering economizer	105	104
Feed water leaving economizer	208	213
Temperature of acid entering acid absorber, °C	82	80
Temperature of acid leaving acid absorber, °C	120	92
Strength leaving acid absorber	99.17%	99.1%
Acid level tank-o-meter, in.	$8\frac{1}{4}$	8.2
Temperature of acid entering dry tower, °C	68	60
Temperature of acid leaving dry tower, °C	66	—
Acid level tank-o-meter, in.	7	7
Temperature of acid leaving pumps, °C (66°Bé)	110	98

additional data to be collected can be obtained from an approximate simulation. (The importance of the approximate simulation in other process problems is discussed in the problems in this chapter and in Chapter 13.)

Collecting information to obtain the process matrix

It is now time to leave the preliminary simulation and devise a simple version of the information flow diagram to answer the needs of the problem. From the preliminary study it can be seen that the energy balances are important.

This more sophisticated information flow diagram includes only equipment for those calculations that affect the stream variables. For example, because the hot gas filter changes only one significant variable, the pressure,

TABLE 4.8
FINAL PROCESS MATRIX TO SATISFY REQUIREMENTS FOR C.I.L. ACID PLANT

Unit number	Unit Computation name	Stream numbers				
1	BLOW1	1	65	−2	−66	0
3	DRIES1	2	39	−4	−40	0
4	MIXER3	4	−5	−25	−26	0
5	BURN2	5	7	−6	0	0
6	MIXER3	6	−8	−27	0	0
7	BOILS1	8	56	−9	−54	0
8	MIXER3	9	26	27	−10	−70
9	MIXER3	25	−28	−29	−30	0
10	REACT4	10	−11	0	0	0
11	BOILS1	11	57	−12	−55	0
12	MIXER3	12	28	−13	0	0
13	REACT4	13	−14	0	0	0
14	COOL4	14	61	−15	−63	0
15	MIXER3	15	29	−16	0	0
16	REACT4	16	−17	0	0	0
17	COOL4	17	62	−18	−64	0
18	MIXER3	18	30	−19	0	0
19	REACT4	19	−20	0	0	0
20	COOL4	20	52	−21	−53	0
21	MIXER3	21	−22	−48	0	0
22	MIXER3	22	50	−23	0	0
23	SORB2	23	33	−24	−34	0
24	MIXER4	41	34	40	42	−31
25	PUMPS1	31	−32	−35	0	0
26	COOL3	32	−33	0	0	0
27	COOL3	35	−36	0	0	0
28	PUMPS1	36	−37	−38	−39	0
29	COST2	37	−69	0	0	0
31	MIXER3	38	43	−45	0	0
32	COOL3	48	−49	0	0	0
33	SORB2	49	45	−50	−46	0
34	COOL3	46	−47	0	0	0
35	PUMPS1	47	−44	−43	−42	0
36	PUMPS1	51	−52	0	0	0
37	DRUM1	53	54	55	−60	−58
38	PUMPS1	58	−59	−56	−57	0
39	MIXER3	60	−61	−62	0	0
40	MIXER3	63	64	−65	0	0
41	SETSP1	59	−67	0	0	0
42	PTRACE	67	−68	0	0	0
43	PRESEQ	68	−69	0	0	0
44	CONT01	69	−51	0	0	0

this calculation can be combined with some adjacent Unit Computation, resulting in an overall simplification in information handling. That is, as far as the information flow diagram is concerned, the hot gas filter does not exist. (If the particle size distribution had been a significant variable, then the hot gas filter would have been a Unit Computation because it would have been cumbersome to incorporate its effect on both pressure and percent solids into another Unit Computation.) For this simulation, the demister (24), the dust filter (8), and pressure drop in the pipelines added physical complications that were not considered in the simulation because they affected a single variable and the effect could be easily included in another Unit Computation.

A second consideration in selecting the Unit Computations to be included in the information flow diagram is the function of the equipment unit. It is useful to separate the units according to function. For the variables of the acid plant and in particular because of the importance of the energy balances, it is easier to divide the reactor into an integrated collection of MIXERS, COOLERS, and individual REACTOR beds. The dilution tank is a MIXER and a PUMP. This is a separation of one equipment unit into several Unit Computations according to function, and will be elaborated on in Chapter 5.

Furthermore, because the Unit Computations are *information modifiers*, some Unit Computations are not related to a physical piece of equipment; for example, a useful calculation is to add the operating costs for all the units, or to set variables at different locations throughout the information network. Such calculations are included in special Unit Computations on the information flow diagram. For convenience this collection of Unit Computations is called the supervisory section.

Information can flow in the streams in any direction. In the acid plant the pressure moves backwards through the system because the stack pressure of the gas leaving the absorbing tower is known and the pressure at the blower exit is unknown, whereas the temperature and concentration information flows forward through the system in the same direction as the physical material.

Figure 4.6 is the information flow diagram for the acid plant of Fig. 4.1, subject to the criteria and variables chosen for the system. This information flow diagram shows the simplest form of Unit Computations and expands the flow diagram of Fig. 4.5 to include the energy balance. Note that extensive use is made of a Unit Computation called SETST1. This merely sets the temperature in any line. The use of this model eliminates the need for complex heat exchange models and usually eliminates some recycle that complicates the system at the beginning. Technically more sophisticated Unit Computations are gradually substituted for the simple Unit Computations.

This progressive substitution yielded the information flow diagram shown in Fig. 4.7. This later arrangement satisfied the purpose of the simula-

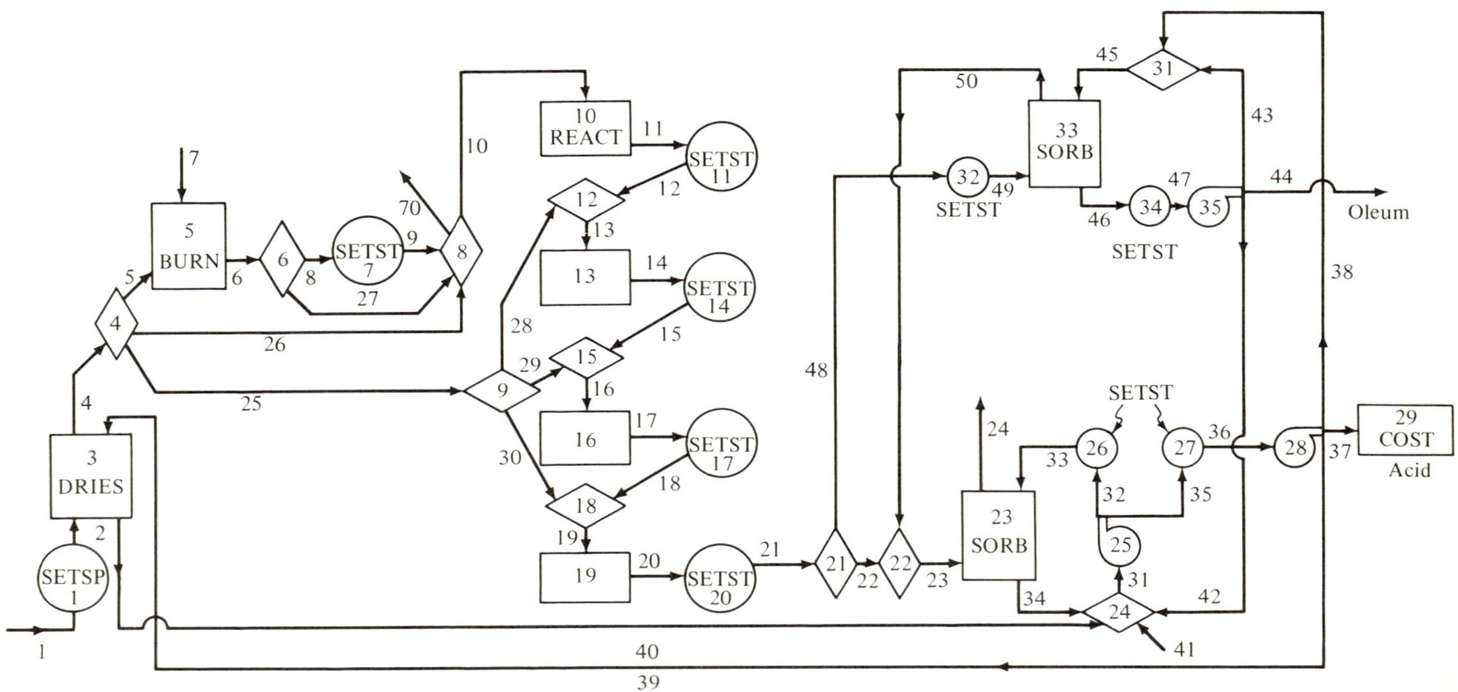

Figure 4.6. Preliminary information flow diagram for the sulfuric acid plant.

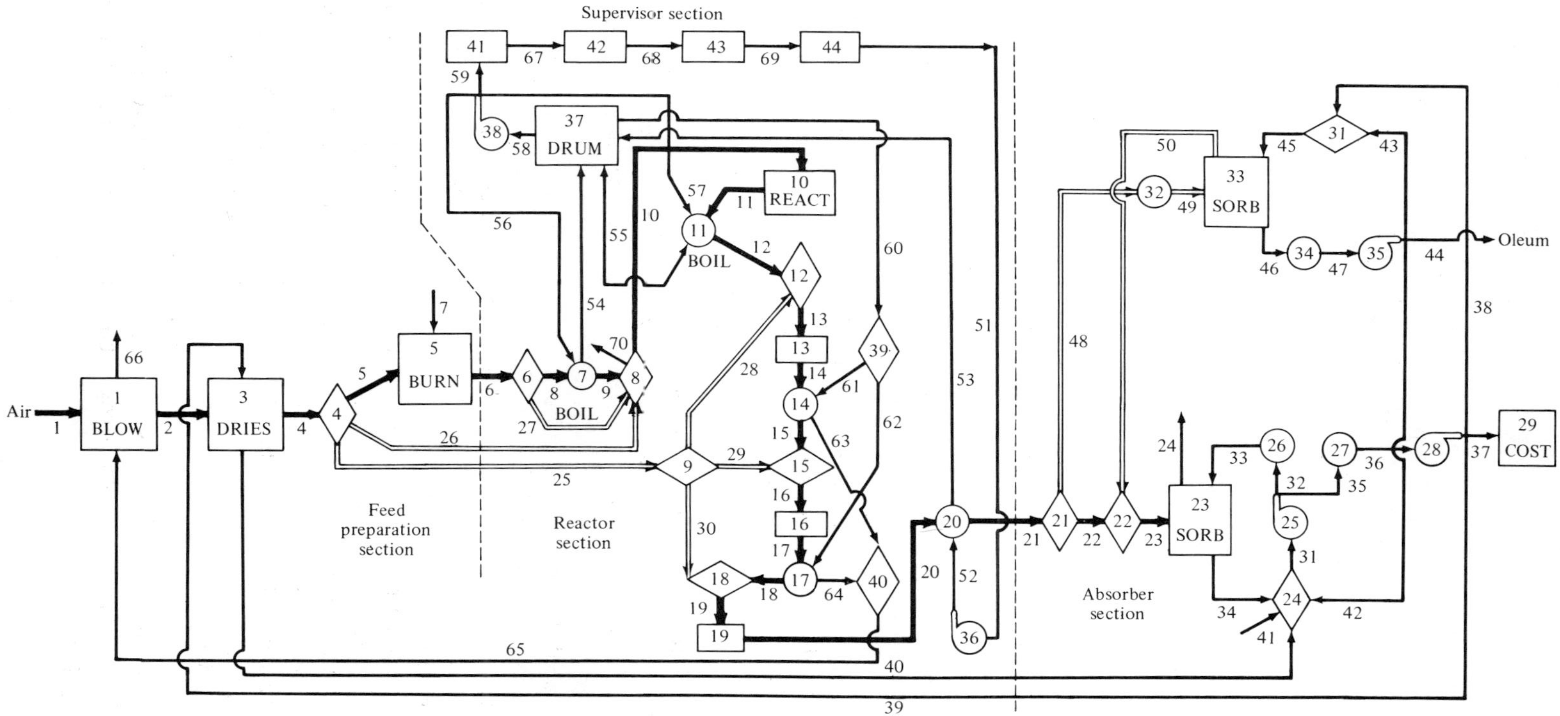

Figure 4.7. Final information flow diagram for the sulfuric acid plant.

tion. Thirteen mixer-splitters, a reactor of four beds, (10, 13, 16, 19) a dilution tank of a mixer and a pump (24 and 25), and five units in the supervisory section (29, 41, 42, 43, 44)—these are the Unit Computations for the information flow diagram.

A process matrix for the information flow diagram of Fig. 4.7 is shown in Table 4.8.

Collecting additional corroborating plant data

Two forms of additional data are needed: those required to complete mass and energy balances for the basic operating conditions, and those for conditions different from the basic conditions.

The importance of these additional plant data should not be underestimated, and simulations that lack sufficient corroborating data should be treated with skepticism. Plant data are often difficult to obtain, but if any simulation work is anticipated for future plants, sufficient provision must be made in the design of new plants for the required instrumentation.

The accuracy of the measurement method should be checked, and an evaluation of the location of the sensing elements should be made.

A careful study of the important mass balances around the feed preparation to the reactor section in the plant being studied indicated that seven additional Pitot traverse points were needed. More data were preferred because some flows (especially around boiler #1 bypass) had to be estimated by difference, but it was physically extremely difficult, if not impossible, to do this because special allowances were not made in the original design. The accuracy of the resultant flow rate predictions varied between 5% and 30%. The Pitot traverses were difficult to interpret because of back flow and skewed profiles.

Detailed mass balances around the absorber-drier system indicated that the flows of acid needed to be known. A radioactive manganese (Mn^{56}) tracer technique was used to measure the flow rates of the hot concentrated liquid sulfuric acid flowing to the absorber-drier system. Reproducible and accurate results were obtained.

Useful corroborating data would be the concentrations of O_2, SO_2, and SO_3 in between the beds in the reactor. Great care and effort were spent using wet chemistry and gravimetric determinations to analyze the results. However, reliable data were not obtained.

This case history of the difficulties encountered for the acid plant is given here to emphasize the difficulty in obtaining reliable, accurate corroborating data.

The reliable data obtained from the additional plant measurements are summarized in Table 4.9. Routine data for half-capacity operation were available to give some knowledge about plant behavior at conditions other

TABLE 4.9
ADDITIONAL DATA MEASURED FOR 316 TON PER DAY RATE

(a) Summary of air flow measurements determined from Pitot tube traverse

	Flow rate		Temp., °C	Pressure, psig
	lb moles/hr	scfm		
Blower exit	3,370 {+ 10%, − 0%}	21,300 {+10%, − 0%}	29.2	1.23
Total secondary air	430 ± 20%	2,720	61	1.195
Dilution to boiler #2 exit	142 ± 5%	900	58	1.190
Dilution to reactor	211 ± 5%	1,330	58	1.182
Absorber vent	3,820 ± 30%	24,200	83	1.01
Net boiler #1 bypass calculated by difference	77 ±130%	—	—	—

(b) Recirculation acid rates determined from radioactive tracer measurements

Acid to absorber		880,000 lb/hr
Acid to drier	Low flow rate	357,000 lb/hr
	High flow rate	392,000 lb/hr

than the standard 316 ton per day data. These are later referred to as the 150 ton per day operation for October 20, 1965. The data were checked for consistency and it was found that the major inconsistencies occurred in establishing the total gas flow rate of the cold dry air that bypassed the burner, the gas flow rates to the reactor bed, and the concentrations between the bed. Some data therefore had to be accepted as being 100% accurate; the inconsistent data were rejected. The data most confidently accepted were the total sulfur input, the gas concentrations at the exit of boiler #1 and in the feed to the reactor, the air to the reactor beds and the inter-bed temperatures. Some details as to why these were chosen will be presented in subsequent chapters. A brief summary of the data to illustrate the inconsistencies is given in Fig. 4.8. This type of consistency check and establishment of the "most reliable data" is an inevitable task in establishing the operating data that is to be used to corroborate the simulation models. Other data besides those mentioned were also established as key check data, but the details are not essential here.

4.4 Summary

A description was given of the contact sulfuric acid plant used in this text as an example of a plant to be simulated. This description included details of the flow diagram, the sizes of the major pieces of equipment, and some

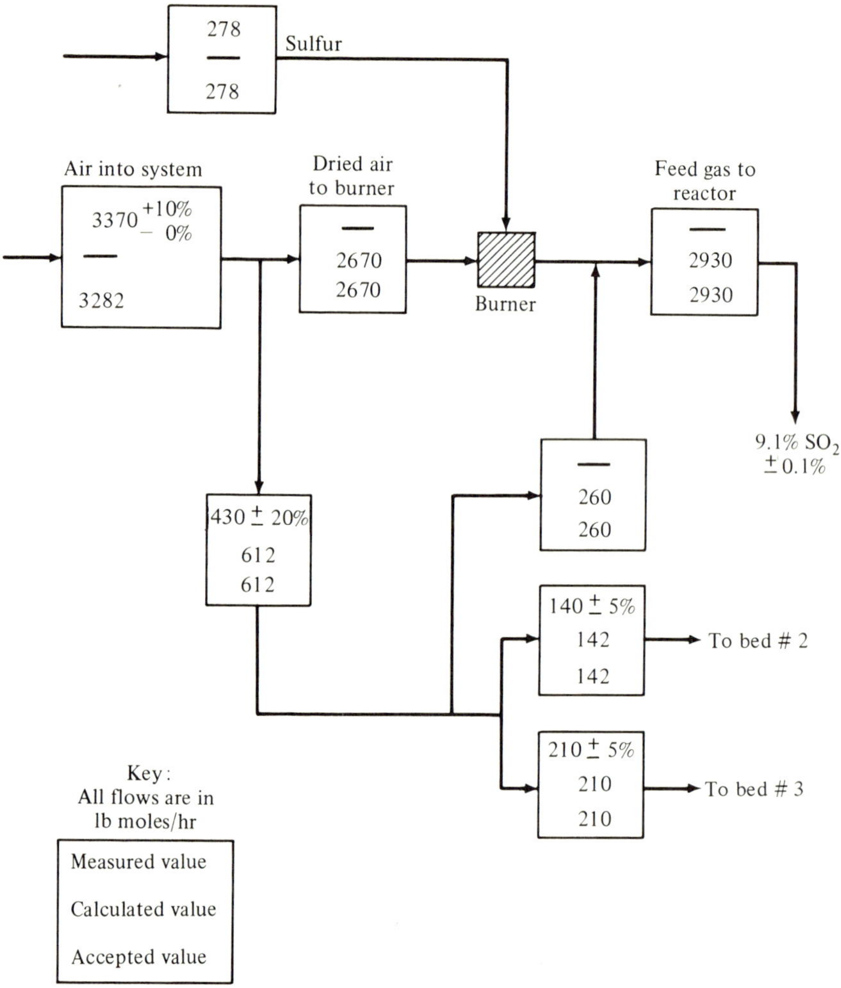

Figure 4.8. Comparison of data measured and calculated in feed preparation and reactor.

pertinent operating conditions. Typical economic conditions and some of the constraints and specifications were presented.

It is useful (before the details are given of the example of PACER as a simulation technique applied to an acid plant) to see the overall strategy for solving the problem. Furthermore, it is beneficial to realize where simulation fits into the strategy for improving plant operation. First, the stategies for problem solving, for design, and for plant improvement follow a similar four-step procedure of define, plan, carry out, and look back. Simulation is an important part of the definition step in the strategy for plant improvement.

The important stages in the definition step are: define the need or questions to be answered, select criteria to judge when the questions have been answered, choose the variables, collect information, devise models or Unit Computations to simulate information-changing units on the information flow diagram, and simulate sections of the plant, and the complete plant. Two decisions that influence the strategy are the level of modeling chosen and the method of handling the physical properties. The time and personnel available need to be carefully allocated. As a general principle, it is advisable to start simply, say with a simple mass balance simulation, and gradually increase the technical sophistication as the need demands.

PROBLEMS

4.1. What major differences would occur between summer and winter operation of this plant? How does this affect the simulation?

4.2. To understand the structure of the contact sulfuric acid process, answer the following questions.

About the reactions

(a) What are the four main reactions?

(b) What can we establish from thermodynamic considerations about each reaction? Factors we should bear in mind are whether the reaction is exothermic or endothermic, the quantity of energy involved, equilibrium conversion, Le Chatelier's principle, effect of state variables on conversion, the temperatures, pressures and concentrations for all reactors, and constraints on operation.

(c) Have we learned enough about these reactions from thermodynamics or should we consider the rate? If we elect to find rate data, where would we look and what information would we be looking for? Assume the process uses vanadium pentoxide as the catalyst.

(d) Consider the reaction $SO_2 + \frac{1}{2}O_2 \rightleftharpoons SO_3$.

(i) Plot on linear-linear paper mol percent conversion to SO_3 versus temperature (°C) for equilibrium conversion assuming initial sulfur dioxide and oxygen concentration of 10.0%. The relationship is

$$K_p = \frac{x(b - x/2)^{1/2}}{(1 - x)(a - x/2)^{1/2}\pi^{1/2}} \text{ (atmos)}^{-1/2}$$

where K_p = equilibrium constant
x = moles SO_3 formed
a = moles O_2 initially present
b = total moles initially present
π = total pressure, atmos

(ii) Common catalysts used are platinum or vanadium pentoxide. Assume this plant uses the latter. To prevent excessive degradation of the catalyst, the temperature should not exceed 600°C [Duecker and West (1959), p. 184]. Burkhart (1968) suggests that the useful range of the catalyst is 430° to 600°C although some are active from 404° to 650°C. Include this information on the graph if you think it is important. What is the significance of the lowest temperature?

(iii) A boiler is used between the sulfur burner and the reactor. Why is the gas cooled before it enters the reactor? Suggest a reasonable inlet temperature for the gas entering the reactor. What information is most useful in helping you answer the question?

(iv) There are many routes that could be followed to go from an inlet of 0% SO_3 to a higher concentration of SO_3. On the equilibrium graph, plot lines of adiabatic reaction, isothermal reaction, and cooling with no reaction. On the equilibrium graph draw the lines for at least six different combinations of reaction and cooling; for example, alternating sequences of adiabatic reaction and cooling with no reaction. Sketch the physical layout of both the catalyst beds and the coolers and the gas flow for each of these operations. For the adiabatic reaction, assume gases enter the reactor at 400°C and calculate the temperature of the gas leaving a catalyst bed in which equilibrium has been achieved. [See Hougen, Watson, and Ragatz, (1954) Vol. I, p. 352].

(v) A sketch of a commercial reactor is given in Fig. 9.1. Plot the temperature conversion path assuming the inlet temperature to each bed is 430°C. [See Duecker and West (1959), p. 252, or Shreve (1967), p. 336, for information that may save you calculation time].

(vi) What radically different methods of running the reactor could be considered?

About the processing

(e) Consider the drier. Sulfur trioxide and sulfuric acid vapor in contact with water vapor rapidly form mists of $\frac{1}{2}\mu$ to 10μ diameter droplets. The mist is hazardous, corrosive, unsightly, and difficult to remove [see Duecker and West (1959), p. 353]. From a materials of construction viewpoint the corrosiveness is the main reason why every attempt must be made to prevent mist from forming inside the process. The air entering the sulfur burner is dried by contacting the air with sulfuric acid. The desirable moisture content entering the burner is ≤ 5 mg water per standard ft^3 air [Industrial and Engineering Chemistry (1950), p. 10]. Does this put any constraints on the operation of the acid drier?

(f) To prevent mist formation within the system, we must ensure that no water vapor contacts the SO_2.

(i) Identify means by which water vapor could enter the gas system.

(ii) Since hydrocarbons can oxidize to produce water vapor, how much hydrocarbon contamination in the sulfur can be tolerated if a maximum of 10 mg water per standard ft^3 air is permitted? How could this hydrocarbon be removed?

(g) The unit usage of process water per ton of 100% sulfuric acid is reported as:

0.2 ton by Connor (1968)
0.14 ton by Industrial and Engineering Chemistry (1950), p. 13

0.15 ton by Duecker and West (1959), p. 215
From the stoichiometry determine how much process water is needed. Why are these three values different?

(h) How many horsepower are required to supply air to the system if the pressure drop through the complete plant is 50 in. water gauge? What is the best location for the compressor?

(i) What would be the advantages and disadvantages of recycling the exit gas from the absorbers to the sulfur burner?

(j) Sulfur is pumped from the sulfur pits to the sulfur burner in a molten state.
   (i) How much heat is required to maintain the sulfur pits if solid sulfur is supplied?
   (ii) How much sulfur is lost if the pits are maintained at 135° to 140°C by manual adjustment [Industrial and Engineering Chemistry (1950), p. 9]?
   (iii) The viscosity of molten sulfur is given in Duecker and West (1959), p. 463. What does this tell us about the operation?
   (iv) The ignition temperature of sulfur is 230°C. Of what significance is this?

(k) If the gas from the sulfur burner is to contain 10.0 mole % SO_2, what is the percentage of excess air and the theoretical adiabatic flame temperature in °C?

(l) How many PCU are theoretically available per ton of 100% sulfuric acid
   (i) From the combustion of sulfur?
   (ii) From the reaction of SO_2 to SO_3?
   (iii) From the absorption of SO_3 in dilute sulfuric acid?
   (iv) From the absorption of water vapor in the drier?

Examples of analysis of structure for other processes are given by King (1967) and by Woods (1969).

4.3. Perform a functional analysis for the major sections of this plant. That is, define the major function of each piece of equipment, generate five alternative ways to do each function, and evaluate these alternatives based on technical feasibility. [See Woods (1968) and Elliott and Owen (1968).]

4.4. The blowdown from the steam drum is not shown on Fig. 4.1. Should it be included in the information flow diagram? Why and under what conditions?

4.5. Calculate the total amount of heat available from the heat of combustion of sulfur and the heat of reaction to produce SO_3. How much of this heat is eventually used in producing steam? Does this suggest an area for probable optimization studies? How does this affect the choice of stream variable?

4.6. Develop an overall algorithm or flow chart of the steps and decisions needed to produce a simulation for *any* process based on the principles outlined in this chapter. While you study succeeding chapters, polish and modify the algorithm.

4.7. Common units of measurement for water and waste-water treatment processes are mg per liter or parts per million for the concentration of contaminants, and millions of US gals per day for the total flow. However, in the solids disposal portion of the treatment process the units are tons per day or cu ft per day. What should we use as the units for the SN list for a waste treatment process?

4.8. Consider the demands of structuring an SN list for both approximate and rigorous simulations. The variables that can be of interest for water and waste treatment processes are the 24-hour average total volumetric flow rate, the daily and hourly average minimum and maximum volumetric flow rates, the mg per liter of total suspended solids, solid organic carbon, solid nonbiodegradable carbon, solid organic nitrogen, solid organic phosphorus, solid fixed matter, dissolved organic carbon, dissolved nonbiodegradable carbon, solid biochemical oxygen demand, volatile suspended solids, dissolved phosphorous, dissolved fixed matter, dissolved biochemical oxygen demand, alkalinity, concentrations of heavy metals, pesticides, coliforms, and the particle size distribution.
   (a) Are there any variables that should be added to this list?
   (b) Should any be eliminated from the list? Why?
   (c) How can we sequence the list so that a short list of components used for approximate simulations is consistent with a long list required for rigorous simulations?

4.9. In the sequence of the SN list, does it make much difference whether the solid and dissolved components are side by side, separated by a fixed number of elements, or located arbitrarily? An example of part of an SN list is given in Table 4.10.

4.10. In the Bayer process, bauxite is digested in caustic to dissolve alumina; the gangue is filtered off and the alumina trihydrate is precipitated from the filtrate liquor. The alumina trihydrate is washed and calcined to produce

TABLE 4.10
PARTIAL STREAM LIST FOR AN ALUMINA PROCESSING PLANT

Element	Meaning	
10	Fe_2O_3	(solid)
11	P_2O_5	(solid)
12	CaO	(solid)
13	TiO_2	(solid)
14	Fe_2O_3	(dissolved)
15	P_2O_5	(dissolved)
16	CaO	(dissolved)
17	TiO_2	(dissolved)
10	Fe_2O_3	(solid)
11	Fe_2O_3	(dissolved)
12	P_2O_5	(solid)
13	P_2O_5	(dissolved)
14	CaO	(solid)
15	CaO	(dissolved)
16	TiO_2	(solid)
17	TiO_2	(dissolved)

crystals of alumina. Details of this process are given in Appendix A. In particular, to satisfy the questions asked for a given simulation, the following variables were selected: water, Na_2CO_3, NaOH, $3(Na_2O \cdot Al_2O_3 \cdot 2SiO_2) \cdot 0.6\,Na_2O \cdot Al_2O_3 \cdot 0.45\,Na_2CO_3$, $Al_2O_3 \cdot 3H_2O$, $Al_2O_3 \cdot 2SiO_2 \cdot 2H_2O$, Fe_2O_3, CaO, TiO_2, $CaCO_3$, carbon compounds, unknown inert solid, $NaAlO_2$, Fe_2O_3, Na_2SO_4, $Na_2C_2O_4$.

Some of these compounds are insoluble, some are soluble, and some are distributed between the phases. Important plant units of measurement were weight percent of solids, the ratio of Al_2O_3/Na_2CO_3, equivalents of NaOH, cfh, cfm, gm per liter, grain per 10 ft^3, and causticity. Causticity is defined as the weight ratio of sodium hydroxide in liquor expressed as sodium carbonate to the total sodium carbonate equivalent in the liquor as determined from a titration with acid. What units of measurement would you select for the variables in the SN list?

4.11. Consider the problem of developing an information flow diagram for the filtration section of the Bayer process as shown in Fig. 4.9. Mother liquor containing dissolved alumina and the insoluble gangue is pumped to a battery of pressure-leaf filter presses operating at constant pressure for the complete wash-drain-dump cycle. The total cycle is about 240 min with about 200 min of filtration per cycle. This amounts to about 140 dumps per 8-hour cycle. The filtrate from the first 3 to 5 min is reprocessed. The clear filtrate of about 10 to 12 cfm per cycle per press goes to the precipitation circuit. At the end of the filtration cycle, steam is injected into the enclosed filter, and the condensate and the pregnant liquor slurry drain from the press, and this slurry is reprocessed. Then wash liquor flows through the cake for 5 to 15 min; the time depends upon the demands downstream.

Finally, the press is opened and the solids are washed free of the leaves by manually hosing the leaves with water.

How can we use the simple models given in Table 4.3 to represent this time-dependent cyclic process for a steady-state simulation?

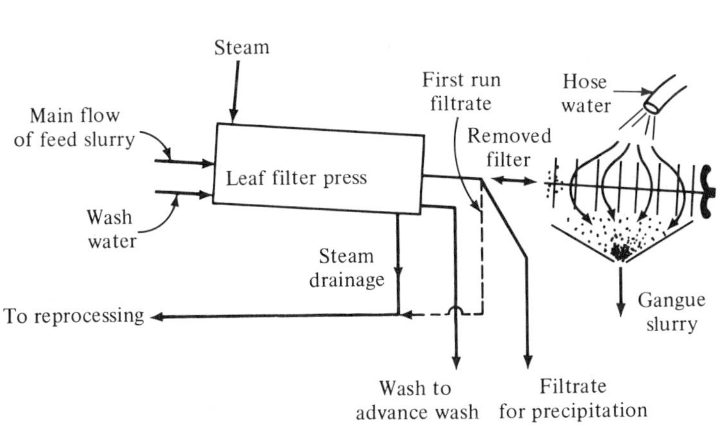

Figure 4.9. Process flow diagram of a filtration process.

Sample Solution:

The whole process could be represented on a time-average basis by a single Unit Computation MIXER1 as shown in Fig. 4.10. However, this model would be limited in its applicability for more rigorous simulations.

Simple models SEPA03 and MIXER1 from Table 4.3 could be used as shown in Fig. 4.11, 4.12, and 4.13 to yield a simulation system that is easier to extend to more accurate simulations. Hence, it is preferable to reject the single MIXER1 system shown in Fig. 4.10 in preference to the schemes given in Figs. 4.11, 4.12, or 4.13.

Some care should be taken in arranging the simple models so that their eventual replacement by rigorous models least alters the information flow diagram. Although we do not discuss modeling until Chapter 6, the way we *plan* to model the filtration system affects which of the three schemes shown in Fig. 4.11, 4.12, and 4.13 we select. We return to this problem after we have completed Chapter 6.

4.12. Assume the operating procedure described in Prob. 4.11 is changed so that

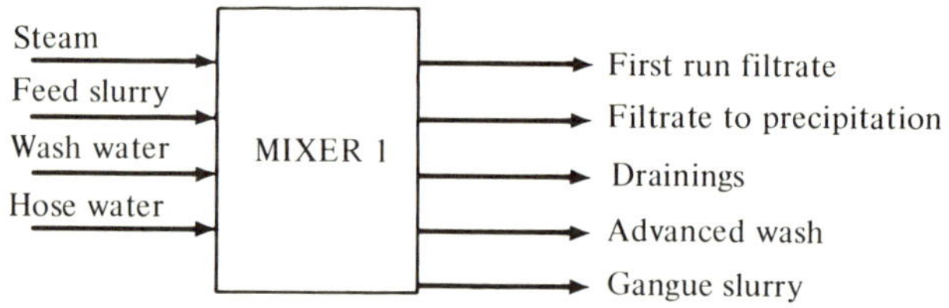

Figure 4.10. A simple filtration Unit Computation.

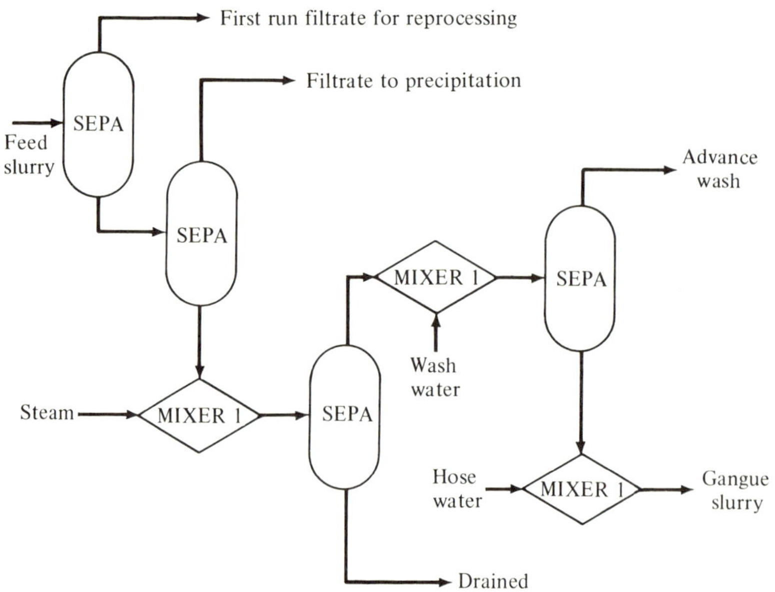

Figure 4.11. An information flow diagram for filtration.

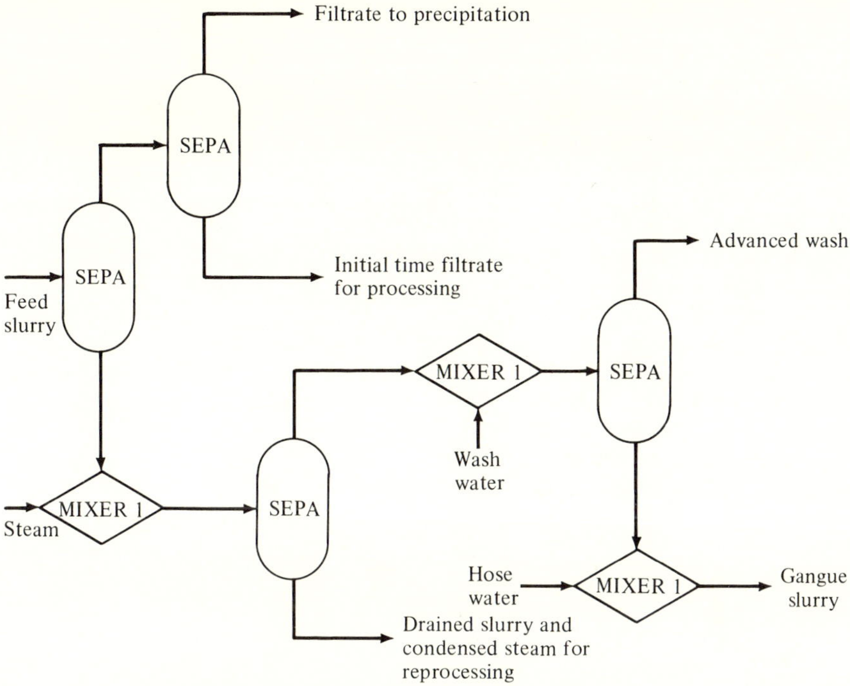

Figure 4.12. An information flow diagram for filtration.

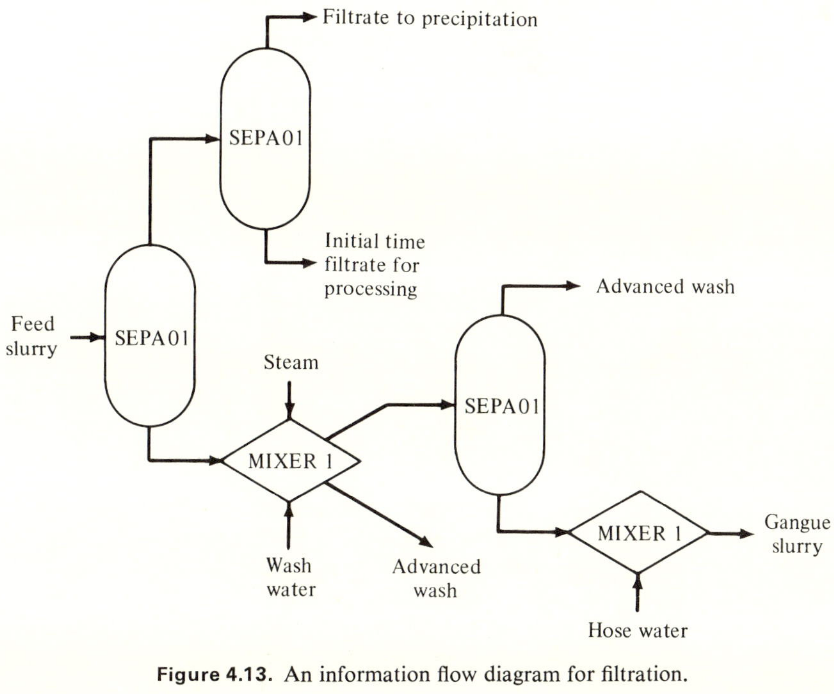

Figure 4.13. An information flow diagram for filtration.

after the backwash is complete, blow-off steam is sent into the closed press, the condensed steam and interstitial fluid are drained, and then the press is opened so that the solids are hosed off the leaves. Modify the information flow diagrams in Figs. 4.11, 4.12, and 4.13 to account for this new operating procedure.

4.13. Figure 4.14 shows part of the heat exchange-digestion circuit of the Bayer

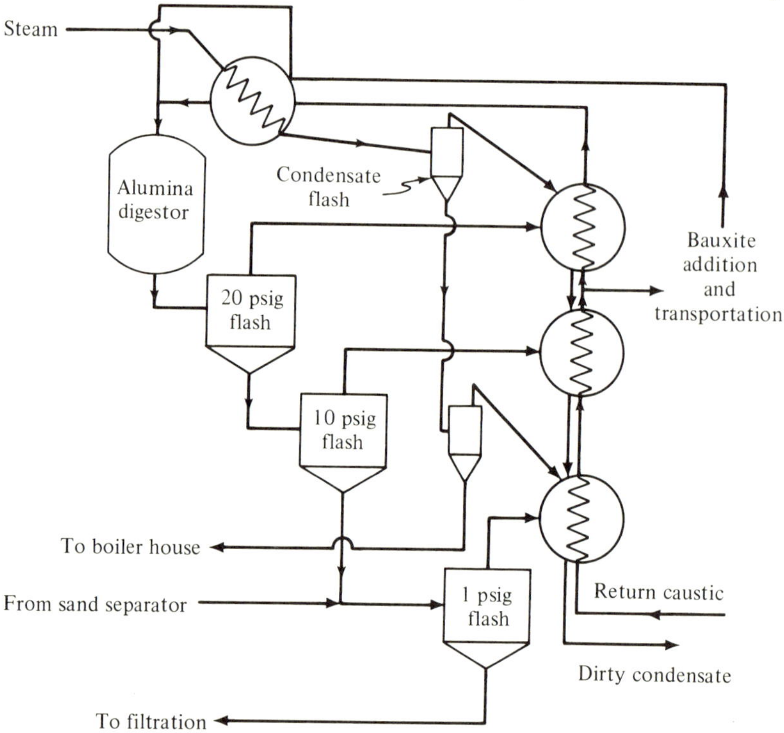

Figure 4.14. Heat exchange-digestion circuit.

process. Details are given in Appendix A. Convert this into an information flow diagram to be used for

(a) A simple mass balance in terms of the models in Table 4.3;

(b) A simple mass balance including SETST models described in Prob. 3.3 to represent the first step in the more technically sophisticated simulation where temperature effects are included in the models but heat balances are not done.

4.14. A flow diagram for the production of alkylate is given in Appendix A. Convert this into an information flow diagram to answer the same questions as posed in Prob. 4.13 using the simple models listed in Table 4.3.

4.15. We have devised a unit converter to convert internal units of measurement, say °C, to external units desired by the engineers, say °F. The model uses the elements in the EN list to identify those streams whose answers are to be reported in the new units. The EN list has 8 elements that can be so specified. We want to use this model UCONV1 by placing it in the process matrix as the Unit Computation following the converged answers.

(a) If the process matrix is shown in Table 4.8, what would be the last entry in the new process matrix to include UCONV1?

(b) We have 100 streams in the process; there are only 8 elements in the EN list. We want the results reported in the *new* units for all 100 rather than just $\leqslant 8$ streams. What alternative ways do we have of accomplishing this? Which one do you think would be the easiest?

4.16. Select a suitable SN list for a simulation of a polymer reactor system which involves bulk polymerization and an initiator. The polymer is soluble in the monomer. Some background information is given by Levenspiel (1962), Chapter 8, and by Hui and Hamielec (1969).

(a) For an approximate simulation, use the moments of the molecular weight distribution for characterizing the polymer produced. How can these become part of the complete SN list?

(b) For a comprehensive simulation consider the flow of polymer species of various chain length and branching. For copolymerization, chemical composition should also be included. How many species need to be accounted for to satisfactorily approximate the molecular weight distribution? How can the length of the SN list be reduced and yet still carry the molecular weight distribution information?

(c) What units are appropriate?

Two Key Decisions About Modeling and Physical Properties

5

Before Unit Computations themselves are created, some guidelines on the decisions to be made are needed. These decisions appear simple, but they should not be made without understanding the ramifications. There are two basic questions to decide:

1. How accurate, how general, and how reliable for extrapolation should the Unit Computations be made?
2. Should the physical property information be supplied within each Unit Computation or from a physical property package that will be developed in addition to the Unit Computations?

Before examining these decisions below, the collection of physical property, equipment, and cost information and their implications on the use of the EN list are discussed. Finally, the problem of the organization of resources to handle the overall simulation problem is briefly considered.

5.1 Collection of the Input Data

Physical properties, equipment parameters (such as efficiencies, sizes, and configurations of equipment), and cost information are needed for most simulation problems.

Obtaining acceptable physical properties

Physical properties of pure components or mixtures can be measured in the laboratory or in the plant, found in the literature, or estimated from known data. The method chosen depends upon the time available and the accuracy required in the final answer.

The ability to measure the data depends primarily on the type of equipment readily available, the manpower, the peculiarities of the material in question (for example, ketene is a difficult substance to handle), and the existing system of information retrieval in the company in which the simulation is being carried out. References for locating data are given in a list at the end of this chapter in order of decreasing usefulness.

Data found in the literature are not necessarily reliable because the accuracy and interpretation usually depend on the experimental method used, because the data may not be given for the conditions of interest so that some extrapolation is needed, or because typographical errors frequently occur.

For example, the surface tension may be needed for the simulation of boiling phenomena or for mass transfer in a packed column. The value of the surface tension reported in the literature depends on whether an oscillating jet technique or a Du Noüy ring Tensiometer was used. The physical data accepted should have been measured with the technique that most closely resembles the phenomenon under study. The hazard of extrapolating data is illustrated by the following example. Suppose the latent heat of vaporization of acetic acid is needed at 93°C. Data from Perry (1950), plotted on Fig. 5.1(a), indicate that the range of interest is not covered. This function seems to be smooth but an extrapolated value for the latent heat of 102 is about 10% higher than a value reported by Weltner (1955) indicated by △ in Fig. 5.1(b), Ramsay, et al., (1886) indicated by □, and the International Critical Tables indicated by ▽.

Typographical errors are fairly common for physical property data. For example, the latent heat of vaporization of acetic anhydride is given as 92.2 PCU per lb by Perry (1950). This is a typographical error; the accepted value is 66.2 PCU per lb as checked from International Critical Tables and experimental laboratory tests.

The data should also be consistent. If the consistency cannot be checked, only one source of data should be used when possible. For example, vapor-liquid equilibrium data from two sources were used to cover the complete range of composition for the plate-to-plate calculation of a multicomponent distillation column. Anomalous behavior was found in the region of composition where the sources of data overlapped.

Data which can be neither measured nor located can perhaps be estimated. For a mixture, this is normally a two-step process: the properties of the pure components are estimated and then the mixture properties are computed.

TABLE 5.1

CURRENT STATE OF DATA AND CORRELATIONS OF IMPORTANT ENGINEERING PROPERTIES OF FLUIDS
(BY PERMISSION OF R. C. REID AND *Chemical Engineering Progress*)

Property	Condition	Available data			Estimation techniques		
		Usually Sufficient	Could use more	More definitely needed	Satisfactory for engineering use	Fair	Poor; better ones needed
Critical properties	......	–	x	–	x	–	–
Pure component constants (boiling point, freezing point, etc.)	......	x	–	–	–	–	x
Volumetric properties	Gas, pure component	–	x	–	x	–	–
	Gas, mixture	–	–	x	–	x	–
	Liquid, pure component	x	–	–	–	x	–
	Liquid, mixture	–	x	–	–	x	–
Vapor pressures	Pure components	x	–	–	x	–	–
Latent heats	Vaporization, pure component	x	–	–	x	–	–
	Vaporization, mixture	–	–	x	–	–	x
	Fusion, pure component	–	x	–	–	–	x
Heat capacities	Gas, low pressure	–	x	–	–	x	–
Heats of formation	......	x	–	–	x	–	–
Free energies of formation	......	x	–	–	–	–	x
Thermodynamic properties	Gas, high pressure, pure component	x	–	–	–	x	–
	Gas, high pressure, mixture	–	–	x	–	–	x
	Liquid, pure component	–	x	–	–	x	–
	Liquid, mixture	–	–	x	–	–	x

Viscosity	Gas, low pressure, pure component	x	–	–	x	–	–
	Gas, low pressure, mixture	–	x	–	x	–	–
	Gas, high pressure, pure component	–	x	–	–	x	–
	Gas, high pressure, mixture	–	–	x	–	–	x
	Liquid, pure component	x	–	–	–	–	x
	Liquid, mixture	–	–	x	–	–	x
Diffusion coefficient	Gas, low pressure	–	x	–	x	–	–
	Gas, high pressure	–	–	x	–	–	x
	Liquid	–	x	–	–	x	–
Thermal conductivity	Gas, low pressure, pure component	–	x	–	x	–	–
	Gas, low pressure, mixture	–	x	–	x	–	–
	Gas, high pressure, pure component	–	x	–	–	x	–
	Gas, high pressure, mixture	–	–	x	–	–	x
	Liquid, pure component	–	x	–	x	–	–
	Liquid, mixture	–	–	x	–	–	x
Vapor-liquid equilibrium constants	......	x	–	–	–	–	x
Surface properties	Nonaqueous	x	–	–	x	–	–
	Aqueous mixtures	–	x	–	–	x	–

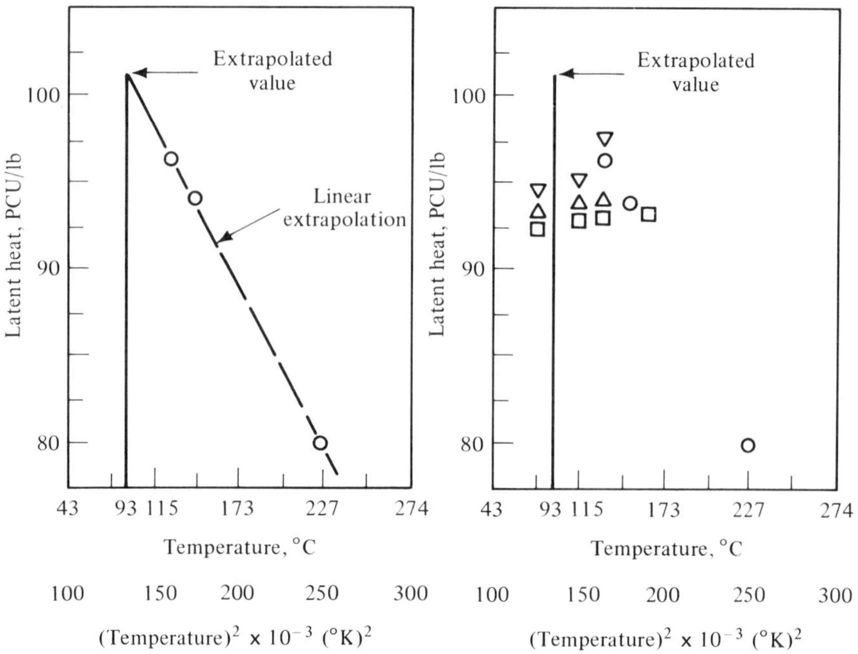

(a) Latent heat of vaporization of acetic acid: experimental data

(b) Latent heat of vaporization of acetic acid: experimental data

Figure 5.1.

Reid (1965) summarizes in Table 5.1 the present availability of data in the literature and the state of estimation techniques for various properties. For more details see Reid and Sherwood (1966). Many papers, monographs, and books have been presented on the estimation techniques available.

To incorporate estimation formulas in a simulation, those techniques should be chosen which require a minimum of storage and computation time and yet supply the desired accuracy. This favors estimation methods which require the same input data (say, critical constants). Finally, no matter how data are obtained, the source should be recorded with the data for future reference.

Three general methods for presenting technical data are graphs, tables, and equations. At present, most digital computers do not accept graphical information so that graphical data must be recast in the form of either tables or equations. Tables are not ideal because the data needed must often be interpolated between discrete values of the independent variable or parameter. Thus, data will ultimately be correlated with an equation. This is a three-step process involving the choice of function, the determination of the constants, and testing the significance.

The form of the function may be suggested by the theory of the phenomenon. The theory may also reduce the number of variables required, often by indicating dimensionless combinations of variables. Once the variables are selected, the data are plotted on linear-linear paper to observe the trends of each variable. A form of the function is then either chosen from a study of the background theory or selected empirically by trial and error plotting or by finite difference analysis. Rowe (1963) has some useful comments to make about selecting the functional form.

The constants can be determined by regression analysis or by finite difference interpolation techniques. Regression analysis is preferable to finite difference polynomials because, in general, the overall residual error is less and the significance of the regression can be tested. Regression analysis programs are readily available at most computation centers.

It is wise to compare the observed values with the values predicted from the correlation. Furthermore, if more than one functional form or method has been used to generate a correlation, the simpler and more significant correlation is preferred. In general, however, the major problems in regressing data are reducing the number of variables and selecting the form of the equation.

Collecting information about the equipment

Two types of information about equipment are needed. These are the physically measurable characteristics and the derived information. The only precaution in collecting the former is to ensure that the blueprints and plant records are correct. The measurable characteristics may include tube sizes, pitch, baffle spacing and type of heat exchanges, types and height of column packing, the number of plates, the horsepower and rpm of the motors, type and characteristics of catalyst, and geometrical configurations and height of interfaces between immiscible liquids.

Derived information about the equipment includes efficiencies, head capacity data for pumps, fans, and blowers, packing porosity, fouling factors, and reliability numbers. To a limited extent such equipment data may be available from correlations. However, it is advisable to corroborate the information from the operating data collected.

Collecting cost information

Since simulations are often used to suggest ways of increasing profits, reliable cost data are needed. Although most of these should be available from within the company, reasonable order-of-magnitude values and sources of information are summarized in Table 5.2.

TABLE 5.2
ORDER-OF-MAGNNITUDE VALUES OF USEFUL COST DATA

Information	Price in 1967	Other sources
Utility*		
Electricity (AC)	\$0.54/100 kwh	
Purchased water for process steam	\$0.20/1000 US gal	Peters (1958), Peters and Timmerhaus (1968)
Pump water for cooling	\$0.02/1000 US gal	
Bituminous coal (7,400 PCU/lb)	\$8.00/ton	Lawrence (1954)
Raw Materials*		
Sulfur	\$40.00/ton	*Oil, Paint, and Drug Reporter*, Happel (1958), Aries and Newton (1955)
Maintenance		
$3\frac{1}{2}\%$ to 8% of fixed capital investment		Lawrence (1954)
Capital Costs		
Update in years by the Marshall-Stevens index		Bauman (1964)
Scale-up for plant size using $C_1/C_2 = (g_1/g_2)^n$ where C = capital cost, g = characteristic parameter, n = value depends on plant and equipment; average values are 0.6 to 0.7		Dryden and Furlow (1966)

* Indices for capital costs, such as the Marshall-Stevens, *Chemical Engineering*, or *Engineering News Record*, must never be used to update material and utility costs.

5.2 The Modeling Decision

A process simulation consists of a set of Unit Computations that represents a process. The characteristics (such as accuracy, generality, and ability to extrapolate) of the simulation of the complete process can be distinguished from the characteristics of the individual Unit Computations that form part of the simulation. The modeling decisions about the complete process may differ from those about individual Unit Computations. For example, there can be a very accurate process simulation for the purpose of minimizing cost, which uses an inaccurate individual Unit Computation to represent an equipment unit whose behavior does not significantly affect the overall cost. That is, the unit itself makes a small contribution to the direct cost, and it does not cause large changes in the cost of other units or streams through changes in any of its equipment parameters or output stream variables. Hence, when discussing accuracy a distinction must be made between the accuracy of the overall simulation and that of an individual Unit Computation representing a unit. In this section decisions are made to define an overall consistent policy that applies to the construction of the Unit Computations. The main policy decisions are:

1. The overall accuracy of the complete simulation;
2. The accuracy of the Unit Computations for each equipment unit;
3. The generality of the individual Unit Computations;
4. The ability to extrapolate the individual Unit Computations.

The decisions are tempered by ability to test the models.

Overall accuracy of the complete simulation

For the complete simulation the overall accuracy is affected by the questions to be answered by the simulation, the time and money available, and the accuracies possible for the key individual Unit Computations.

The questions to be answered by the total plant simulation fix the desired accuracy. This can be achieved if the time and the money are available and if sufficiently accurate models of the sensitive units can be devised. For example, the simulation of the alkylation plant described in Appendix A could not achieve the desired accuracy because there was insufficient background knowledge about the reactor system. Thus, the accuracy of the overall simulation was limited by the accuracy possible for the Unit Computation representing the reactor.

As the project progresses, there must be a reevaluation of the original simulation objectives subject to these constraints of time, money, and accuracy of individual Unit Computations.

Accuracy of the individual Unit Computations

Perhaps the first and the greatest temptation in simulation work is to devise exceedingly accurate models for every equipment unit. Such enthusiasm should be restrained. Since the required accuracy for each unit cannot be known at the beginning of the study, it is wise to start with very simple models such as given in Table 4.3 and do an approximate simulation, as has been described in Sec. 4.3.

The development of more accurate models needed for some equipment units depends on the data available, the level of final simulation desired, and the number of times the model is used in the computation.

Sometimes physical properties, plant corroborating data, or mechanistic information are so inaccurate that accurate models are not warranted because insufficient knowledge is available.

Unit Computations representing much-used calculations should be designed to minimize computing time. Sometimes this means using a comprehensive model to generate a simple statistical model. This example is discussed in Sec. 14.3 for a distillation column encountered in the alkylation plant simulation.

Generality

Generality has two meanings. Models can be general with respect to physical arrangement or with respect to materials handled. For example, Table 5.3 illustrates four levels of generality for each. The model that one might choose to develop could be specific with respect to physical arrangement (level 1) and general within a given chemical compound family (level 3). An example of this would be a model of the condenser on the batch still AX-5 in a resin plant modeled so that the behavior for any hydrocarbon charge could be predicted. General models have several advantages. They require less total storage space because they avoid duplication of similar computations. The general model is adaptable to many processes and units. Development of general models builds a library of models which can, at a moment's notice, be used to simulate virtually any process. However, the advantages of general Unit Computations are achieved by removing equipment and physical property parameters from inside the Unit Computation program. That is, to move from the specific (level 1) to a model independent of equipment configuration (level 2) requires that the equipment parameters be stored outside the Unit Computation program, say, in the EN list. This often requires a long, cumbersome EN list and internal switching to minimize computation time. To make the model independent of the chemicals handled, the physical properties need to be removed from the Unit Computation program. This normally requires the establishment of a physical property block separate from the Unit Computations. Depending on the type of calculation done, this may be awkward and even wasteful. Often the generality depends on the system and the type of unit. For example, REACT models are usually

TABLE 5.3
LEVELS OF GENERALITY

Generality level	With respect to physical arrangement	With respect to materials handled
Specific 1	Feed preheater on column 1 (Size, configuration, stream condition, duty specified)	Feed to column 1 (Compositions and species specified)
2	Any preheater of a given type (Configuration and stream condition, duty specified)	Materials in a process (Range of chemicals specified)
3	Any preheater (Stream condition, duty specified)	All chemicals within a chemical compound family
4 General	Any heat transfer device (Condenser, boiler, furnace heat exchanger, evaporator)	Any chemical

specific with respect to kinetic models, flow configurations, and heat exchange details; MIX and PUMP models can easily be general.

Ability to extrapolate

The usual application of simulation is to predict the behavior of a process under conditions different from the present. This requires that the Unit Computations be accurate enough to be extrapolated to the new conditions. This accuracy at extrapolated conditions is usually easier to obtain from fundamental equations (analytical models) rather than statistical descriptions of the macroscopic responses of the output streams to inputs (statistical models). In between these two extremes is the range of mixed models that depends partly on fundamental concepts and partly on statistical analysis of responses. When there is a choice of using either fundamental equations or statistical correlations to describe the behavior to the required accuracy, the fundamentals should be used.

In summary, the modeling decision concerns accuracy, generality, and

TABLE 5.4
MODELING DECISIONS

Modeling decision	Depends on	Affects
Overall simulation accuracy	Questions to be answered Time available Money available Accuracy possible in key individual Unit Computations	Accuracy required of individual Unit Computations
Accuracy of individual Unit Computations	Accuracy needed to match overall accuracy as demonstrated by sensitivity analysis Data available Number of times model is used in simulation	Number of different versions of models developed Accuracy required of physical property data Research and plant studies needed
Generality:		
For equipment	Equipment type; length of EN list	EN list
For chemicals	Existence of storage in executive program	How and where physical properties are manipulated
Ability to extrapolate	Fundamental knowledge available	Time spent to obtain data required through plant tests or theoretical research

ability to extrapolate. These decisions are summarized in Table 5.4. Now consider the implications.

Implications of the modeling decisions for individual Unit Computations

The accuracy needed for the overall simulation and, as discovered from the sensitivity analysis on the approximate simulation, for the individual Unit Computations, will suggest the number of different levels of models needed for each Unit Computation. This will be illustrated beginning in Chapter 6. Sensitivity analysis shows when plant tests or research work needs to be done to obtain better information. For example, in the alkylation plant simulation, enough was known to write a very accurate simulation of the distillation columns; however, the reaction kinetics of alkylation are unknown and information about acid losses in the decantation-wash system was not available. Since the main objective was to minimize costs, knowledge of the acid losses and utilization within the reactor were needed. Better information had to be obtained experimentally.

The accuracy demanded by the model also affects the accuracy demanded in the physical properties handled in each model. If simple models are sought, then constant values of the physical properties may be sufficient. On the other hand, accurate models may demand that the property be known as a function of temperature, pressure, and composition. The decision of the model accuracy affects the number and character of the physical properties and equations which have to be manipulated. The implications, though, are straightforward; once an accuracy is specified, the accuracy required for the physical properties is specified.

The generality decision affects the EN list and the manipulation of physical properties. The decision on the generality of equipment models dictates whether the equipment parameters are incorporated in the Unit Computations or fed into the Unit Computations through the EN list. Unit Computations which are specific to one piece of equipment require that few, if any, parameters be fed into the program; there may not even be a need for an EN matrix. On the other hand, Unit Computations which are general with regard to equipment require a large EN list to accommodate all the required input information. The problem of length of the EN list can be dealt with by:

1. Choosing a standard length which is longer than necessary for some equipment models but is perhaps too short for very general models;
2. Having an EN list of variable length;
3. Having auxiliary EN (AEN) list that can be used where needed.

Since the costs are most closely associated with equipment, additional elements in the EN list are needed to store such input information as unit costs, and such calculated cost information as hourly operating cost, capital cost, or depreciation and total available energy as a measure of efficiency. If the general model is sought, the demand for elements in the EN list can be high enough that it is convenient to establish a separate cost matrix.

Models that are general with respect to chemicals require that the physical property correlations specific to chemical species be stored outside the Unit Computation. On the other hand, models specific to the process can include physical properties within the Unit Computation. A decision is reached based on the plans for future use of the models in other processes, the amount of project time available, the types of innovations envisaged, and the existence of a reliable property handling system.

5.3 How To Handle Physical Properties

Unless a physical property handling system has been developed as part of the executive system, which was not the case with PACER, a method of handling properties needs to be devised.

Five characteristics should be incorporated into the handling system. First, the system should be modular and open-ended so that both future developments can easily be incorporated and new processes can be handled. Second, data should be stored in accessible common locations for rapid retrieval because iterative calculations can demand exorbitant amounts of time if the property is recalculated at each loop. An example is the estimation of the vapor-liquid equilibrium compositions. Third, the system should allow for easy updating and permit the user to keep track of the literature sources of data. Fourth, the user should be able to understand and use it easily. Finally, both tables and correlations of data should be acceptable input information for the system.

Apart from these, the system should combine the correlating constants, the correlating equations, and the values of the state variables to yield a calculated property that is transmitted to the place where it is needed. It is here that the implications of model generality are important. If models specific to a process are to be developed, then physical properties can be stored and calculated within the Unit Computation. In Fig. 5.2, this would imply that all of the operations below line 1 would be within the Unit Computation. On the other extreme, completely general models with respect to the material handled would demand that the physical property storage and calculation be outside the Unit Computation. This is shown on Fig. 5.2 as having all the operations above line 3 with only the answer being used within the Unit Computation, as indicated below line 3. In between these two extremes are several combinations of those operations that are included within the Unit Computation and those without. A common alternative is indicated by line 2 where the constants are stored outside the Unit Computation and

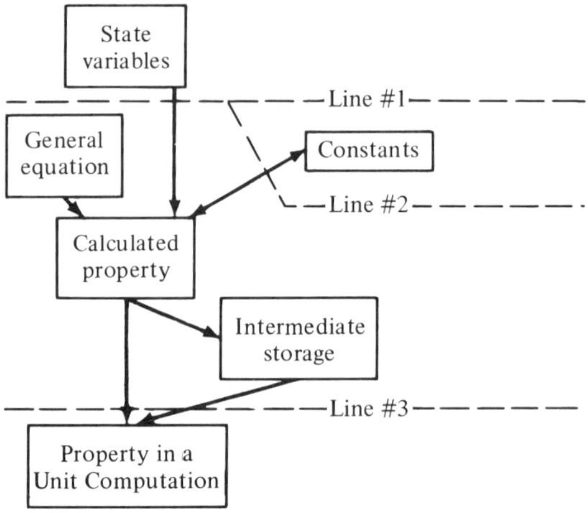

Figure 5.2. Operations needed to calculate a physical property (the dashed lines indicate possible groupings of the operations inside or outside the Unit Computation).

the general equations, the calculation of the answer, and any intermediate storage are within the Unit Computation.

This brief enumeration of useful characteristics, mode of operation required, and implications of generality indicates that the decision on how to handle physical properties may dictate that a miniature executive program be devised to handle physical properties. Since most of the problem revolves about how much of the physical property handling is done within the Unit Computation, it is instructive to consider some examples of possible methods of handling properties as characterized by the numbered boundaries indicated in Fig. 5.2.

Constants stored outside Unit Computation

Generalized expressions for all properties as functions of the state variables are stored within the Unit Computation. The constants, however, are stored in a data block and identified through coded numbers in the EN list. The crux of this system is a simplified code that rapidly brings the correct constants into the right locations within the Unit Computation. This example method does not accept tabular data; it requires correlations.

Calculations done outside Unit Computation

An example system in this category could store the sets of constants for each property and component in a data block. A calculational routine could then match the appropriate constants with the generalized form of correlating equation and calculate the pure component properties. It might be useful to have an intermediate storage matrix for the calculated pure component properties. Finally, a routine to estimate mixture properties is needed. It might also have an intermediate storage. Again, the major problem in making such a system work is in devising a simple coding system to direct the flow of information. A simplified sketch of the general concepts is shown in Fig. 5.3.

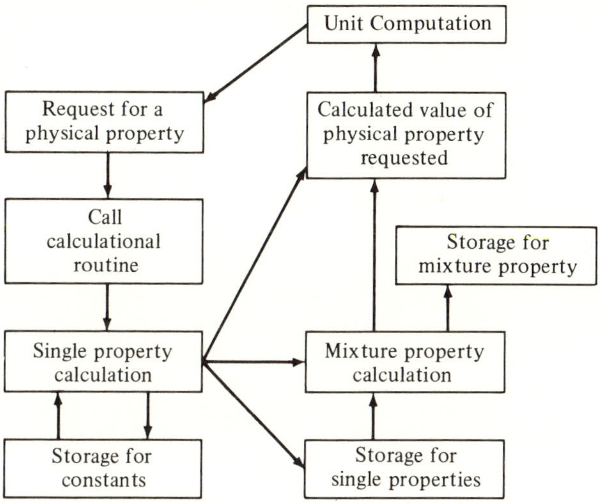

Figure 5.3. Information flow diagram for physical property calculation outside the Unit Computation.

A request for a calculated physical property at given values of the state variables is directed to the calculational routine. The form of correlating equation (and hence the accuracy) is specified. First the pure component calculations are done, the answers are passed to the calculational procedure for the mixture, and the resulting answer is transmitted back to the Unit Computation. When this procedure has been done once, if the accuracy warrants it, the properties can be lifted directly from either of the pure component or mixture storage matrices.

5.4 Decisions Made for the Acid Plant Study

Modeling decision and the EN list

The modeling decision made for the acid plant work was that two types of models would be developed. The first models devised were simple and specific, to be used in the preliminary simulation shown in Fig. 4.6. These Unit Computations are identified by a suffix of unity in the following chapters.

This type of model quickly pointed out the interactions in the information flow diagram and showed areas in which care had to be taken. The other type of model prepared—and it is the development of this type of model that is emphasized in the following chapters—was relatively specific, yet as accurate and as analytical as was warranted. Nevertheless, attempts were made to make models that were general. For example, the calculation of the gas flow rate for a given exit pressure for the steam-driven blower 1 of the flow diagram, Fig. 4.1, and EN 1 of the information flow diagram, Fig. 4.7, requires:

1. Rpm for the blower;
2. Head capacity data for the blower as a function of rpm.

A completely general Unit Computation would require that both be fed in through the EN list. For this simulation, the head capacity information was incorporated in the Unit Computation; the rpm was supplied through element 4 in the EN list for the BLOW Unit Computation. The EN list for a COOL Unit Computation is shown in Table 5.5. This illustrates that the design equation constants are all kept within the Unit Computation and that primarily specific physical dimensions are fed in through the EN list. (The information in elements 13, 14, and 15 is, however, the result of calculations that could have been performed within the Unit Computation but were not.)

TABLE 5.5
EN AND ENC LISTS FOR COOL UNIT COMPUTATION

EN list	ENC list
1. Equipment number	1. Equipment number
2.	2. Pressure drop 1, psia
3. Total number of tubes	3. Pressure drop 2, psia
4. Number of tubes in parallel	4.
5. Tolerance on temperature calculation	5. Energy available, PCU/hr
6. Tube length, ft	6. Operating cost, \$/hr
7. Inside diameter of exchanger tube, ft	7. Capital cost, \$/hr
8. Diameter of tubes at fin contact, ft	
9. Tube wall thickness, ft	
10. Fouling transfer coefficient for inside of tubes	
11. Fouling transfer coefficient for outside of tubes	
12. Contact transfer coefficient for fins	
13. Minimum area available for gas flow, sq ft	
14. Total surface area of finned tubes, sq ft	
15. Effective diameter of tubes, ft	

Physical property decision and details of methods

When the acid plant study was done, a generalized physical property handling system was not available. The decisions described here, then, represent those which are made for these conditions. Nevertheless, if a handling system is available, or is to be developed as described in Chapter 13, then this section supplies background information on some problems in handling physical properties.

Two approaches were taken that directly reflected the modeling decisions. For the simple models, constant values of the physical properties adequately matched the accuracy demanded. Any manipulations were done within the Unit Computation; i.e., a division 1 method of Fig. 5.2 was used. Values of heat capacity, molal volume, and viscosity of the stream mixture were fed into the program, stored, and transmitted through the process as part of the stream list. At each unit the appropriate values of the properties for the leaving streams were calculated from equations incorporated within the Unit Computations. This method of storing and transmitting physical properties in the stream list is not recommended because the properties had to be recalculated within the Unit Computation in any event.

For the accurate and specific models, the most accurate physical properties available were used. The properties were no longer constant but were functions of temperature, pressure, and concentration. Because the development of a complete physical property handling system could not be done in the time available, and because specific models were developed, arithmetic statement functions and subroutines for the physical properties were stored in or with the Unit Computations. These programs calculated mixture properties for the compounds, and for the range of temperature, pressure, and concentrations used in the acid plant. This system worked reasonably well.

Programs developed

Wherever possible, a simple arithmetic statement function was preferred to a subroutine. The arithmetic statement functions used in this case study are listed in Table 5.6, and a sample is given in Fig. 5.4. The expressions for each component were put on separate cards, a reference source was included with each set of data for quick identification and updating, and the accuracy of each was checked against known data.

Subroutines were necessary when the desired accuracy could not be

TABLE 5.6
ARITHMETIC STATEMENT FUNCTIONS DEVELOPED FOR THE ACID PLANT STUDY

Name	Calculation	Mixing law	Functional dependence
CPGMX(T)	Heat capacity gas mixture	Molal weighted	Temperature
CPGAV(T)	Molal average gaseous heat capacity relative to 298°K	Molal weighted	Temperature
VISGM(T)	Viscosity of gas mixture	Molal weighted	Temperature
TCGMX(T)	Thermal conductivity of gas mixture	Molal weighted	Temperature
VGM(T)	Molal volume	Molal weighted	Gas law

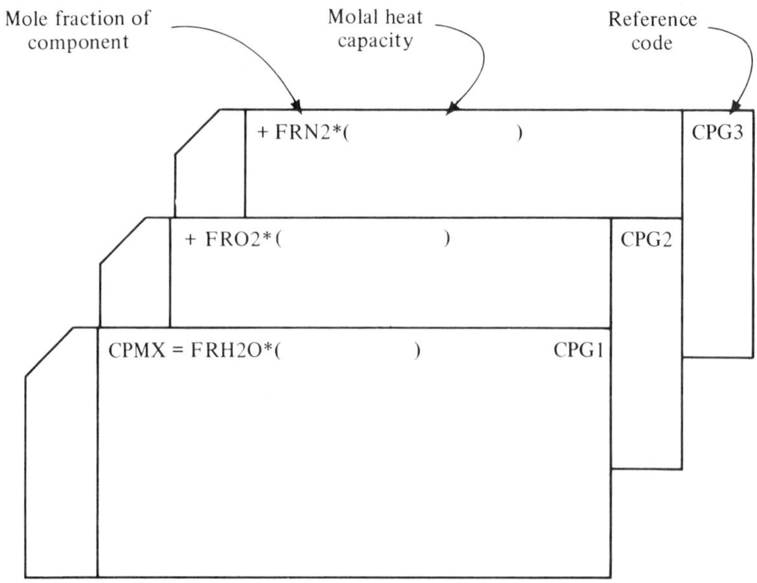

Calculation of molal heat capacity of a mixture from the equation

$$CPMX = \sum_i Y_i C_{P_i}$$

where Y_i = component mole fraction
C_{P_i} = component molal heat capacity

Figure 5.4. Sample arithmetic statement function set.

obtained through the use of a simple arithmetic expression. Three subroutines were developed.

A subroutine was developed to predict the vapor pressures of water vapor, sulfur trioxide, and sulfuric acid vapor over sulfuric acid, correlated as a function of temperature, and concentration. Data were taken from Gmitro and Vermeulen (1964).

Subroutine SUPER was developed for superheated steam by Martin (1961) as an improvement on the earlier work of Steltz and Silvestri (1958). A nonlinear regression was used to calculate the enthalpy, entropy, and specific volume of steam in the superheated region as functions of both pressure and temperature. The data used to prepare the correlation were from the Keenan and Keyes Steam Tables (1963). The form of the equations used was based on specific volume expressions quoted by Keenan and Keyes.

Subroutine STABLE was a combined subroutine for saturated steam; it was used to calculate for a given pressure the saturation temperature, the specific volume, and the enthalpy of the saturated liquid and then, for the same pressure, to calculate the specific volume and the enthalpy of the saturated vapor.

Guidelines and comments

Great care must be taken in collecting and evaluating physical property data. This is such a monumental task that it can easily be the sole or major

TABLE 5.7
SUGGESTED UNITS AND NOMENCLATURE

Units		
Temperature		°C or °K
Pressure		psia
Mass		lb moles
Heat		PCU
Nomenclature		
Temperature	T	°K
	TEMP	°C
	TEMM	°F
Pressure	P	psia
	PRES	atm
Mole fraction	FR	
Molar flow	F	moles/hr
Mass flow	W	lb/hr
Specific heat	CP	PCU/lb mole °K
Viscosity	VIS	lb/ft hr
Thermal conductivity	TK	PCU/ft^2 hr (°K/ft)
Molar volume (density)	VOL	ft^3/lb mole
Critical temperature	TC	°K
Critical pressure	PC	psia
Reduced temperature	TR	
Reduced pressure	PR	
Molecular weight	Z	
Subscripts		
Gas	G	
Liquid	L	
Solid	S	
Water	W	
Molal average above a reference datum of 298°K	AV	
Mixture	MX	
Components	H20	
	02	
	N2	
	SO2	
	SO3	

Examples: VISGMX viscosity of gas mixture
CPGAVW average molal specific heat for water vapor above 298°K

responsibility of one of the members on the simulation team. It is worthwhile to prepare a physical property data book with an index, pertinent data, and photocopies of important articles. This can be extended to supply tested data packs to each engineer. These should all be written with a consistent system of units (those chosen in Sec. 4.3) and with consistent nomenclature for common terms. The units and nomenclature should be specified at the beginning of the project so that each engineer can make his programs compatible with any physical property routines that may later be developed. Some suggested symbols are given in Table 5.7.

Some typical problems were encountered in collecting and evaluating the physical properties. A few properties, especially for sulfur trioxide, were unknown; several simple estimation techniques were used to obtain the desired quantity. For example, the method of Bromley and Wilke was used to estimate the viscosity of gaseous sulfur trioxide; the thermal conductivity was estimated from Eucken's equation for the Prandtl number as suggested by Reid and Sherwood (1958). For the development of SUPER and STABLE, the nine-place accuracy of the regression constants reported in the original articles was reduced to eight places to avoid time-consuming double precision accuracy where it was not required. Another difficulty was that most correlations for the heat capacity of steam, including the ones used in this

TABLE 5.8
COMPARISON OF TEST CORRELATIONS FOR THE HEAT CAPACITY OF WATER VAPOR (PCU/LB°C)

Temperature	Estimated value from correlation		Tabulated data
°C	Kelley (1960)	Spencer (1945)	Eckert and Drake (1959)
27	0.447	0.443	
77	0.453	0.450	
107			0.492
127	0.460	0.457	0.481
177	0.467	0.464	0.473
227	0.473	0.472	0.474
277	0.481	0.479	0.477
327	0.488	0.486	0.484
377	0.494	0.493	0.491
427	0.501	0.500	0.498
477	0.508	0.508	0.506
527	0.515	0.515	0.514
577	0.522	0.524	0.522
627	0.528	0.531	
677	0.535	0.538	
727	0.542	0.547	
827	0.556	0.563	

work, are based on spectrographic data that suggest a decrease in heat capacity with a decrease in temperature below 200°C. Experimentally measured heat capacities of steam increase with a decrease in temperature for temperatures below 200°C. Table 5.8 illustrates this and also indicates the differences in results calculated from different correlations.

5.5 Organization of Resources

There are essentially four activities to be coordinated: the problem definition, the organization and collection of data, the systems work, and the development of Unit Computations. The problem is usually defined by all associated with the project. The remaining three activities are often assigned to two or three separate groups of people as shown in Fig. 1.6. Possible variations in the distribution of these activities among the team members are as follows.

The organization and collection of data can be assigned to one group as its sole responsibility, can be assumed by the coordinator, or can be distributed between the systems and Unit Computations groups. The responsibility of the systems group is to develop an initial simple simulation, to study convergence promotion, and to ensure the smooth insertion of new Unit Computations into total simulation. The responsibility of the Unit Com-

TABLE 5.9
CHECKLIST AND DISTRIBUTION OF ACTIVITIES

Activity	Total resource team
1. Determine need and criteria. 2. Select variables; decide on units and format. 3. Prepare information flow diagram and process matrix. 4. Organize resources to match objectives. Define responsibilities and lines of communication. 5. Understand format and requirements of executive program. 6. Develop consistent format for handling physical properties. 7. Decide on the generality of the Unit Computations.	10%
8. Collect physical property data and develop system for handling physical properties.	5%
9. Collect data: costs, equipment parameters, corroborating plant data.	15%
10. Develop Unit Computations.	35%
11. Perform systems analysis.	20%
12. Communicate results.	15%

TABLE 5.10
APPROXIMATE DISTRIBUTION OF ACTIVITIES
WHEN THE RESOURCES ARE DIVIDED

Activity	Organization and collection of data	Develop Unit Computations	Study system
1 to 7	10%	10%	10%
8	30%	—	—
9	30%	10%	5%
10	—	50%	20%
11	—	15%	50%
12	30%	15%	15%
	100%	100%	100%

20% for the organization and collection of data
60% for the development of Unit Computations
20% for the systems study

TABLE 5.11
APPROXIMATE DISTRIBUTION OF ACTIVITIES
WHEN THE RESOURCES ARE DIVIDED

Activity	Develop Unit Computations	Study system
1 to 7	10%	10%
8	—	20%
9	20%	—
10	45%	10%
11	10%	45%
12	15%	15%
	100%	100%

75% for the development of Unit Computations
25% for the systems study

putation development group is to devise the models for the equipment units, consistent with the overall objectives. The various Unit Computations may be grouped by section of the plant or by research interest of the participants. For the acid plant study, the latter was selected. Tables 5.9, 5.10, and 5.11 provide a summary of the activities needed for a simulation and some estimates of the percentage of the total time required by each team for each activity. In Table 5.11 it is assumed that a separate team organizes and collects the data; in Table 5.12 it is assumed that the collection and handling of physical properties are done by the systems group and the collection of plant data is done by those developing the Unit Computations.

5.6 Summary

To completely define the problem, the strategy toward modeling and the method of manipulating physical properties must both be decided. But before the decisions can be discussed, the information available for physical properties, equipment, and costs should be appreciated.

Accurate, reliable, consistent physical properties are needed for any simulation or design project. Properties can be measured, extracted from the literature, or estimated. Whether constant values or values as functions of temperature, pressure, and concentration are needed depends on the degree of technical rigor sought in the mathematical modeling. But the interplay between physical properties and mathematical modeling does not stop at that point. The reverse may be true, i.e., the rigor of the modeling may depend on the physical properties available. The generality of the modeling also hinges upon the availability and form of physical properties.

The physical properties have to be represented as equations (or to a lesser extent, in tables) so that the computer can process the information. Generating equations requires selecting the appropriate functional form for the equation, evaluating the constants, and checking the significance of the fit.

Data about the equipment and cost information need to be collected and assessed.

The first decision that has to be made is a choice of modeling approach. This means that the accuracy sought for both the overall simulation and the individual models, the generality of the model, and the ability to extrapolate the model to a range of conditions have to be specified. This decision affects the development of the EN matrix, and it also interacts with the decision about the treatment of the physical properties.

The second decision on the manipulation of the physical properties determines whether the Unit Computations will be general or specific as far as physical properties are concerned. Once this decision is reached, care must be taken in the development of the mechanics for handling the information. Several systems are presented.

Finally, the modeling and the physical property decisions made for the acid plant study are described. Two types of models were developed: simple, specific models and accurate, specific models. The physical properties were also specific in that they were stored and manipulated within the Unit Computations. Two methods were used. The SN list was a storage point and transmitter of constant physical properties for preliminary, simple mathematical modeling, and later in the project, physical property arithmetic statement functions and subroutines were included in the Unit Computation. The former method is not recommended whereas the latter was satisfactory.

Some suggestions for overcoming the challenge of handling physical properties and some interesting side notes that illustrate typical problems of handling physical properties for digital simulation are included. The method chosen for distributing resources was described.

PROBLEMS

5.1. The data for the viscosity of gaseous SO_3 reported by Amelin are correlated by $\log \mu = -3.065 - 630.86/T + 305559.0/T^2$, where μ is the viscosity in centipoise and T is in degrees Kelvin for the range 283° to 323°K. Generate an expression for the viscosity (in lb mass/ft hr) as a function of absolute temperature for the range 283° to 900°K. Note that an extrapolation of Amelin's correlation to 900°K is risky and great care is needed to avoid excessive error. Carefully elaborate all the steps you took to minimize the uncertainty in your results.

5.2. The table below lists the viscosity of gaseous SO_2 for various temperatures. The data for °K were reported by Bromley (1949) and those for °C were reported by Reid and Sherwood (1958).

Temperature	Viscosity, centipoise
273°K	0.0117
288	0.01235
10°C	0.0120
40	0.0135
100	0.0163
300	0.0246
500	0.0315
700	0.0376
900	0.0432

(a) Generate an appropriate correlation of viscosity as a function of temperature, both by multiple regression and by Newton's forward difference interpolation equation.

(b) From an analysis of Table 6.2 in Reid and Sherwood (1958) and from the estimation techniques available, generate an equation to estimate the viscosity of SO_2 given the critical temperature, the critical pressure, and molecular weight. Check the validity of this expression by comparing it with the above data.

(c) Estimate the viscosity for gaseous SO_3 given the molecular weight is 80.0, the critical temperature is 491°K, and the critical pressure is 83.6 atmospheres. Compare your result with the answer to Prob. 5.1.

5.3. Pure water, acid from the absorber, and acid efflux from the drying tower are mixed together in the dilution tank. We want to do a heat balance over the dilution tank where there is a temperature rise because of the heat of dilution, using data from an enthalpy concentration diagram for this heat balance such as that reproduced in Fig. 21, p. 277 of Perry (1950). Prepare an algorithm and a working program that will calculate the exit temperature of the mixture from the dilution tank. Use standard nomenclature and units suggested on Table 5.8.

5.4. The calculation of the pressure drop through the reactor beds requires a knowledge of the viscosity of the mixture. Take a representative composition between the 2nd and 3rd beds in the reactor and determine the percentage difference in viscosity that results if we use the simple molar weighted mixing law as compared with the Wilke equation.
  (a) In doing this problem, develop an algorithm and a working program for the Wilke mixing equation.
  (b) Choose the viscosity correlations for the individual components that you feel are most reliable by writing a simple program to print out a table of calculated pure component viscosities as a function of temperature and by comparing the calculated with the known values found in the literature. Values for SO_2 and SO_3 are given in Probs. 5.1 and 5.2. Values for O_2 and N_2 are given in Eckert and Drake (1959).

5.5. How realistic is it to try to develop models with a generality of level 4 in Table 5.3 (as presented in Sec. 5.2) for pumps, heat exchangers, distillation columns, crystallizers, decanters, and ejectors? What are the problems with each?

5.6. Devise several division 3 (from Fig. 5.2) models for manipulating physical properties. Discuss the advantages and disadvantages of each system.

5.7. Cost information must also be manipulated.
  (a) Should it be stored and treated as a stream variable or as an equipment parameter? Does it make any difference?
  (b) The costs are normally not computed until the material and energy balances have been completed. How could this be worked into the information handling part of the system?
  (c) If the cost data are treated as stream parameters, what problems arise because of recycle, how does the present PACER system handle this, and what could be done to minimize the problem?
  (d) When a process makes valuable by-products, such as in the sulfuric acid plant where steam, acid, and sulfur dioxide are sold as products, how can a cost be assigned to each? Is there any simple way that this can be incorporated into the cost handling scheme?

5.8. Chapters 4 and 5 define the problem. List the uses you could make of a mathematical model for plant improvement (as briefly outlined in Fig. 4.4). Some of the uses suggest aspects of the problem definition that have not been discussed in Chapters 4 and 5. How would these be incorporated in the simulation? For example, a model may be used to test the safety of the plant or to investigate trouble in the plant. If this is the "need" for the simulation, what should be incorporated into the simulation work? Background for reliability studies is given by Rudd and Watson (1968).

Sources of Physical Properties for Fluids

These are arranged in order of decreasing generality.

General Compounds

1. Landolt, H. H. and Bornstein, R. (1950), *Zahlenwerte und Funktionen aus Physik, Chemie, Astronomie, Geophysik und Technik.* 6th ed. Springer, Berlin. Best source of indexed physical property data.

2. Perry, J. H., ed. (1962), *Chemical Engineer's Handbook.* 4th ed. McGraw-Hill, New York.

3. Lange, N. A., and Forker, G. M. (1961–62), *Handbook of Chemistry.* 10th ed. McGraw-Hill, New York.

4. Weast, R. C., ed. (1966), *Handbook of Chemistry and Physics.* 47th ed. The Chemical Rubber Co., Cleveland, Ohio.

5. *International Critical Tables* (1926–33). Voluminous; watch the units, beware of misprints, and try to get more recent data.

6. Seidell, A., et al. (1952), *Solubilities of Inorganic and Organic Compounds.* 3rd ed. Van Nostrand, Princeton, N.J.

7. Kaye, G. W. C., and Laby, T. H. (1966), *Tables of Physical and Chemical Constants.* 13th ed. Wiley, New York.

8. Touloukian, Y. S., ed. (1960), *Retrieval Guide to Thermophysical Properties Research Literature.* vol. 1. McGraw-Hill, New York. Review since 1940 of thermal conductivity, specific heat, viscosity, emissivity, diffusion coefficient, thermal diffusivity, Prandtl number.

9. Dreisbach, R. R. (1955, 1959), *Physical Properties of Chemical Compounds.* Advances in Chemistry Series, nos. 15, 22. American Chemical Society, Washington, D.C.

10. Din, F. (1956), *Thermodynamic Functions of Gases.* vol. 1, ammonia, carbon dioxide, carbon monoxide; vol. 2, air, acetylene, ehylene, propane, argon; vol. 3, methane, nitrogen, ethane. Butterworth Scientific Publications, London.

11. Matheson Co. Inc. (1966), *Matheson Gas Data Book.* 4th ed. East Rutherford, N.J. Vapor pressure and physical constants for about 100 compressed gases.

12. Carnegie Institute (1956–), *Selected Values of Properties of Chemical Compounds.* Dept. of Chemistry, Carnegie Inst. of Technology, Pittsburgh.

13. Emmett, P. H. (1954–60), *Catalysis.* 7 vols. Reinhold, New York. Source of kinetic data for heterogeneous organic and inorganic reactions.

14. (1948), *Advances in Catalysis and Related Subjects.* Academic Press, New York. Source of kinetic data for heterogeneous organic and inorganic reactions.

15. Sax, N. I. (1963), *Dangerous Properties of Industrial Materials.* 2nd ed. Reinhold, New York.

16. Stephen, T., and Stephen, H. (1963), *Solubilities of Inorganic and Organic Compounds.* 2 vols. Macmillan, New York.

17. Thermophysical Properties Research Center. *Data Book*. vol. 1, Metallic Elements and their Alloys; vol. 2, Nonmetallic Elements, Compounds, and Mixtures, Liquid and Gaseous State at NTP; vol. 3, Nonmetallic Elements, Compounds, and Mixtures, Solid State at NTP. Purdue, Lafayette, Ind. Gives data on thermal conductivities and viscosities.

18. Bondi, A. (1968), *Physical Properties of Molecular Crystals, Liquids and Glasses*. Wiley, New York.

Organic Compounds

19. Fenske, M. R., Braun, W. G., and Thompson, W. H. (1966), *Technical Data Book-Petroleum Refining*. American Petroleum Institute, New York. Physical constants, critical properties, vapor pressure, density, enthalpy, entropy, heat capacity, heat of vaporization, fugacity, vapor-liquid equilibrium, hydrocarbon-nonhydrocarbon equilibria, surface tension, viscosity, thermal conductivity, diffusivity, combustion characteristics.

20. Richter, F., ed. *Beilsteins Handbuch der organischen Chemie*. Springer, Berlin. See Bottle, R. T. (1962), *Use of the Chemical Literature*. Butterworth, Toronto.

21. Rossini, F. D., Pitzer, K. S., Armett, R. L., Braun, R. M., and Pimentel, G. C. (1953), *Selected Values of Physical and Thermodynamic Properties of Hydrocarbons and Related Compounds*. American Petroleum Institute, Project 44. Carnegie Press, Pittsburgh.

22. Rossini, F. D., Wagman, D. D., Evans, W. H., Levine, S., and Jaffe, I. (1952), *Selected Values of Chemical Thermodynamic Properties*. National Bureau of Standards Circular C500. U.S. Government Printing Office, Washington D.C.

23. National Bureau of Standards (1951), *Table of Chemical Kinetics of Homogeneous Reactions*. Circular 510. U.S. Government Printing Office, Washington D.C.

24. Maxwell, J. B. (1950), *Data Book on Hydrocarbons*. Van Nostrand, Princeton, N. J. Charts on thermodynamic properties.

25. Jordan, T. E. (1954), *Vapor Pressures of Organic Compounds*. Interscience, New York.

26. Chu, Ju-Chin, et al. (1950), *Distillation Equilibrium Data*. Reinhold, New York.

27. Chu, Ju-Chin, et al. (1956), *Vapor Liquid Equilibrium Data*. J. W. Edwards, Ann Arbor, Mich.

28. Horsley, L. H., et al., eds. (1952, 1962), *Azeotropic Data*. vols. 1,2. Advances in Chemistry Series, nos. 6, 35. American Chemical Society, Washington, D.C.

29. Hala, E., Pick, J., Fried, V., and Vilim, O. (1958, 1968), *Vapor-Liquid Equilibrium*. Pergamon, London. Detailed bibliography.

30. Natural Gas Processors Suppliers Association (1957), *Engineering Data Book*. Tulsa, Okla.

31. Natural Gasoline Association of America (1957), *Equilibrium Ratio Data Book*. Tulsa, Okla.

32. Lenoir, J. M. (1959), *Equilibrium Ratios of Hydrocarbons.* C. F. Braun, Alhambra, Calif.

33. Edmister, W. C. (1961), *Applied Hydrocarbon Thermodynamics.* vol. 1. Gulf Publishing Co., Houston, Tex.

34. Marsden, C., ed. (1963), *Solvents Guide.* 2nd ed. Cleaver-Hume, London.

35. Scheflan, L. and Jacobs, M. B. (1953), *The Handbook of Solvents.* Van Nostrand, Princeton, N. J.

36. Spiers H. M. (1958), *Technical Data on Fuel.* 6th ed. The British National Committee, World Power Conference, London.

37. Dreisbach, R. R. (1952), *Pressure-Volume-Temperature Relationships of Organic Compounds.* 3rd ed. Handbook Publishers, Sandusky, Ohio.

38. Hougen, O. A., Watson, K. M., and Ragatz, R. A. (1961), *Chemical Process Principles.* vol II. Wiley, New York. Corrections based on the law of corresponding states.

39. Timmermans, J. (1959), "*The Physico-Chemical Constants of Binary Systems in Concentrated Solutions.* vol. 1. Interscience, London.

40. Timmermans, Jean (1950), *Physico-Chemical Constants of Pure Organic Compounds.* Elsevier, New York.

41. Francis, A. W. (1963), *Liquid-Liquid Equilibriums.* Interscience, New York. Extensive tabulation of references for approximately 4,000 immiscible systems.

42. Egloff, G. (1939, 1940, 1946, 1947), *Physical Constants of Hydrocarbons.* American Chemical Society Monograph Series, no. 78, 2 vols. Reinhold, New York.

43. Coffey, S., ed. (1964), *Rodd's Chemistry of Carbon Compounds.* 2nd ed. Elsevier, New York.

44. Canjar, et al., in a series of articles on thermoproperties of hydrocarbons and nonhydrocarbons appearing in *Hydrocarbon Processing and Petroleum Refiner.* Individual references are tabulated below.

Vol	No	Page	Part	Title
41	8	121	1	Universally Consistent Reference State for Thermodynamic Properties
41	9	263	2	Methane
41	10	149	3	Ethane
41	11	203	4	Propane
41	12	115	5	*n*-Butane
42	1	129	6	*n*-Pentane
42	8	127	7	*i*-Butane
43	6	177	8	*n*-Hexane
44	9	219	9	Ethylene

Vol	No	Page	Part	Title
44	10	137	10	Propylene
		141	11	1-Butene
	11	293	12	Acetylene
		297	13	Benzene
45	1	135	1	Ammonia
		139	2	Carbon dioxide
	2	158	13	Carbon monoxide
		161	4	Hydrogen
	3	137	5	Nitrogen
		141	6	Oxygen
	4	161	7	Sulfur dioxide
		165	8	Water

45. Holland, C. D. (1963), *Multicomponent Distillation.* Prentice-Hall, Englewood Cliffs, N.J. In Appendix, constants for regressions of half a dozen light paraffins for equilibrium constants and enthalpy as functions of temperature and pressure.

46. Prausnitz, J. M., Eckert, C. A., Orye, R. V., and O'Connell, J. P. (1967), *Computer Calculations for Multicomponent Vapor-Liquid Equilibria.* Prentice-Hall, Englewood Cliffs, N.J.

47. Prausnitz, J. M. and Chueh, P. L. (1968), *Computer Calculations for High-Pressure Vapor-Liquid Equilibria.* Prentice-Hall, Englewood Cliffs, N.J.

48. Stull, D. R., ed. (1965), *Joint Army-Navy-Air Force Thermochemical Tables.* Issued as PB-168370.

49. Gallant, R. W. (1968), *Physical Properties of Hydrocarbons.* part I. Gulf Publishing Co., Houston, Tex. Extensive graphs of vapor pressures, heat of vaporization, density, heat capacities, viscosities, surface tension, and thermal conductivities for a variety of hydrocarbons.

50. Gallant, R. W., in a series of articles on physical properties of hydrocarbons appearing in *Hydrocarbon Processing and Petroleum Refiner.* Typical references are tabulated below; any annual index since 1962 for this journal will refer the reader to all pertinent articles.

Vol	No	Page	Part	Title
44	9	225	3	C_2 to C_4 alkynes
46	9	155	17	C_4-C_5 branched hydrocarbons
46	10	135	18	C_6-C_8 branched hydrocarbons
46	12	119	19	Chlorinated C_2's
49	1	137	41	Cyclic hydrocarbons

Inorganic Compounds

51. Gmelin, L. (1924–), *Handbuch der anorganischen Chemie.* 8th ed. Verlag Chemie, Weinheim. See Bottle, R. T. (1962), *Use of the Chemical Literature.* Butterworth, Toronto.

52. Kelley, K. K. *Contributions to the Data on Theoretical Metallurgy*. See *List of Publications Issued by the Bureau of Mines*, Bureau of Mines Special Publications, U.S. Government Printing Office, which provides brief abstracts listed by report number with author, and subject indices. For example, see reports 406 (1937) and 476 (1949).

53. Mellor, J. W. (1922–37, 1956–), *Comprehensive Treatise on Inorganic and Theoretical Chemistry*. Longmans, Green, London.

Estimation Techniques

If the property has not been reported, an estimate of the property might be sufficient.

54. Reid, R. C., and Sherwood, T. K. (1966), *Properties of Gases and Liquids*. 2nd ed. McGraw-Hill, Toronto.

55. Gambill, W. R. (1960), How to Estimate Engineering Properties. *Chemical Engineering*. See vol. 67, 1, p. 109 for a description of all the articles in the series.

56. Janz, G. J. (1958), *Estimation of Thermodynamic Properties of Organic Compounds*. Academic Press, New York.

57. Bird, R. B., Hirschfelder, J. O., and Curtiss, C. F. (1954), Theoretical Calculation of the Equation of State and Transport Properties of Gases and Liquids. *Trans. ASME* **76**, 1011–1088.

The Simulation of the Fluid Handling Equipment 6

6.1 Introduction to and Strategy for Modeling Equipment

The strategy of simulating the acid plant has now reached the stage at which the Unit Computations need to be developed. That is, the input variables have been specified and the process matrix and the SN and EN lists have been established to serve the needs of this problem. The complex total problem as exemplified by the information flow diagram is now reduced to a series of tractable problems of modeling the equipment units. This chapter describes the development of the Unit Computations for the fluid handling equipment. This is a natural first choice for discussion because:

1. Some of the models are so simple that the principles of modeling within the constraints of the PACER framework can easily be illustrated;
2. These modeling problems emphasize the importance of the decisions discussed in Chapter 5;
3. They demonstrate the need for accurate data and extensive fundamental knowledge of relatively simple equipment;

4. The reader is familiar with the simple MIXER that has been used as an example in Chapters 2 and 3.

The fluid handling equipment in the acid plant includes pipe tees, mixing tanks, circulating pumps, and the main blower. No matter how simple these appear, care must be taken in developing models that will transmit information of the required accuracy. To develop the Unit Computation the programmer must:

1. Decide what types of models are required, choose the physical phenomena to be included in the model, and select the assumptions needed.
2. Formulate the equations of the model under the assumptions made. The equations must be mathematically consistent so that the number of output variables equals the number of independent equations.
3. Obtain the data required for solving the equations. Physical properties, operating curves, operating data, and perhaps cost data are needed.
4. Choose the method of solving the equations, prepare the program, and minimize the time of calculation. If there are several streams entering or leaving an equipment unit, the right information must be stored with the right stream. The order in which the streams are given in the process matrix determines the order in which the streams will be assigned by PACER to STRMI and STRMO lists. Realizing this, the programmer should check to see that stream consistency is maintained within the Unit Computation. The program should include many comment cards to clearly document the steps in the calculations, and the program should print warning comments when process conditions are beyond the range of the model.
5. Determine values of any remaining unknown parameters to match the model to the plant and test the agreement between predicted and measured plant data. At this stage the programmer should also check the mass balance around each Unit Computation. Each Unit Computation must give reasonable outputs.

These steps are illustrated in this chapter for the simulation of the fluid handling equipment and in the subsequent four chapters for the other equipment.

6.2 Choice of Models and Physical Phenomena to be Included

Based on the statement of need, the criterion selection, and the variables chosen in Sec. 4.3, the function of each piece of fluid handling equipment is studied for similarities in operation. The different fluid dynamical functions for the acid plant are:

1. Combine streams;
2. Separate streams or distribute streams;
3. Control the concentration by the addition of water;
4. Specify the total flow rate;
5. Adjust the flow rate and pressure drop across the blower;
6. Extract energy from the steam to run the blower.

To fulfill these functions, three groups of models were chosen: MIXER, PUMP, and BLOW. The physical phenomena included and the calculations envisaged for these general models are listed in Table 6.1.

After it has been decided that three groups of models are to be developed, the next decision is how many different models are needed within each group. Recall that the modeling decision was to make two types of models: simple and specific, and accurate and specific. Wherever possible, models are to be based on fundamentals rather than on statistical correlations. Since every pipe junction in the plant is a MIXER, 13 specific models would be required.

TABLE 6.1
GROUPS OF MODELS CHOSEN

Mnemonic name of model	Physical phenomena included	Calculates
MIXER	Mixes streams together and/or splits a uniform mixture into different streams, all of the same composition and temperature. It does *not* control the total flow in any line, but it specifies the fractional split. It does not calculate the energy required for the mixing, the residence time in any vessel, the pressure drop, or the costs.	Unknown flow rates, concentrations, and temperatures in all of the streams; transmits pressure.
PUMP	Splits a uniform mixture into different streams and controls the total flow (not the fractional split) of one stream. It does not calculate energy, cost requirement, pressure drop, or temperature rise because of pumping. Head capacity data were not used.	Unknown flow rates and concentrations in the streams; transmits temperature and pressure.
BLOW	Supplies energy to convey gas. Energy-capacity relationships are used to determine the pressure rise across a blower from the flow through the unit. An energy balance can be done; a mass balance is superfluous because only one stream is blown. For a steam driven blower, the energy extracted from the steam can be used to estimate steam back pressure conditions. Costs were not included.	Unknown pressure rise across the blower and exit steam conditions for turbine-driven blower.

A more detailed analysis suggests that four units split one stream so that an enthalpy balance is not needed. One mixer model is needed to do a heat balance which includes the heat of mixing, and one to add water to achieve a specified concentration. Furthermore, pumps and blowers are described by statistical equations rather than fundamental relationships. Based on such an evaluation of the model requirements for this plant, the models listed in Table 6.2 were developed.

Having decided the type and variety of models to be developed and the

TABLE 6.2
FLUID HANDLING UNIT COMPUTATIONS DEVELOPED FOR THE ACID PLANT CASE STUDY

Name	Flow control	Heat effects	Description	Modeling characteristic
MIXER1	no	no	Sum inputs and equally divide among the NOUT output streams	simple general
MIXER2	no	no	Sum inputs and divide total among the NOUT output streams according to fractions given in the EN list	simple general
MIXER3	no	yes	MIXER2 modified to include an enthalpy balance (excluding heat of mixing) based on constant heat capacity	simple general
MIXER4	no	yes	MIXER3 modified to include a control on the concentration of exit acid by adding water	simple specific
MIXER5 (for acid-water)	no	yes	MIXER4 modified to include heat of mixing in the enthalpy balance for acid-water mixtures	accurate specific
MIXER6 (for gases)	no	yes	MIXER3 modified to include gaseous heat capacities as functions of temperature	accurate specific
PUMPS1 (for acid-water)	yes		MIXER3 modified to preset the flow in the first output line and control the split to the other streams by EN list	simple general
BLOW1	no	no	Calculate energy required to force the air through the blower	simple general
BLOW2	no	yes	Predict pressure drop across blower for given volumetric throughput and rpm. Calculate temperature rise across the blower for the gas and predict, for total steam flow through the turbine, the temperature drop on the steam side	accurate specific

physical phenomena to be included, the development of two typical Unit Computations is now discussed. The steps of formulation, information collection, model preparation, and matching the model to the plant are illustrated with these models.

First, the structure of PACER suggests that the calculations should follow the three-step pattern of transfer in, calculate, and transfer out. The model should also include a comment block to clearly identify the program. Furthermore, throughout the program, comment cards should identify the different steps. Therefore, any computer program would have the blocks shown in Fig. 6.1. Figure 6.2 gives an example of a complete program for MIXER1 written in Fortran IV to illustrate these principles. Second, the assignment of the EN list for each model is dictated by how general it is to be. BLOW1 and MIXER6 are described below because BLOW1 clearly illustrates the three-step process suggested in Fig. 6.1, whereas MIXER6 shows the assignment of the EN list.

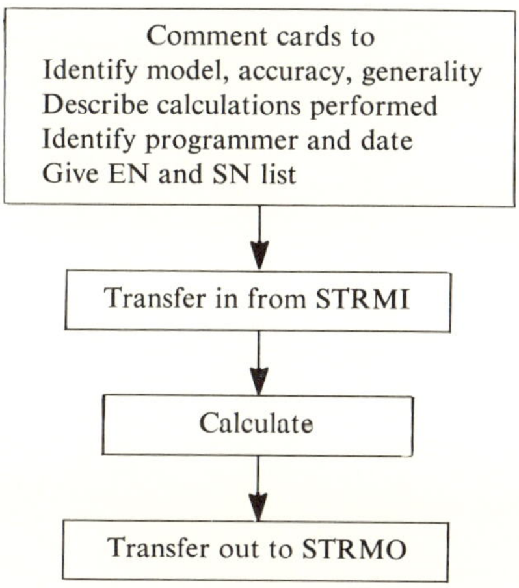

Figure 6.1. Blocks of instructions and information for a typical Unit Computation program.

6.3 The Development of BLOW1 as an Example

The five-step approach of Sec. 6.1 is followed in devising the model.

Choice of physical model and assumptions

This model, as outlined in Table 6.2, is simple and general. It calculates the theoretical energy required to compress a given quantity of air to a given pressure. Such a model may be useful for cost estimation but does not alter the material flow. It does illustrate simply the principles of modeling within PACER. It assumes 100% energy conversion, ideal gas law behavior, no condensation problems, and no temperature rise.

Formulation of the equations

BLOW1 calculates the energy needed to force the air through the blower which equals, by assumption, the reduction in energy of the driving steam. All the values of the stream variables except pressure are the same for the input streams and the output streams; only the available energy in the steam and the pressure change. This is calculated assuming 100% energy conversion from the relationship:

$$\text{energy} = 144(\text{POUT} - \text{PIN})\frac{\text{CFHIN}}{778 \times 1.8} \tag{6.1}$$

where POUT = outlet air pressure from the blower, psia
PIN = inlet air pressure, psia
CFHIN = volumetric gas flow, cfh

One additional calculation is required, however, because the gas flow, FA, is given in moles per hr, and Eq. 6.1 requires volumetric flow in cfh. That is,

$$\text{CFHIN} = (FA)359\left(\frac{T + 273}{273}\right)\frac{14.67}{P} \tag{6.2}$$

where T = temperature at inlet, °C
FA = flow of air at inlet, moles/hr
P = pressure, psia

Collection of data

In the acid plant, the inflow of ambient air and hence the gas flow rate through the complete process are dictated by the operation of the blower, which is located at the start of the process.

For the present BLOW1 model, the head capacity data are not needed per se but they do yield reasonably consistent values for the flow rate FA and pressure drop across the blower. These are difficult to obtain because of the recycle. That is, the gas flow rate into the plant, FA, is known and the pressure at the exit of the plant is atmospheric. However, the pressure at the exit of the blower, POUT is not known; it must be such that the pressure drop through the complete plant is consistent with the gas flow rate. For a fixed flow rate of gas through the plant, it is possible to either

1. Assume a pressure at the exit of the blower, calculate the pressure losses through all the equipment, and compare the calculated stack pressure with the actual atmospheric pressure, or
2. Start with atmospheric pressure at the stack and let the pressure feed

```
$ I B F T C MIXER1
C
C             SUBROUTINE MIXER1 (MIXER—MASS BALANCE)
C             REVISED BY T. TOONG FEB. 20, 1968
C
C             *MAXIMUM VALUE OF NOCOMP, NUMBER OF FLOW COMPONENTS, IS  26
C             *        "              NIN,     "       INPUT STREAMS,       5
C             *        "             NOUT,     "      OUTPUT STREAMS,       5
C
C             EN LIST—
C             1. EQUIPMENT NUMBER
C             2. TEMPERATURE OF OUTPUT STREAMS, DEG C, IF O., OUTPUT TEMPERATURE
C                ARE SET EQUAL TO TEMPERATURE OF 1ST INPUT STREAM
C             3. PRESSURE OF OUTPUT STREAMS, PSIA, IF O., OUTPUT PRESSURE
C                ARE SET EQUAL TO PRESSURE OF 1ST INPUT STREAM
C             4. (BLANK)
C             5. (BLANK)
C             6. FRACTION OF TOTAL INPUT IN  1ST OUTPUT STREAM
C             7.              "              2ND        "
C             8.              "              3RD        "
C             9.              "              4TH        "
C             10.             "              5TH        "
C
C             ENC LIST (OPTIONAL)—
C             1. EQUIPMENT NUMBER
C             2. FLAG
C             3. LENGTH OF EN LIST (FOR PRINTING)
```

Figure 6.2. Sample program to illustrate the use of comment cards, lay-out, and error messages (courtesy T. Toong).

```
C
C          SN, STREAM LIST—
C          1. STREAM NUMBER
C          2. STREAM FLAG
C          3. TOTAL FLOW IN LBMOLE/HR
C          4. TEMPERATURE IN DEGREE C
C          5. PRESSURE IN PSIA
C          6. FLOW OF COMPONENT 1  IN LBMOLE/HR
C          7.         "          2       "
C          8.         "          3       "
C          9.      ETC.
C
C          SNC, LIST—NOT REQUIRED
C
C
           SUBROUTINE MIXER1
C          ******* MACSIM COMMON DECK (REDUCED). T. T. FEB. 20, 1968. ********
           COMMON NOCOMP, KSETS, NIN, NOUT, NE, STRMI (5,31), STRMO (5,31), EN (50,20)
C          ***************************************************************************
C        TO CHECK IF INPUT STREAMS ARE NOT MORE THAN 5
           IF (NIN.LE.5) GO TO 10
           WRITE (6,90) NE
           NIN = 5
     10    JJ = NOCOMP+5
```

Figure 6.2. (cont.). Sample program to illustrate the use of comment cards, layout, and error messages (courtesy T. Toong).

```
C
C M A S S     BALANCE
C ( I N P U T )
            DO 1 J = 6, JJ
            STRMO (NOUT,J) = 0.
            DO 1 I = 1, NIN
        1   STRMO (NOUT,J) = STRMO (NOUT,J) + STRMI (1,J)
C (  O  U T P U T )
            DO 2 I = 1, NOUT
            STRMO (I,4) = EN (NE, 2)
            IF (EN(NE,2). EQ.0.) STRMO (I,4) = STRMI (1,4)
            STRMO (I,5) = EN(NE,3)
            IF (EN(NE,3).EQ.0.) STRMO (I,5) = STRMI (1,5)
            STRMO (I,3) = 0.
            DO 2 J = 6,JJ
            STRMO (I,J) = STRMO (NOUT,J)*EN (NE,I+5)
        2   STRMO (I,3) = STRMO(I,3) + STRMO (I,J)
C
C F O R M A T S
     90     FORMAT(////1X, 25H**ERROR**IN MIXER1 MODULE, 13,
          1 73H, NIN EXCEEDS 5, CALCULATION CONTINUES USING THE FIRST FIVE INP
          2 UT STREAMS////)
C
            RETURN
            END
```

Figure 6.2. (cont.). Sample program to illustrate the use of comment cards, layout, and error messages (courtesy T. Toong).

backwards through the plant to the blower to compare the calculated blower inlet pressure with atmospheric pressure.

Regardless of which system is selected, and despite the simplicity of the BLOW1 model, it is useful to obtain reasonable values of the pressures and flows. Figure 6.3 presents the original operating curve supplied by the manu-

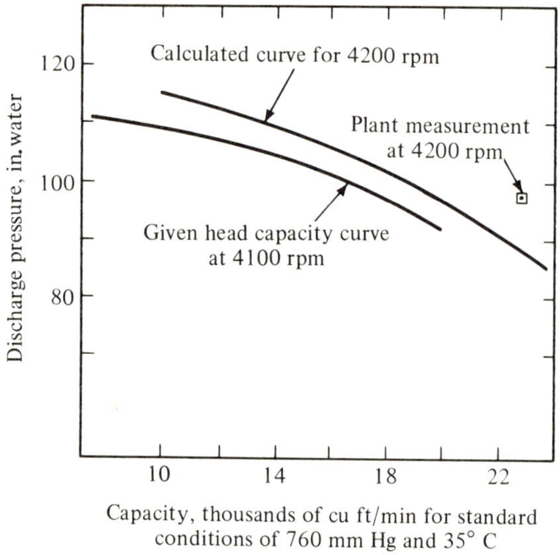

Figure 6.3. Head capacity information for the blower.

facturer. However, the blower has been modified since installation. Since the present blower operates at 4,200 rpm, a more reasonable operating curve can be calculated from the idealized relationships, given for example by Ellis (1947), as

$$\frac{\text{CFHIN(1)}}{\text{CFHIN(2)}} = \frac{\text{rpm}_1}{\text{rpm}_2}$$

$$\frac{\text{(POUT—PIN)(1)}}{\text{(POUT—PIN)(2)}} = \left(\frac{\text{rpm}_1}{\text{rpm}_2}\right)^2$$

This theoretical curve, shown in Fig. 6.3, can give trial values of FA, POUT, and PIN for the model.

Preparation of the model

The program, written in Fortran IV, is given in Fig. 6.4. Only one input stream—the air—is used. The effect of the energy change in the steam on steam temperature and pressure is not included. Two STRMI and two STRMO lists are assigned by PACER with the first one of each referring to the air. First, the information from the stream lists is transferred into the program variables and then Eq. 6.2 and 6.1 are solved. Finally, the calculated loss in steam energy is transferred out of the program to its allocated storage element in the ENC list. The ENC list was used because its elements were the same for all models in order to locate easily the energy and cost values. All other information is transferred between input and output streams, since all information received must be passed on, even if no calculation is done.

Comparison with plant operation

This model was so simple and was of such little use that detailed comparisons with plant data were not made. It was included here to illustrate how easily models can be developed to fit within PACER, and how the information is manipulated. Incidentally, the head capacity data measured on the plant did not agrcc with the estimated curves shown in Fig. 6.3. A modification of the theoretical relationships so that they did match the measured data was necessary in the preparation of BLOW2 and would have been helpful in suggesting reasonable trial values of pressure for the BLOW1 model.

6.4 The Development of MIXER6 as an Example

MIXER6 is a good illustration of the use of the EN list of parameters (but is not a good illustration of the three blocks listed in Fig. 6.1 because the

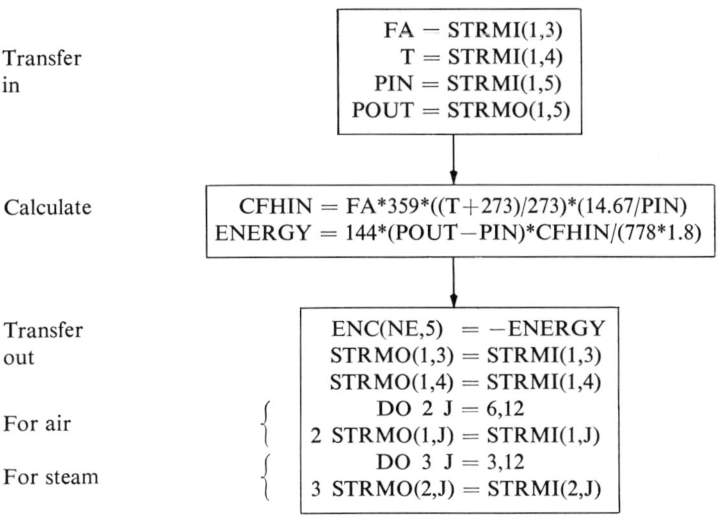

Figure 6.4. BLOW1 Unit Computation program illustrates the major program blocks.

transfer in and calculate blocks are combined). The five steps in developing the model are discussed in turn.

Choice of assumptions and physical model

This model is accurate and specific for compounds but is general for the fractional split of the output streams. This model assumes that:

1. The total number of input and output streams is less than 6, i.e., NIN + NOUT $\leqslant$ 6;
2. The components are the gases water vapor, oxygen, nitrogen, sulfur dioxide, and sulfur trioxide;
3. The molal heat capacity of the mixture is the sum of those of the pure components, weighted by mole fraction. This implies that there is no heat of mixing. The arithmetic statement function used is described in Chapter 5 and shown in Fig. 5.4;
4. There is complete mixing;
5. No work is required or done in the device;
6. There are no heat losses;
7. There is no vaporization or condensation, i.e., that the compounds remain as gases.

The equations can now be formulated.

Formulation of the equations

MIXER6 combines the input streams:

Total molal flow: $$\text{ALL} = \sum_{i=1}^{\text{NIN}} F_i \tag{6.3}$$

Total component flow: $$\text{TOTAL(J)} = \sum_{i=1}^{\text{NIN}} F_{ij} \tag{6.4}$$

where F = molal flow/hr
$_i$ = input stream
$_j$ = species of component

Then the total input flow and the total flow of each component are divided among the exit streams according to values in the elements of the EN list. Through the EN list, the model is general in that the engineer can later specify the proportions in all output streams even though the model is specific to the compounds used. The equations are

$$\text{F(K)} = \text{ALL*EN(NE, K + 5)} \tag{6.5}$$

$$\text{F(J,K)} = \text{TOTAL(J)*EN(NE, K + 5)} \tag{6.6}$$

where NE is the equipment number and
K refers to exit stream; it can be seen from Fig. 6.2 that the assigned

meanings of the elements of the EN list important in Eqs. 6.5 and 6.6 are in element 6 and onwards. Hence, the subscript (K + 5) is used. The enthalpy balance to establish the exit temperature T_2 is based on molal average heat capacities of the gaseous components for the acid plant. Hence, the model, as far as constituents are concerned, is specific. The equation is:

$$\text{THENT} = \underbrace{\sum_{i=1}^{\text{NIN}} F_i C_{pav_i}(T_i - 298)}_{\substack{\text{enthalpy in} \\ \text{relative to } 298^\circ\text{K}}} = \underbrace{\text{ALL}\, C_{pav}(T_2 - 298)}_{\substack{\text{enthalpy out} \\ \text{relative to } 298^\circ\text{K}}} \tag{6.7}$$

where F_i = lb moles/hr of input stream i

C_{pav_i} = molal average heat capacity of the mixture in stream i relative to 298°K

C_{pav} = molal average heat capacity for a gas mixture (CPGAV)

$$\sum_{\substack{\text{all components} \\ j}} y_j \frac{\int_{298}^{T} C_{pj}\, dT}{(T - 298)}$$

T_i = inlet temperature of stream i, °K

y_j = mole fraction of component j

Collection of data

The data needed are concentrations, flows, temperatures, and physical properties. Most data were measured during the plant trial described in Sec. 4.3 and are summarized in Table 4.7 and 4.9. Since this model was to be as accurate as possible, expressions for the molal average heat capacities were calculated as functions of temperature. The fractional split was determined strictly from plant data.

Preparation of the model

All incoming streams are mixed so that the input streams can be in any order in the process matrix. However, each value stored in the EN list must correspond to the correct output stream. From Eqs. 6.5 and 6.6 the fraction split for the first exit stream in the process matrix is in element EN (NE,6). Since the first exit stream corresponds to STRMO (1,__), care must be taken to ensure the split refers to the correct stream on the information flow diagram. STRMO (1,__) means the stream that has the first negative number as shown on the process matrix. For example, from Table 4.8, STRMO (1,__) for equipment #12 is stream 13.

The algorithm of the program is given in Fig. 6.5. The separate blocks of transfer in, calculate, and transfer out are not as apparent as in the BLOW1

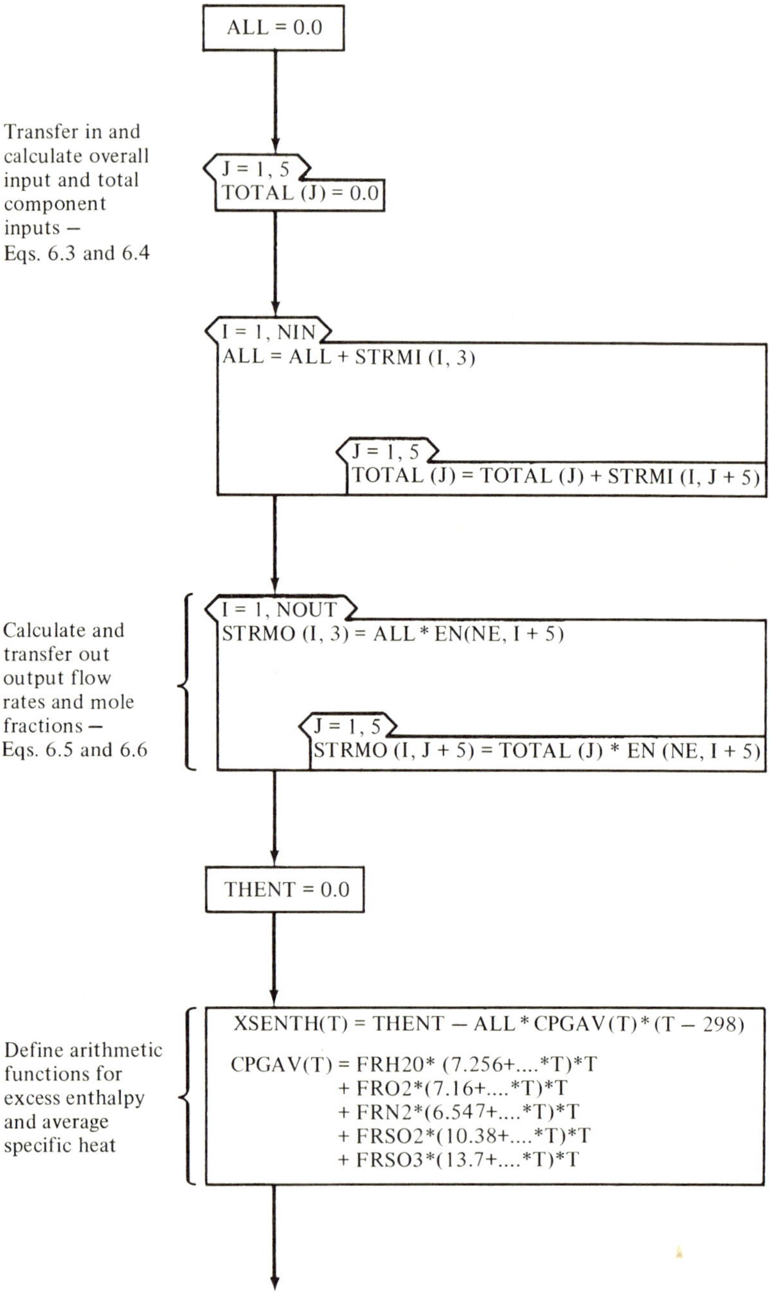

Figure 6.5. Algorithm for MIXER6.

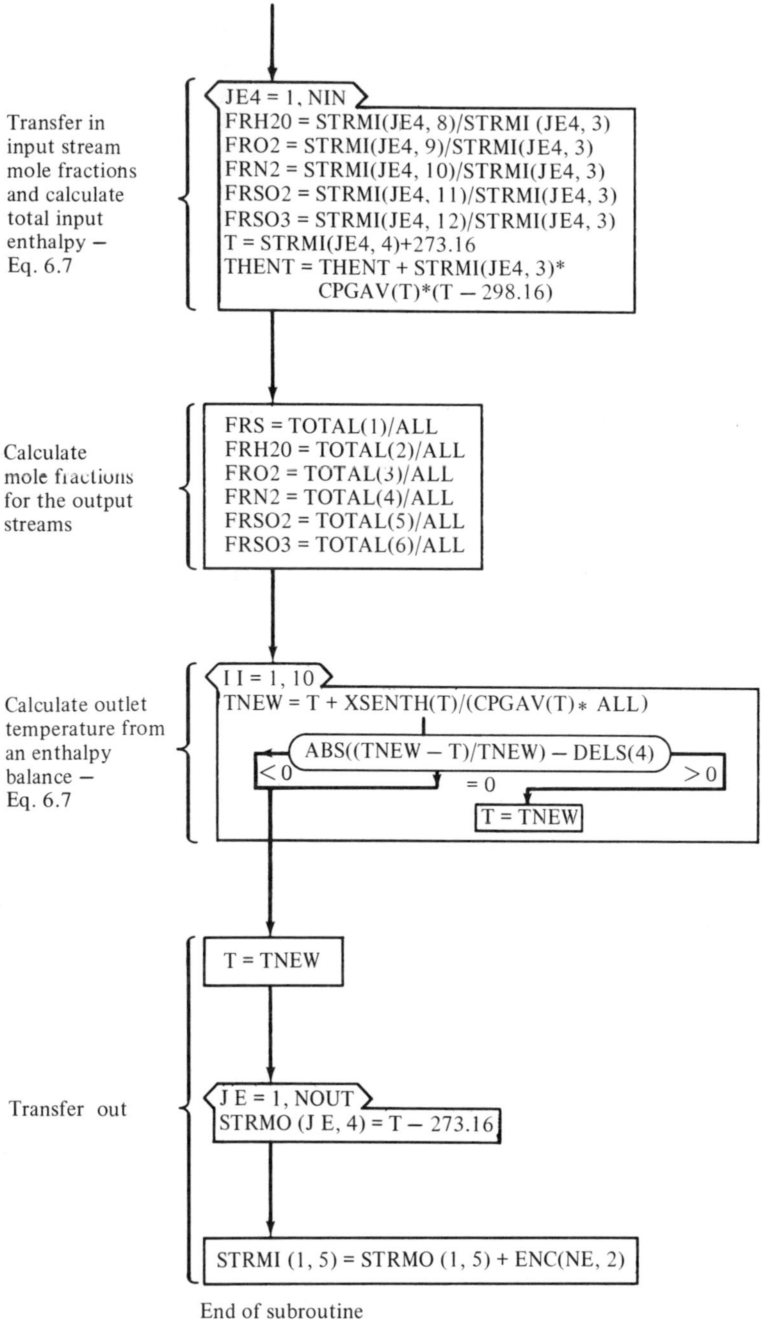

Figure 6.5 (**cont.**). Algorithm for MIXER6.

because the calculation is combined with the transfer in. Within the program, a column list TOTAL(J) and a single location ALL are established and set to zero. From the input STRMI (I,3) the number of total moles in each stream I is transferred to ALL and the molal flows of components J of each stream are systematically transferred from the respective elements of STRMI into TOTAL (J). Thus, the transfer in process and the calculations of Eqs. 6.3 and 6.4 are done concurrently.

The next calculation block considers Eqs. 6.5, 6.6, and 6.7 together with further transfers. To facilitate the estimation of the exit temperature T_k, Eq. 6.7 is rearranged to give the excess enthalpy, XSENTH, as the difference between the calculated entering enthalpy, THENT, and the exit enthalpy evaluated at a guessed temperature T′:

$$\text{XSENTH} = \underbrace{\text{THENT}}_{\text{Entering}} - \underbrace{(\text{Total molal flow})\begin{pmatrix}\text{Exit mixture}\\ \text{average molal}\\ \text{heat capacity}\\ \text{for T}'\end{pmatrix}(\text{T}' - 298)}_{\text{Leaving}} \tag{6.8}$$

To start the calculation, the first guess for the exit temperature is taken as the absolute temperature of the NINth stream into the mixer T′. Then an improved estimate of the exit temperature is found by the Newton-Raphson Method:

$$T_{\text{NEW}} = T' + \frac{\text{XSENTH}}{(\text{Total molal flow})\begin{pmatrix}\text{Exit mixture}\\ \text{average molal}\\ \text{heat capacity}\\ \text{for T}'\end{pmatrix}} \tag{6.9}$$

This calculation is repeated until $|T_{\text{NEW}} - T'|/T_{\text{NEW}}$ is less than the restriction in DELS(4). Finally the calculated results are transferred out of the program into the appropriate locations. This MIXER6 example illustrates the way in which the EN list can be used to generalize a model. The split of the total moles entering the system is not specified within the Unit Computation. Instead, the general program is made specific by the EN list. This example also shows that although the three steps of transfer in, calculate, and transfer out *always* occur, it is sometimes convenient to combine some of them.

Comparison with plant operation

No direct comparison was possible because complete data about inputs and outputs to a mixer were not known. Nevertheless, this model was often used in the complete plant simulation and the quality of the results obtained depended to a limited extent on its reliability. This is indicated later in Sec. 9.4 when the simulation of the reactor system is discussed.

6.5 Summary

The development of the models for the equipment is facilitated if a logical approach is taken. A suggested method is to:

1. Decide on the assumptions and the physical phenomena to be included in the model;
2. Formulate the equations and check for mathematical consistency;
3. Obtain required data;
4. Prepare the program and check for stream consistency;
5. Relate the model to the plant operation.

This strategy was applied to the fluid handling equipment not only to introduce the strategy but to illustrate how the models are developed so that they fit within PACER.

A survey of the fluid handling equipment suggested that three different basic types would be useful: MIXERS, PUMPS, and BLOWERS. Furthermore, because of the particular needs of this simulation and because of the modeling and physical property decisions described in Chapter 5, more than one variety of each type was developed.

The development of BLOW1 and MIXER6 was described. BLOW1 showed the three-block process of information handling between the PACER executive program and the models. The information is first transferred from STRMI into the Unit Computation, the calculations are performed, and the results are transferred out into STRMO.

MIXER6 illustrated how the EN list could be used to make the Unit Computation general. In this case, a mixer is a unit that combines and later divides material among outgoing streams. The percentages of the total flow which leave in various streams are stored in the EN list so that general equations can be used within the Unit Computation. Nevertheless, the heat capacities required for the calculation of the temperature of the mixture are stored within the Unit Computation and not in the EN list. Hence, the Unit Computation is general for the proportions of the splits but specific to the species chosen. Illustrative algorithms are included.

PROBLEMS

6.1. Table 4.3 gives a list of simple models useful for approximate simulation and the preliminary mass balances for design.
   (a) Are these all the simple models that are needed?
   (b) Prepare algorithms for these Unit Computations and FORTRAN IV programs for these such that they can be used by PACER or by whatever executive program you are using.

6.2. It is good practice to include in your programs for Unit Computations messages indicating when the model is being applied to conditions that exceed the region of applicability. One such constraint is shown in the program in Fig. 6.2. Examples of fluid dynamical constraints on centrifugal pumps might include NPSH and a minimum flow of fluid through the pump.

(a) List common technical constraints that you should consider, including those in models for pipelines and ducts, pumps, blowers, fans, compressors, ejectors, gas lift systems, settling basins, cyclones, centrifuges, filters, decanters, mechanical mixers, fluidized beds, gas spargers, and jet mixers.

(b) The equipment listed in (a) is mostly classified as fluid dynamical equipment. What fluid dynamical constraints should be considered for the models of distillation columns, heat exchangers, reactors, absorbers, and extraction columns?

6.3. In this chapter, details of only two Unit Computations are given. Devise algorithms for the remaining Unit Computations listed in Table 6.2. Prepare FORTRAN IV programs for these such that they can be used by PACER or by whatever executive program you are using.

(a) MIXER3.
(b) MIXER4.
(c) MIXER5 with heat of mixing information given by Porter (1917).
(d) BLOW2. Correlate the head capacity information reported in Fig. 6.3 as a function of rpm.
(e) PLINE1, which is a model to estimate the pressure drop in a pipeline.
(f) PUMPS1.

6.4. Extend the Unit Computations in Prob. 6.3 by developing algorithms and working programs for the following Unit Computations:

(a) PUMPS2 to accept one input stream and up to five output streams; it specifies the total flow in the first output stream but otherwise is the same as MIXER1.
(b) PUMPS3 model to calculate the utility cost for running the pump.
(c) PUMPS4 model to handle only one input and one output, transfer the temperature from input to output, and calculate the pressure drop, given the flow as specified is STRMI (1,3). Decide how you want to feed in the head capacity information.
(d) MIXER7 to dilute a gas stream with air to yield a specified SO_2/O_2 ratio. The temperature is to be accurately predicted.
(e) MIXER8 to do everything that MIXER5 does except that it calculates the heat of dilution from an enthalpy concentration diagram such as that programmed in Prob. 5.3.
(f) Generalize MIXER6 to apply for any gas mixture.
(g) BLOW3 to consider the performance of the steam side as well as the air side. Generalize the model to allow for variations in the rpm.

6.5. The alkylation process is described in Appendix A and the information flow diagram was constructed in Prob. 4.14.

(a) Do an analysis similar to that described in Sec. 6.2 and Tables 6.1 and 6.2 for this process. What functionally different types of rigorous models might you have to develop?

(b) Can you model the compressors, coalescers, decanters, the water wash system, and pumps based on fundamentals and empirical correlations, or are you required to statistically correlate the output responses to inputs for this equipment?

6.6. Consider the compressor-condenser combination for the alkylation process. This consists of the product separator which is connected to the compressor by a pipeline with a flow control valve regulated by the pressure in the product separator, a water-cooled condenser at the exit of the compressor, and a receiver for the condensate. The prime function of this system is to liquefy the gas. The compressor flow depends on changes in upstream and downstream pressures. The system was *designed* so that the throttling valve and the condenser cooling water flow could be adjusted automatically to control the pressure in the product separator. In *operation*, the condenser water flow is at its maximum, and an increase in the vapor flow results in a higher pressure and temperature driving force in the condenser; that is, the outlet pressure of the compressor increases. All the total exit flow from the compressor must be condensed; the backpressure adjusts accordingly. The throttling valve is wide open normally so that an increase in flow of vapor increases the inlet pressure to the compressor. Devise an algorithm and a working program to simulate the compressor-condenser system. Use typical head capacity data, energy data, and temperature rise data. For example:

HP = 0.88*TLBS + 252.0

PC = 77.8 + TLBS*[−0.2712 + TLBS*(0.372 × 10^{-3} − TLBS* 0.1589 × 10^{-6})]

POUT = (95.0/PC)PIN

TLBS = lbs/hr

PIN = inlet pressure, psia

POUT = outlet pressure, psia

6.7. Head capacity and efficiency data for a pump are given by the third order polynomial correlations in terms of the volumetric flow rate. Prepare an algorithm and a working computer program to calculate the horsepower, pressure drop, and operating cost for such a pump.

6.8. For the Bayer process given in Appendix A, identify the fluid dynamical models that probably need to be devised. This analysis is similar to Prob. 6.5 and to that described in Sec. 6.2 and in Tables 6.1 and 6.2.

6.9. (a) Devise algorithms and a working program to simulate the filter press operation outlined in Prob. 4.11. Assume the cake is compressible but that the pertinent coefficients will be available through the EN list. Attempt several levels of rigor for each part of the operation.

(b) Based on this set of Unit Computations, should we have used Figs. 4.11, 4.12, or 4.13 for the original information flow diagram composed of simple models?

(c) During the washing, the rate of removal of interstitial fluid may depend on the rate of backwashing and the elapsed time of backwashing. How can this be accounted for in the modeling for steady-state simulation?

6.10. (a) Devise a Unit Computation for a rigorous representation of a thickener or settler to be used in the simulation of the Bayer process.

(b) In one settler, the settling solids are countercurrently washed. Allow for this behavior. As a start, see Scandrett (1963).

(c) Modify these rigorous Unit Computations so that they can be used to simulate the thickener in a water and waste treatment process.

6.11. Prepare an algorithm and a working Unit Computation to rigorously simulate the two-phase flow of a gas-liquid mixture in a circular conduit. This topic is considered in detail in Chapter 7, but it provides insight into how well we have grasped the principles of modeling fluid dynamical behavior to solve the problem without reference to Chapter 7.

Simulation of Heat Exchangers

7

7.1 Statement of the Problem

In general, the problem of simulating a heat exchanger can be stated as follows: given a heat exchanger of fixed geometry and area, involving any one or a combination of the mechanisms of heat transfer under various conditions of flow and temperature, and also involving gases, liquids, or solids or a combination of phases, what are the temperatures (or enthalpies) of the outlet streams?

7.2 The Approach to the Simulation Problem

In any simulation problem it is important to begin with the simplest models for all equipment units and to ascertain the sensitivity of the results of the simulation to the parameters of each model. If this sensitivity analysis suggests a more accurate model, a way to refine that model must be chosen.

There are two different approaches to a given modeling problem. One approach is to accumulate large amounts of input and output data and to analyze them statistically to provide an equation which describes the operation of the particular equipment unit over the range of variables investigated. The other approach is to start from the basic mechanisms and to describe mathematically what is actually occurring in the equipment. In some cases the programmer's knowledge is not advanced enough for the fundamental mathematical model and the first approach is necessary. However, there are many advantages in using fundamental models since they provide a much better understanding of the operation of the equipment and also allow confident extrapolation to areas outside the usual range of operation.

The mathematical model based on fundamental knowledge is the recommended approach in the formulation of Unit Computations. However, since the use of rigorous fundamental models can lead to an involved mathematical description, the computer programs may become very intricate and time-consuming. In these instances, after a program for a single unit has been tested by itself, the numerical results may be found to be insensitive to some parameter of the model. The model could be accordingly simplified by eliminating such a parameter. For example, an accurate calculation of the boiling heat transfer coefficient may not improve the accuracy of the overall heat transfer rate from gas to boiling liquid. Indeed, the programmer must strive at all times to reduce the computing time of any given model by eliminating any part of the program which is not necessary for the accuracy of the overall simulation.

Whenever the fundamental models lead to excessive computer time, they may be used to generate the statistical models which in turn can be used in the overall plant description.

Research in heat transfer has provided sufficient knowledge to allow the fundamental mathematical model to be formulated for most of the phenomena in conventional process equipment. At the same time, some information will be required in these models which only experience can reliably dictate. It is the purpose of this chapter to show how a fundamental knowledge of the three basic mechanisms of heat transfer—conduction, convection, and radiation—and their relative roles in a particular heat exchanger system, can be used to write mathematical models of what appear to be quite complex processes.

It is assumed that the reader is familiar with the fundamentals of the heat transfer mechanisms so that the discussion here will be restricted mainly to their application to the problem at hand. References will be noted for those who wish more detail; their translation to a working simulation should follow the approach outlined here.

7.3 Formulation of the Model Equations

Equations related to heat exchanger performance

The overall heat transfer coefficient dictates the performance characteristics of any convective heat exchanger. It is defined by the equation:

$$dq = U\,dA\,(\Delta T)_{\text{local}} \tag{7.1}$$

Equation 7.1 implies that the overall coefficient applies to a given point in the exchanger where the difference between the two fluids which are exchanging heat is $(\Delta T)_{\text{local}}$.

The magnitude of this coefficient is a measure of the ease with which heat can be transferred from one fluid to another in a given exchanger. It is in turn determined by the resistance to heat transfer offered by the fluids and by the material of the heat transfer surfaces. The overall resistance (which is the inverse of the overall coefficient or conductance) is the sum of the individual resistances when they are in series. This allows the overall coefficient to be evaluated from the well-known equation:

$$\frac{1}{UA} = \frac{1}{h_i A_i} + \frac{\Delta r}{k A_{av}} + \frac{1}{h_o A_o} \tag{7.2}$$

The use of A_{av} implies that the difference between the outside and inside radius of a pipe, Δr, is quite small. Clearly, the evaluation of the overall coefficient requires knowledge of individual coefficients h_i and h_o, which are known to be complex functions of fluid properties, flow conditions, and geometry. Numerous textbooks, for example, Kreith (1965), Eckert and Drake (1959), and Grober, Erk, and Grigull (1961), indicate how the basic flow and energy equations can be reduced to actual design equations for h_i and h_o or recast to indicate the important dimensionless groups for their empirical correlation. The evaluation of these individual coefficients for the present equipment will be discussed.

In exchangers in the chemical industry, foreign material or scale can form deposits which add resistance to the transfer of heat. Scale coefficients h_{fi} and h_{fo} must therefore be included in Eq. 7.2, which are at present evaluated on the basis of experience. The *Chemical Engineers' Handbook* [Perry (1963)] lists recommended approximate values for a variety of environments and conditions.

The preceding discussion has pointed to the two main problems in predicting the performance of an exchanger, namely:

1. Proper integration of Eq. 7.1 over the entire exchanger;
2. Evaluation of the overall heat transfer coefficient from the individual or film coefficients.

The integration problem

For any exchanger, the overall heat transfer rate may be obtained by integrating Eq. 7.1 to yield:

$$A = \int dA = \int \frac{dq}{U\Delta T} \tag{7.3}$$

Since the fluids will be either heated or cooled over the differential area, *dq* may be expressed mathematically in terms of this sensible heat change of the hot or cold fluid, namely:

$$A = \int_{T_{in}}^{T_{out}} \frac{(WC_p)_c dT_c}{U(T_h - T_c)} \tag{7.4}$$

Most standard texts, for example, Kreith (1965) and McAdams (1954), indicate the details of the analytical integration when the following assumptions are made:

1. The overall heat transfer coefficient is constant throughout the exchanger;
2. The specific heats of the two fluids can be assigned an average constant value;
3. The mass flow rates of the fluids are constant;
4. The exchanger is adiabatic with respect to the surroundings.

These assumptions hold fairly well for a low viscosity, turbulently flowing fluid when the temperature differences are not extreme.

With these four assumptions, the heat transfer rate for the exchanger with total heat transfer area A is given by:

$$q = UA\,\Delta T_{lm} \tag{7.5}$$

where

$$\Delta T_{lm} = \frac{\Delta T_1 - \Delta T_2}{\ln(\Delta T_1/\Delta T_2)} \tag{7.6}$$

The temperature differences ΔT_1 and ΔT_2 are those existing between the two fluids at each end of the exchanger, with the same expression valid for either cocurrent or countercurrent flow.

If the exchanger is not truly countercurrent or cocurrent but rather of a multipass or crossflow nature, the correct mean temperature difference is not the logarithmic mean. If the above four assumptions are made together with assumptions concerning the mixing or nonmixing of the shell-side fluid and the distribution of the heat transfer area among passes, expressions for the correct mean temperature differences may be derived [Bowman, et al. (1940)]. These expressions allow a correction factor to be calculated for use in Eq. 7.5 which is defined by

$$Y = \frac{\Delta T_{\text{true mean}}}{\Delta T_{\text{log mean}}} \tag{7.7}$$

and hence

$$q = YUA\,\Delta T_{lm} \tag{7.8}$$

In this case the log mean temperature is evaluated as if the exchanger were a countercurrent one. These factors Y, which are always less than unity, are usually presented graphically as a function of the inlet and outlet temperatures and thermal capacities (the product of weight flow and specific heat) of the fluids.

Exchanger effectiveness

Unfortunately, the use of Eq. 7.8 requires a knowledge of the inlet and outlet temperatures of both streams. Other disadvantages have been cited by Cary (1959). In simulation work, it is the outlet temperatures which usually are to be calculated, given the surface area and configuration of the exchanger, the rates of flow, and the fluid properties. The use of the heat exchanger effectiveness factor alleviates many of the difficulties associated with the use of the log mean temperature driving force.

The definition of the heat exchanger effectiveness is based on the hotter or colder fluid, whichever has the smaller thermal capacity. Thus, when $(WC)_{\text{hot}} = (WC)_{\text{smaller}}$

$$E = \frac{(WC)_{\text{hot}}(T_{h,\text{in}} - T_{h,\text{out}})}{(WC)_{\text{smaller}}(T_{h,\text{in}} - T_{c,\text{in}})}$$

and when $(WC)_{\text{cold}} = (WC)_{\text{smaller}}$ (7.9)

$$E = \frac{(WC)_{\text{cold}}(T_{c,\text{out}} - T_{c,\text{in}})}{(WC)_{\text{smaller}}(T_{h,\text{in}} - T_{c,\text{in}})}$$

Thus, the effectiveness E is the ratio of the heat actually transferred to the maximum heat which could be transferred in an infinitely long exchanger.

If one makes the same assumptions as were used in deriving the equation for the log mean temperature difference, it can be shown [Kays and London (1964)] that the effectiveness is a function of the thermal capacity ratio,

$$R = \frac{(WC)_{\text{smaller}}}{(WC)_{\text{larger}}} \tag{7.10}$$

and the number of transfer units,

$$NTU = \frac{UA}{(WC)_{\text{smaller}}} = \int_{T_{\text{in}}}^{T_{\text{out}}} \frac{dt_{\text{fluid}}}{t - t_{\text{fluid}}} \tag{7.11}$$

Effectiveness also depends on the flow arrangement in the exchanger. For example, the expression for the effectiveness of a countercurrent exchanger is given by:

$$E = \frac{1 - \exp[-(NTU)(1 - R)]}{1 - R\exp[-(NTU)(1 - R)]} \tag{7.12}$$

TABLE 7.1
SUMMARY OF HEAT TRANSFER CORRELATIONS USED IN HEAT EXCHANGER SIMULATIONS

System	Equation	Eq. no.	Reference	Properties evaluated at
1. Flow of fluid inside tubes without change of phase	Dittus-Boelter Equation $(Nu)_{DB} = 0.023(Re)^{0.8}(Pr)^n$	(7.14)	Kreith (1965)	Average bulk temperature
2. Flow of fluids inside tubes with boiling	Chen Equation $h = h_{mac} + h_{mic}$; $h_{mac} = \left(\frac{k}{D}\right)F'(Nu)_{DB}$; $h_{mic} = \frac{(0.00122)(S)k_L^{0.79}C_L^{0.45}\rho_L^{0.49}g_c^{0.25}\Delta T^{0.24}\Delta p^{0.25}}{\sigma^{0.5}\mu_L^{0.29}\lambda^{0.24}\rho_v^{0.24}}$	(7.15)	Chen (1966)	Saturation temperature except Δp
3. Flow of fluids across a bank of tubes	$\frac{hD_{Eff}}{k} = 0.33C_H\left(\frac{D_{Eff}G_{Max}}{\mu}\right)^{0.6}(Pr)^{1/3}$	(7.16)	Tate and Cartinhour (1945)	Average film temperature
4. Film flow of water over the outside of a vertical tier of horizontal pipes	$h = 65\left(\frac{\Gamma}{D_0}\right)^{1/3}$	(7.17)	McAdams (1954)	Not required

Similar expressions for other flow arrangements have been derived by Kays and London (1964). It is of interest, however, that for other types of exchangers the effectiveness can be obtained from the Y-factor corrections to the log mean temperature difference,

$$E = \frac{1 - \exp[-Y(NTU)(1 - R)]}{1 - R\exp[-Y(NTU)(1 - R)]} \tag{7.13}$$

The effectiveness E measures the performance of a given exchanger relative to the maximum performance possible and as such gives an indication of its relative efficiency. The use of effectiveness avoids the problem of unknown outlet temperatures and allows the relatively easy analysis of complex exchanger arrangements. It can also be used for cases in which the overall coefficient may vary in a prescribed manner [Cary (1959)].

TABLE 7.1 (CONT.)

Limitations	Expected accuracy	Remarks
(i) $n = 0.3$ if fluid being cooled (ii) $n = 0.4$ if fluid being heated (iii) $\mu < 2$ cp Re $> 10^4$ $0.7 < \text{Pr} < 20$	± 10 to 20%	$\text{Re} = \frac{D v_{av} \rho}{\mu}$ $\text{Pr} = \frac{c_p \mu}{k}$ $\text{Nu} = \frac{hD}{k}$
Applies to water in turbulent flow	$\pm 25\%$ claimed	F′ is given as a plot against $\frac{1}{X_{tt}}$ S is given as a plot against (Re) $(\text{F}')^{1.25}$
$\frac{D_{\text{Eff}} G_{\text{Max}}}{\mu} > 6000$ $C_H = 1$ for the number of rows > 10 $C_H < 1$ when number of rows is less or tube arrangement changed ($= 0.75$ for $n = 2$)	$\pm 20\%$	$G_{\text{Max}} = \frac{\text{mass flow rate}}{\text{minimum area for flow}}$ D_{Eff} = diameter of cylinder of equivalent projected area as finned tubes
Streamline flow of water over horizontal tubes $600 < \frac{\Gamma}{D_o} < 3000$	$\pm 25\%$	Γ = water rate from tube per unit width of stream, lb/(hr) (ft) $\Gamma = \frac{W}{2L}$ h is based on the arithmetic average temperature difference

The evaluation of the overall heat transfer coefficient

The role of convection

The key parameters in evaluating the overall coefficient are the individual film heat transfer coefficients. At present, most of these coefficients are evaluated from empirical correlations, (in terms of dimensionless groups) which have been available to the designer for many years. Many of these correlations now have a firm theoretical basis [see, for example, Sparrow, Hallman, and Siegel (1955)] and the errors and limitations can be properly assessed. In either case, the equations apply only to a specific geometry and to known fluids and flow conditions.

For the simulation of heat exchangers in a typical sulfuric acid plant, the following correlations are required:

1. Flow of fluids inside tubes without a change of phase;
2. Flow of fluids inside tubes with boiling;
3. Flow of fluids across a tank of tubes;
4. Film flow of water over the outside of a vertical bank of pipes.

The correlations chosen for this application, along with the limitations in their use, are summarized in Table 7.1.

The role of conduction—fin efficiency

When the outside film coefficient is small and is in fact the controlling resistance to heat transfer, it has proven economical to increase the outside surface area available for heat transfer through the use of extended surfaces known as fins or gills. In the sulfuric acid plant, fins are used in the exchangers which cool the reacting gases.

Clearly the surface area cannot be increased indefinitely since the added surface increases the conduction resistance within the metal, which eventually cancels the advantage of the added area itself. It is possible that the fin surface will not be at a uniform temperature and hence that the temperature driving force from fluid to metal surface will not be uniform over the entire tube wall.

Gardner (1945) analyzed this conduction problem, assuming a constant film heat transfer coefficient over the fin for a number of simple and complicated fin arrangements. These mathematical analyses allow the evaluation of the fin efficiency, defined by the expression

$$\eta = \frac{\text{actual heat transfer rate from fin}}{\text{expected heat transfer rate if fin uniformly at the fin base temperature}} \tag{7.18}$$

Evaluation of the fin efficiency for the specific exchanger tubes will be discussed later. The overall heat transfer coefficient required in Eq. 7.8 is reduced because the outside film coefficient-area product ($h_o A_o$ in Eq. 7.2) is multiplied by the fin efficiency η.

The role of radiation

A considerable amount of heat may be transferred by radiation which, in the absence of interaction, may be added directly to the convective flux. Clearly an evaluation of its importance is necessary here. A brief review of the subject is presented by way of introduction.

The well-known Stefan-Boltzmann equation

$$q/A = \sigma T^4 \tag{7.19}$$

expresses the heat radiated per unit area per unit time from an ideal radiator/

absorber maintained at an absolute temperature T. Since real systems at any given temperature will radiate only a fraction of this maximum radiation, the right-hand side of Eq. 7.19 is multiplied by a factor called the emissivity ($\epsilon_s \leqslant 1$). Moreover, some of this radiation is reflected, the magnitude of which is indicated by the reflectivity ($\rho_s < 1$) of the surface or by its absorptivity ($\alpha_s = 1 - \rho_s$). For very thin layers of solid or for fluid systems, transmission of radiation must also be considered.

Certain gaseous molecules (e.g., CO_2, CO, SO_2) exhibit strong vibrational-rotational energy modes and as a result absorb and radiate appreciable energy in the lower spectral region where thermal radiation is important (wave length <25 microns). Other gases such as oxygen and nitrogen do not radiate and appear transparent to thermal radiation. Again, when a gas does radiate, the radiation per unit volume or per unit area of bounding surface will always be less than that given by the Stefan-Boltzmann equation.

In any real system, an infinite number of rays of radiation will be emitted from any given point over the complete solid angle subtended by that point. Each of these rays will undergo an infinite number of reflections at each surface making up the enclosure. A number of approaches have appeared in the literature to solve this problem of an infinite number of terms in an infinite number of equations; these methods have been summarized by Sparrow and Cess (1966).

In each of these formulations, a factor F, referred to as a view or shape factor, is required. It is defined as the fraction of the radiation which leaves a given surface or gas volume and goes directly to another surface or gas volume. This view factor, which obviously cannot be greater than unity, depends solely on the geometry of the system and the absorbing power of the intervening gas. These factors have been evaluated for various simple geometries and as a function of the absorption strength of the intervening gas. Methods for calculating the radiant interchange in enclosures of various geometries which contain an absorbing real gas are reviewed by Hottel and Sarofim (1967).

In most cases under study here, it can easily be ascertained that either the temperature level is relatively low or the view factor is very small, so that radiation contributes only one or two percent of the overall heat flux. However, in the first waste heat boiler, the temperature level of about 900°C is significant, the view factors approach unity, and the gases contain an absorbing gas, sulfur dioxide. However, since sulfur dioxide is a relatively poor radiator ($\epsilon_G = 0.05 \pm 20\%$ over the temperature and concentration range of interest), the increase in the heat transfer rate resulting from radiation is only about five percent at the highest temperature level. Therefore, considerable error can be tolerated in evaluating emissivities without seriously affecting the simulation.

An approximate equation for calculating the radiant heat flux from a gas to its enclosing surface has been available for some time [McAdams 1954)]:

$$q/A = \left(\frac{\epsilon_s + 1}{2}\right)\sigma(\epsilon_G T_G^4 - \alpha_G T_s^4) \tag{7.20}$$

Since the surface temperature, T_s, will be low relative to that of the gas because of the high thermal resistance on the outside heat transfer surface, little error will be involved in assuming the gas emissivity ϵ_G and absorptivity α_G are equal. The emissivity ϵ_s of the surface will be high because of oxidation and erosion. The insulated brickwork containing the flowing gases will act as a reradiating surface. The net effect will be to direct most of the radiation emitted by the gas to the heat exchanger tubes. The resulting expression

$$(q/A)_R = \epsilon_G \sigma(T_G^4 - T_s^4) \tag{7.21}$$

is relatively simple to use in any program.

This procedure demonstrates the usefulness of an order-of-magnitude analysis in this work.

7.4 Data Required for the Superheater and Economizer Models

Description of the units

It has been shown that the outlet temperature for any heat exchanger can easily be ascertained from its effectiveness. This effectiveness depends on the geometry, the thermal capacity of the flowing fluids, and the product of the overall heat transfer coefficient and the area for transfer. This coefficient depends in turn on the geometry, the physical properties of the flowing fluids, and operating conditions within the exchanger. The information required is therefore related to these factors.

Superheaters

Two steam superheaters are used to provide temperature control of the process gases leaving the second and third beds of the sulfur dioxide reactor. These exchangers are identical, being composed of six 17 ft hairpin sections arranged in parallel (Fig. 7.1). The 2 in. outside diameter steel tubes carrying the steam are equipped with cast iron gilled rings (annular fins) which have a $4\frac{5}{8}$ in. outside diameter to provide greater heat transfer area (Fig. 7.2). A summary of the dimensions and normal operating conditions is shown in Table 7.2.

Hot gases are cooled as they flow across the superheater tube banks.

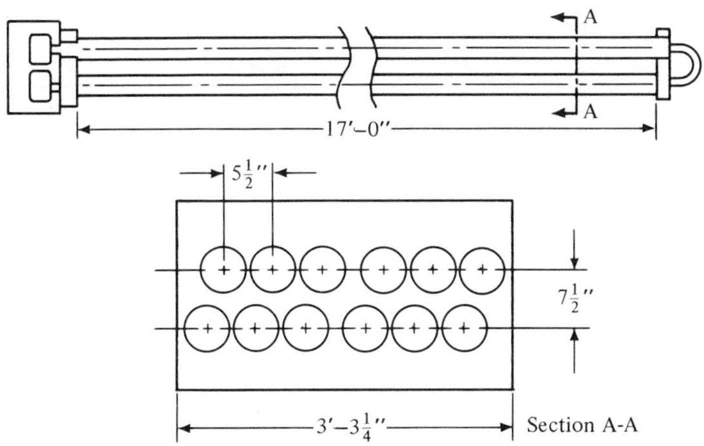

Figure 7.1. Details of the superheater.

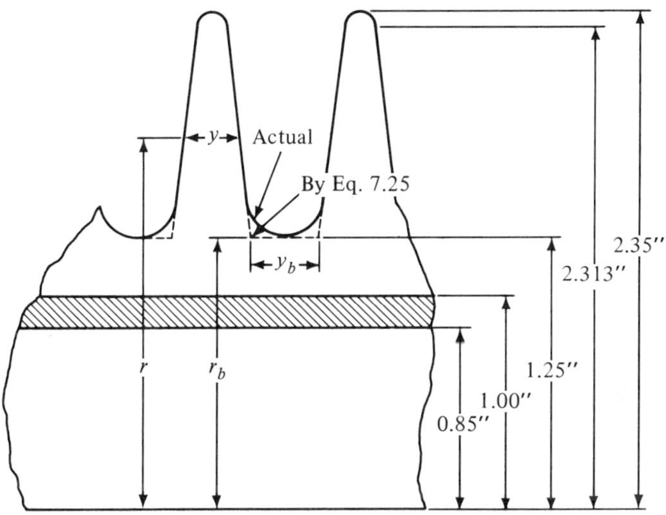

Figure 7.2. Cross section of a finned tube used in the heat exchangers.

Saturated steam coming from the steam drum flows inside the tubes countercurrent to the gas.

Economizer

The economizer not only cools the hot gases leaving the reactor before these gases enter the absorber, but also preheats the water feed to the steam drum of the waste heat boilers. The hot gases enter at the top and are cooled by crossflow over a tube bank which carries water countercurrently.

The economizer is composed of four 7 ft hairpin sections (eight tubes)

TABLE 7.2

DETAILS OF SUPERHEATER SECTION

Shell dimensions	
shell tube length	17 ft.
shell width	3 ft. $3\frac{1}{4}$ in.
Exchanger tubes	
total number of tubes	12
number of hairpins in parallel	6
tube length	17 ft.
outside diameter of tubes	0.167 ft.
inside diameter of tubes	0.139 ft.
tube pitch	$5\frac{1}{2}$ in.
tube clearance	$7\frac{1}{2}$ in.
material of construction	steel
thermal conductivity	29.0 CHU/hr ft °C
Exchanger fins (cast iron gilled rings)	
number of fins per foot of tube	17
outside diameter	$4\frac{5}{8}$ in.
diameter to base of fin	$2\frac{1}{2}$ in.
inside diameter of tube casing	2 in.
material	cast iron

SUPERHEATER OPERATING CONDITIONS

	Tube side	Shell side	
		Superheater 1	Superheater 2
nature of fluid	steam	gas	gas
normal flow	31,800 lb/hr to both superheaters		
inlet temperature	222°C	500°C	448°C
outlet temperature	304°C	439°C	423°C
pressure in	350 psia	33.75 in. H_2O	26.25 in. H_2O
pressure out	340 psia	32.75 in. H_2O	25.50 in. H_2O

TABLE 7.3

DETAILS OF ECONOMIZER

Shell dimensions	
shell tube length	7 ft.
shell width	2 ft. $11\frac{3}{4}$ in.
Exchanger tubes	
total number of tubes	96
number of tows in parallel	6
tube length	7 ft.
tube pitch	$5\frac{1}{2}$ in.
tube clearance	$5\frac{3}{8}$ in.

TABLE 7.3 (CONT.)

Economizer Operating Conditions		
	Tube side	Shell side
nature of fluid	water	gas
flow, lb/hr	31,800 + 11% blow down	59,300
inlet temperature	105°	427°C
outlet temperature	208°	259°C
operating pressure in	375 psia	15.4 psia
operating pressure out	—	15.0 psia

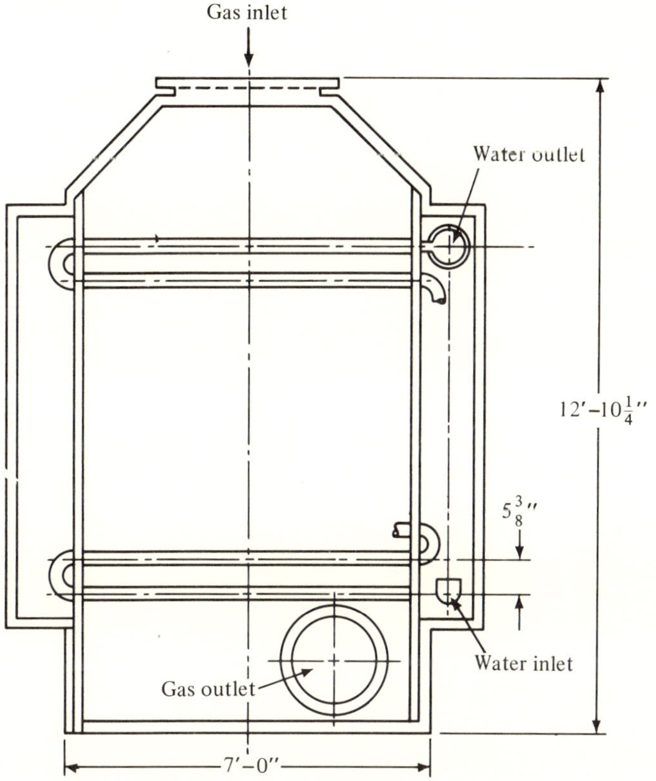

Figure 7.3. Details of the economizer.

arranged as six parallel heat exchangers (Fig. 7.3). The finned tubes are identical to those in the superheaters. The dimensions and operating data for the economizer are shown in Table 7.3.

Calculated information for the exchanger geometries

Finned-tube data

Details of the cast iron gilled rings were obtained from direct measurements of plant equipment and from information given for identical fins in Tate and Cartinhour (1945).

1. Effective diameter (D_{Eff} in Eq. 7.16, Table 7.1)
   The effective diameter in the heat transfer correlation is the diameter of a cylinder with a projected area equivalent to the projected area of the finned tube. The projected area is defined as the cross-sectional area of the tube as seen by the incoming gas. Direct measurement indicated 0.282 in., whereas Tate and Cartinhour suggested a value of 0.275 in.
2. Heat transfer area of fins
   The finned tube area is given by

$$A_o = \int_0^{2\pi} \int_{Z_1}^{Z_2} r dS d\phi \tag{7.22}$$

   where S = the arc length at any axial position Z
   r = the outside radius of the tube at Z
   These values were measured on a dimensioned drawing and the area was evaluated graphically. The calculated area per foot was 3.20 sq ft whereas Tate and Cartinhour suggested 3.10 sq ft.
3. Minimum area for crossflow
   The minimum area available for crossflow was obtained from a detailed scale drawing of the finned tubes and by weighing the paper representing the flow area per unit length. The minimum area was found to be 0.176 sq ft per foot of finned tube.
4. Efficiency of extended fin surface
   Gardner (1945) has derived mathematical expressions for the fin efficiency of various geometries. Among these is an expression for an annular fin whose thickness y varies inversely as the square of the distance from the center, i.e., $n = \frac{1}{2}$ in the expression

$$y = y_b\left(\frac{r}{r_b}\right)^{-2n/1-n} \tag{7.23}$$

   The definitions of the variables are indicated in Fig. 7.2. For this type of fin the efficiency is

$$\eta = \frac{\tanh(u_b - u_c)}{(u_b - u_c)} \tag{7.24}$$

   The variable u is a function of the radial distance from the centerline of the tube to a point in the fin and is given by

$$u = (1 - n)r_b\left(\frac{r}{r_b}\right)^{1/1-n}\left(\frac{h_o}{ky}\right)^{1/2} \tag{7.25}$$

where h_o = the average film heat transfer coefficient over the fin; this is assumed constant and is given by Eq. 7.16

u_b = the value of this function at the base of the fin

u_c = the value at the tip of the fin

For the finned tubes under consideration here, direct measurements indicated that $y_b = 0.21$, $r_b = 1.25$, and $r_c = 1.30$. The dotted line on Fig. 7.2(c) shows how well the empirical expression predicts the actual dimensions.

Heat exchanger effectiveness

Since the superheater is not a truly countercurrent exchanger, an equation similar to Eq. 7.12 may be derived. Alternatively, Eq. 7.12 may be modified by including the factor Y, as in Eq. 7.13, to account for the difference in performance between a truly countercurrent and a crossflow exchanger. Bowman, et al. (1940) present equations for calculating the Y factors for various flow arrangements in heat exchangers. For a two-pass, crossflow exchanger they derive the following expression:

$$\frac{1}{1-p} = e^{Kp/c}\cosh(Kp/c) + (1-K)\sinh(Kp/c) \tag{7.26}$$

where $K = 1 - \exp(-c/2\Delta T_{\text{mean}})$

$Y = \Delta T_{\text{mean}}/\Delta T_{lm}$

ΔT_{mean} = true mean temperature difference for the crossflow (two-pass) exchanger

ΔT_{lm} = log mean temperature difference if the exchanger is operated countercurrently under the same temperature conditions

$p = (T_{c\text{in}} - T_{c\text{out}})/(T_{c\text{in}} - T_{h\text{in}})$

$c = (T_{h\text{out}} - T_{h\text{in}})/(T_{c\text{in}} - T_{h\text{in}})$

Substituting the exponential functions for the hyperbolic functions gives

$$\exp(2Kp/c) = \frac{2 - K(1-p)}{(1-p)(2-K)} \tag{7.27}$$

Note that with this formulation the advantage of the effectiveness approach is lost, since Eq. 7.27 now requires the outlet temperatures for its solution. Fortunately, the solution of the effectiveness factor equation, Eq. 7.13, is not very sensitive to the outlet temperatures used in obtaining Y. Note further that putting the equation in this form isolates the most sensitive term on the right-hand side to ensure that the trial-and-error solution for K (and hence r and Y) will converge rapidly.

In the case of the economizer, the large number of tube rows makes this exchanger approach the performance of a countercurrent exchanger and the Y factor becomes unity.

Other considerations

Fouling coefficients were estimated from the handbook data [Perry (1963)]. The "contact" resistance to heat transfer between the cast iron gilled rings and the steel tube had to be estimated. These coefficients became the adjustable parameters required to fit plant data under a given operating condition. Heat loss and radiation heat effects were assumed to be negligible.

7.5 Programming the Model

The simplified algorithm of the heat exchanger calculations is shown in Fig. 7.4. A number of programming details are discussed below.

1. The superheater or economizer (or the waste heat boiler which will be discussed later) may be simulated by one subroutine (COOL7), since the tube geometry and spacing and exchanger type are essentially the same in each case. The outside heat transfer coefficient can be calculated in the same way each time and will vary according to the process gas composition and average temperature level. The inside coefficient will depend on the fluid (steam, water, or boiling water). The physical properties to be used in calculating the inside film coefficient must be evaluated accordingly. A particular exchanger (i.e., a superheater, economizer, or waste heat boiler) may be simulated from this program simply by providing a numerical switch (1, 2, or 3) within the EN list to indicate which exchanger is to be calculated. The appropriate tests on this switch, at the points in the program where the inside film heat transfer coefficient is calculated, allow the proper calculation to be performed. Similarly, the Y factor calculation can be included or bypassed and, in the case of the boiler discussed below, a very high heat capacity of the boiling liquid can be used to keep the inside temperature constant. All of these details may be easily incorporated into or left out of the calculation by testing the magnitude of this switch.
2. Since the calculation of the heat transfer coefficient for a given exchanger requires the appropriate temperature at which to evaluate the physical properties, the outlet gas temperature must be assumed. The heat transfer coefficient depends only slightly on temperature, so that considerable error can be tolerated in the initial guess. However, if the outlet temperature has been calculated on a previous iteration of the entire plant, it should be used to hasten convergence on the outlet gas temperature. If the outlet temperature has been calculated previously, the magnitude of the number will be within the range of two predetermined limits and this value should then be used as the initial

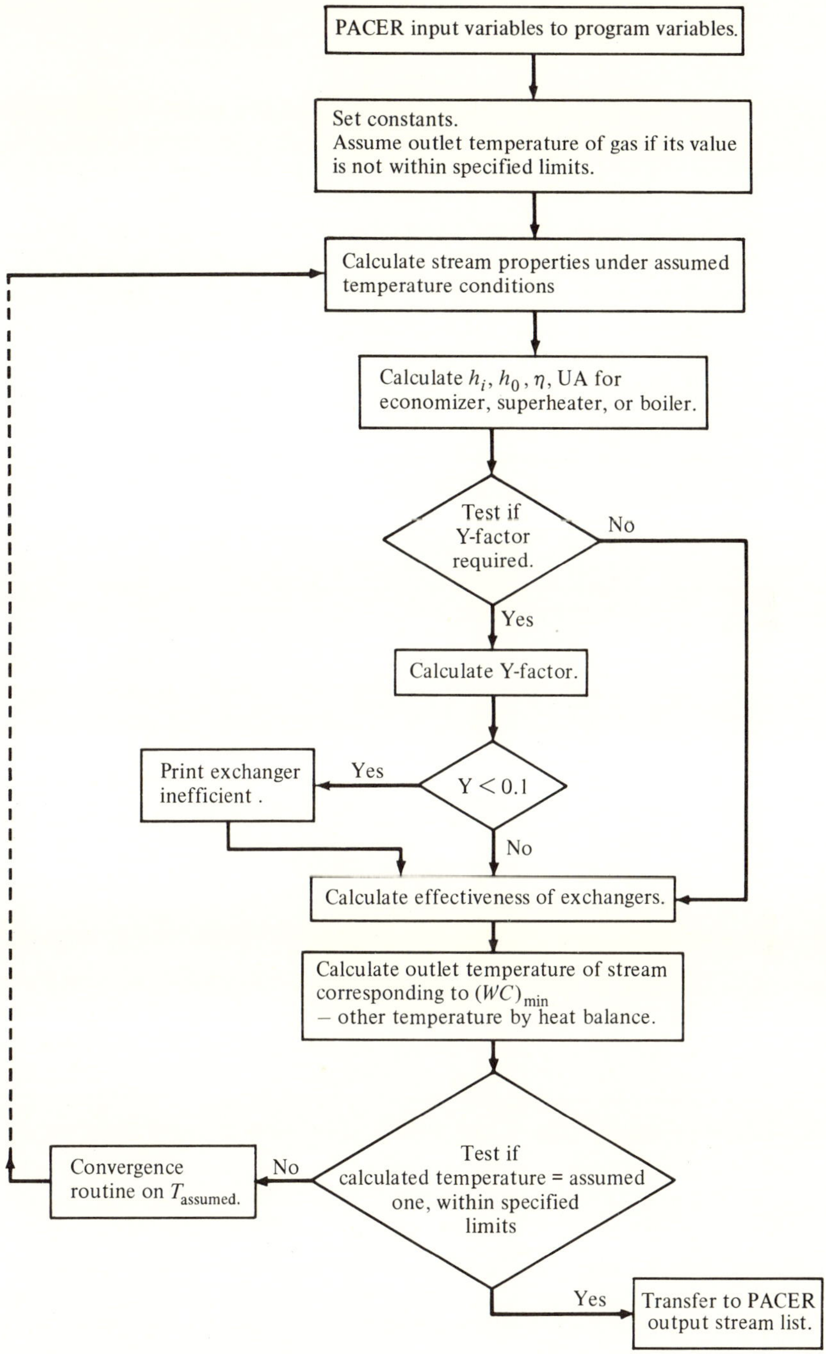

Figure 7.4. Algorithm for superheater and economizer subroutines.

guess of the outlet temperature. If not, a value is assumed as a starting point.

3. Tests must be made throughout the program to avoid encountering a coolant temperature greater than that of the heating fluid, since this can cause computer overflow or other errors. If this circumstance should arise, the cold fluid temperature is set equal to that of the inlet hot fluid, the calculation bypassed, a message to this effect printed, and these temperatures transferred to the appropriate STRMO list. Usually this happens only during the first couple of iterations when the information has not been passed along to all units in the flow diagram.
4. Under some conditions it is possible for the water to begin to boil in the economizer. This condition can be checked if the water pressure within the economizer is known. This in turn requires a pressure drop calculation for the water flow from the pump through the economizer to the steam drum of the boilers, whose pressure is fixed. If boiling does occur, this information should be pointed out since a complete disruption of the water flow rate will result from the interaction of the increased pressure drop for two-phase flow and the head capacity characteristics of the pump.

7.6 Matching the Model to the Plant

Since the coefficients in the correlations represent average values determined under a wide range of experimental conditions, the correlations can only be expected to predict performance within a limited accuracy, as indicated in Table 7.1. Therefore these coefficients can be adjusted somewhat to match the calculated data to plant data.

Some adjustment will be possible through the choice of the fouling coeffi-

TABLE 7.4
ADJUSTMENT OF FOULING COEFFICIENTS TO EQUATE PREDICTED AND PLANT DATA FOR THE ECONOMIZER

Assumed fouling coefficients			Effectiveness	Fin efficiency	Predicted temperatures, °C	
h_{fi}	h_{fo}	h_{CONT}	E	η	$T_{L,\text{out}}$	$T_{G,\text{out}}$
∞	∞	∞	0.52	0.87	210	258
500	∞	500	0.45	0.87	204	281
500	∞	∞	0.48	0.87	209	269
∞	∞	500	0.49	0.87	211	255
Measured temperatures					208	259

cients and the contact resistance offered by the press fit joint between the cast iron fins and the steel tube. Table 7.4 indicates the variation in outlet temperature possible by changing these resistances within the expected limits. It will be noted that if only the contact resistance is included in the economizer model, the predicted and actual temperatures differ only by 4°C. Experience with the superheater was essentially the same. This illustrates how an empirical adjustment using fouling and contact resistances can satisfy the plant data but cannot be justified on the basis of other data, or how important it is to have accurate plant measurements. The correctness of these empirical adjustments can only be tested by operating the plant at another overall condition. In Chapter 9, calculations at half capacity show that the heat exchanger models accurately described the actual performance under these conditions.

7.7 Data Required for the Simulation of the Waste Heat Boilers

This section will develop the simulation of a natural circulation, waste heat boiler in which steam is generated inside the heat exchanger tubes as hot process gas flows across them. A schematic diagram of the waste heat boiler system is shown in Fig. 7.5. The problem is the same as the one stated at the beginning of the chapter: the determination of the outlet temperature of the process gas and the amount of steam generated in the tubes.

The description of the heat transfer on the process gas side will be essentially the same as for the economizer and superheater. However, the phenomena associated with the generation of steam are much more complicated, because of an interaction of two-phase flow and boiling heat transfer.

All the complexities can best be described by analyzing the operation of a boiler, in this case boiler #2, since it is in some ways simpler than the first. Moreover, the capacity of the plant is restricted in part by the cooling capacity of the second boiler, so that its full detailed description becomes more important.

Since all the details of this model would detract from central considerations, only the salient points of the comprehensive model will be included here.

Description of boiler #2

Boiler #2 is a waste heat boiler which acts as an interstage cooler for the process gas as it passes from the first to the second bed in the sulfur dioxide

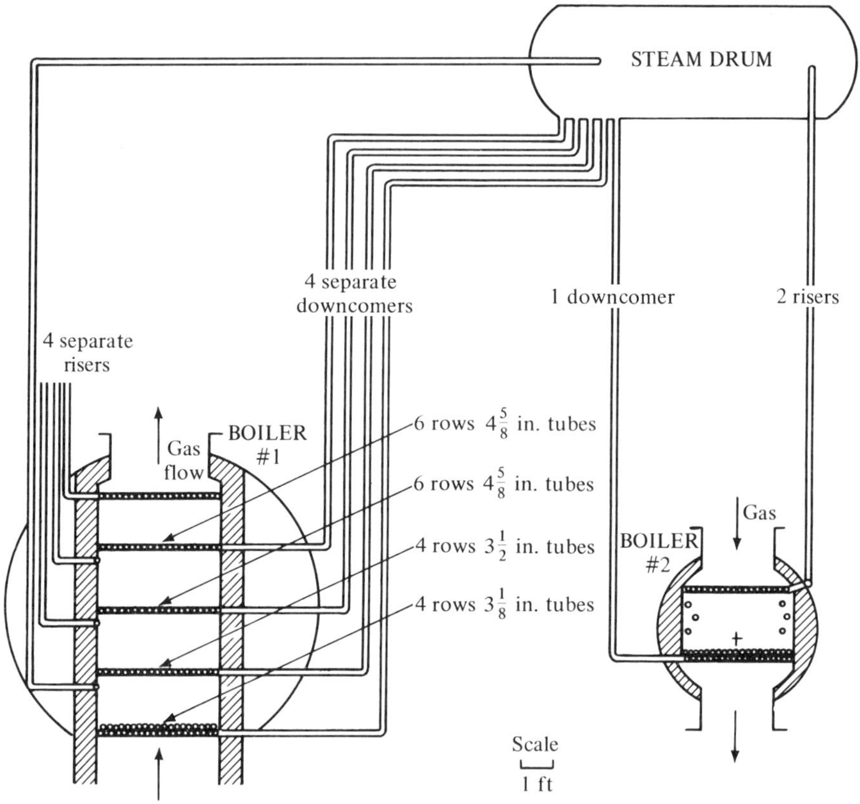

Figure 7.5. Waste heat boiler system.

reactor. Process gas enters at the top at approximately 600°C and flows across the tubes. After cold air is mixed with the exit gas from the boiler, the measured temperature is about 440°C. Steam is generated in the tubes at 350 psia.

The boiler consists of 30 finned hairpin tubes, ten tubes in parallel with six passes in series. These finned tubes are the same as those used in the economizer and superheater; in fact, the clearance and hence the minimum area available for gas flow is the same. The vertical distance between tubes varies slightly from row to row but may be considered constant without serious loss of accuracy.

A constant head of water is maintained at the inlet by a steam drum located approximately 15 ft above the boiler. The steam-water mixture leaving the top of the boiler is returned to the steam drum by two 4-in. diameter risers. Figure 7.5 indicates the arrangement and details of the interconnecting piping between the steam drum and the boiler.

Discussion of the rate processes occurring in the boiler

Water is fed to the boiler via the downcomers and vapor is generated within the boiler and transported by the risers back to the steam drum. The driving force for flow in a natural circulation boiler is the difference in densities between the fluids in the downcomers and in the heat exchanger and risers in the closed system. Therefore, to ascertain the flow rate of the liquid, the effective density of the two-phase mixture in the risers and the heat exchanger must be known. This requires a knowledge of the fraction of the liquid and vapor at every cross section in the boiler and risers, which in turn requires a knowledge of the vaporization or heat transfer rate at every point. Unfortunately this depends directly on flow rate. Moreover, the frictional resistance to flow and other pressure losses will be determined by flow rate and the degree of vaporization. With the flow rate depending upon the heat transfer rate (vaporization) and the heat transfer rate depending upon the flow rate, the solution to this problem will be one of trial and error on the liquid flow rate. The criterion for the correct flow rate will be the equivalence of the pressure increase at the water inlet due to the head in the downcomer with the pressure decrease from this point to the steam drum, i.e., when

$$\int \frac{dp}{dL}\, dL + \sum_i \Delta p_i = 0 \tag{7.28}$$

The integration is performed over the pipe runs in the system; the i indicates all other contributions to pressure change in the flow system. Therefore, the full description of the boiler will require a detailed knowledge of two-phase flow and boiling heat transfer phenomena under these flow conditions. This information will include:

1. Pressure drop in pipes and fittings for single-phase flow [Perry (1963)];
2. Correlation of pressure drop in pipes and fittings for vapor-gas, two-phase flow [Dukler, et al. (1964), Tong (1965)];
3. Correlation of vapor holdup with two-phase flow parameters for a given fraction of liquid vaporized [Hughmark (1962)];
4. Heat transfer coefficient for single-phase flow and for forced flow boiling inside pipes [Chen (1966)].

7.8 Programming the Model BOILS2

The differential change in pressure and vapor concentration could not be easily related to the two-phase flow parameters and physical properties because of the complicated form of the basic correlations for predicting boiling heat transfer, gas and steam properties, and the two-phase flow inside pipes.

Therefore, the vapor quality and flow parameters were assumed constant over short sections of the heat exchanger pipes, so that a constant heat transfer rate and pressure drop over this section could be calculated. The conditions at the start of the short section were assumed to hold over the entire section. Changing the length of the section from six in. to three in. changed the water flow rate by only a negligible amount.

The calculation must start at the entrance to the downcomer with a given flow of liquid and proceed to add and subtract the pressure changes in turn. An additional complication arises when an attempt is made to calculate the heat transfer to the first row of the heat exchanger tubes. Since the model must be formulated with known inlet conditions only, the outlet gas temperature and the effective temperature driving force over these tubes are not known. Therefore, two iterations are required: first, the outlet gas temperature must be assumed, and second, the mixed gas temperature immediately upstream of any tube row must be assumed in order to provide for an effective temperature difference over the tubes. The latter convergence must be achieved on each row in turn after a given temperature has been assumed for the outlet gas. Thus there are three major convergence cycles in the program—the water mass flow rate, the outlet temperature of the gas, and the average temperature driving force over each row of tubes. In addition there are those routines required for calculating the average temperature at which various physical properties are to be evaluated, as well as those for other transcendental equations [for example, the Colebrook equation for the single-phase friction factor, Kaufmann (1963)]. Fortunately, using the reguli-falsi convergence technique on the temperatures, only three or four iterations were necessary on the outlet gas temperature; using the Newton-Raphson method on the water flow rate, four to six iterations were required if the flow was not approximately known from a previous loop; all other iterations also used the Newton-Raphson convergence method. The program required approximately one minute on an IBM 7040 computer to calculate the total pressure change around the loop for one water flow rate trial.

Since this time requirement was clearly disproportionately long for any one Unit Calculation in the PACER simulation of a complete plant, an attempt was made to simplify the program to reduce the time per calculation. Indeed, it was noted that the process gas side resistance to heat transfer was two to three hundred times higher than that on the water side. This meant that considerable error could be tolerated on the inside heat transfer coefficient without seriously affecting the overall heat transfer rate. This led to the modification of the Unit Computation for the economizer and superheater (COOL 7), as has been discussed earlier.

7.9 Matching the Model to the Plant

The agreement between the simplified model and the more detailed one was excellent; for example, the outlet temperature was calculated to be 470.9°C for the simple model and 470.1°C for the more sophisticated one. When 142 lb moles per hr of cold air were mixed with the process gas leaving the boiler, the mixture temperature was 450°C which agrees well with plant measurements. Again, varying the fouling and contact coefficients through the expected limits produced a variation in outlet gas temperature of 10°C so that plant data may easily be fitted in this way.

7.10 Other Heat Exchangers in the Plant

Boiler #1

Boiler #1 is a waste heat boiler which is located after the sulfur burner; it is similar in design to boiler #2 (Fig. 7.5). Four downcomers from the same steam drum supply water to each of the four sections of the boiler. Because of the high temperatures at the inlet, the height of the gilled rings is smaller there than at the outlet, where the tubes are the same as those in the exchangers already discussed.

Although there are these differences in geometry and flow arrangement, the extension of the simple model (COOL7) to this case follows quite easily. Since there are three sizes of tubes, the exchanger must be divided into three parts, each with a given area. This could be done by inserting three consecutive COOL7 routines, each being used as a boiler model, to simulate the one exchanger. The radiation calculation can be included in this simple boiler routine and the additional heat transfer can be calculated at the end of the convection calculation once the average gas temperature and average tube wall temperature are known.

It is difficult to compare the predictions with plant data because a bypass valve in the gas feeding the boiler was always partly open and consequently the gas flow rate through the boiler was unknown. If the prediction of the boiler performance is assumed to be adequate on the basis of previous success, then this simulation can be used to indicate the fraction of the flow which is bypassed to yield the measured temperature of the two gas streams after they are remixed.

Sulfuric acid cascade-drip (horizontal tube film-type) coolers

Cascade-drip coolers are provided to cool the sulfuric acid before it enters the absorption and drying towers and before the 66° Bé product is transferred to storage. In each case the sulfuric acid is pumped from a hold tank through the series of pipes forming the cascade. Water is pumped from a sump to the top of the cooler where it is distributed by a weir arrangement over the entire length of the pipes. The details of the entire cooler are shown in Fig. 7.6. The operating variables and dimensions are given below.

1. Absorption tower circulating acid.
   This cascade consists of twenty-two 48-ft lengths of 8-in. cast iron pipe connected in series as indicated in Fig. 7.6. Water is circulated from the sump at approximately 600 U.S. gal per min to the troughs of this cooler and that of the drying tower acid.

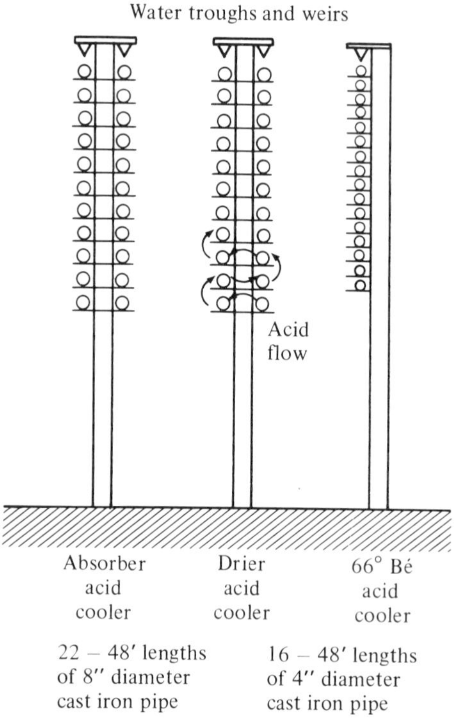

Figure 7.6. Schematic diagram of the sulfuric acid cascade coolers.

2. Drying tower circulating acid.
   This cascade cooler is identical to that for the absorption tower circulating acid.
3. Product acid cooler.
   The product acid is circulated from the bottom in series through sixteen 48-ft lengths of 4-in. cast iron pipe. Water is pumped to the trough at approximately 120 U.S. gal per min.

The simulation of these exchangers is left as an exercise for the reader. Correlations for the film coefficients are available in Table 7.1.

7.11 The Problem of the Oversophisticated Model

The experience with the full-fledged boiler calculation has indicated one of the important questions in plant simulation: how closely does one want to describe the actual process? This question can only be answered by judging the relative importance of the cost and the time available for generating the model. Moreover, one must remember that the accuracy of the model is determined by the accuracy of the information or correlations that are being used in the model. As a general rule, the programmer should strive to make the simplicity of his models commensurate with their intended use. If large extrapolations are contemplated, they should be done with some confidence, or at least one should be able to put confidence limits on the answer.

Often a given parameter within the simple model may be changed deliberately to ascertain the sensitivity of the whole operation to changes in this parameter. Through this type of manipulation, a feel for the importance of certain parts of the model is soon developed.

The long, detailed computer program has a place in a plant simulation only if meaningless results are generated without it. Even in this case, however, this detailed program should be used to generate simpler empirical equations by carrying out statistically designed experiments with the mathematical model.

Indeed, the detailed model may be used in many ways outside of the overall plant simulation. For example, it may be used to check the limits of operation of the simulated equipment. In the present case, the fact that the boiler is now operating with nearly 15 percent vapor in the last tube row indicates that the highest operating temperature has probably been reached, since transition from annular to mist flow with the attendant dry wall formation is expected near this concentration.

On the other hand, if a given piece of equipment becomes the major restriction to increased production, separate studies with the model may show how to improve the design; for example, by arrangement of piping the bottleneck may be removed. Thus the effect of increasing the number of risers and of adding another row of tubes to the boiler was investigated.

In the chapters which follow, attention is directed to the judgment that has been exercised in formulating the reactor and absorber models so that meaningful yet relatively short computer programs will result.

7.12 Summary

This chapter has illustrated how a fundamental knowledge of heat transfer is applied to describe the operation of rather complex heat exchange systems. The models were made to describe the plant exchangers extremely well by adjusting the fouling coefficients within the expected limits. The simulation of one of the waste heat boilers indicated that oversophistication can result in a disproportionate use of computer time. The problem was easily overcome in the present case by introducing an approximation; in other situations, a model based on statistical considerations is suggested.

PROBLEMS

7.1. Develop a Unit Computation COOL7 to simulate the economizer and superheaters.

7.2. Develop a Unit Computation to represent two-phase flows. Combine this with COOL7 to simulate a boiler.

7.3. Develop a simulation for the exchangers described on page 176.

7.4. Devise a general Unit Computation for a shell and tube exchanger, based on effectiveness factors, to handle all of the following configurations: countercurrent, parallel, crossflow—both streams unmixed; crossflow—both streams mixed; crossflow—with the stream having the smaller thermal capacitance unmixed; crossflow—with the stream having the larger thermal capacitance unmixed; parallel—counterflow with N shell passes; and liquid or gas on one side with a condensing or evaporating pure liquid on the other side.

7.5. Develop a Unit Computation for a single-stage flash evaporator.

(a) Assume an adiabatic flash for a given temperature and pressure.

(b) This model can be used to represent the product separator in the simulation of an alkylation plant (described in Appendix A) provided we allow for three sources of heat. These are the enthalpy in the entering feed (assumed to be all liquid), the heat of reaction removed from the reactor, and the heat transfer from the underflow from the depropanizer. Modify the Unit Computation developed in (a) to account for this.

Simulation of Single-Bed Reactors

8

8.1 Statement of the Problem

The problem of simulating a single bed of a catalytic reactor can be summarized as follows: given the bed geometry, catalyst properties, inlet flow rates and temperature, and the outlet pressure, calculate the outlet composition and temperature and the inlet pressure. By contrast, the design problem is to calculate the bed geometry from the inlet and outlet conditions. The approach to the problem is independent of the number of reactions occurring, the detailed kinetics of those reactions, and even of whether it is known which reactions actually occur. The steps to be taken are:

1. Choose the physical phenomena to be included in the model. In other words, establish the assumptions of the model. The rigor of the model will depend on the risks involved in a poor answer, the accuracy of the physical and plant data available, and the cost of the computation. Thus it may be unjustified to include a detailed account of diffusion within the catalyst if the kinetics of the chemical reaction is only approximately known.

2. Formulate the equations of the model under the assumptions made. The equations may be algebraic or differential, depending on the problem, but should be an exact embodiment of the model.
3. Obtain the data required for solving the equations. Only when the equations are written will it be clear what data are needed and to what accuracy. If the kinetics are not well known, there is little advantage in spending time collecting very accurate data on the heat capacity and enthalpy of reaction. It is more important not to use inconsistent thermodynamic data such as a nonzero enthalpy of a reaction together with an equilibrium constant independent of temperature. Similarly, if the flow is turbulent, viscosity need not be known accurately since the friction factor will be independent of Reynolds number.
4. Choose the method of solving the equations and prepare the program. Again, the complexity of the method used to solve the equations should correspond to the accuracy of the data and the degree of approximation in the assumptions.

These steps will be examined in turn, with the reactor for the oxidation of sulfur dioxide by air as an example.

8.2 Choice of the Rate Processes to Be Used in the Model

A rigorous approach demands that all transport phenomena which contribute to bringing together the reactants on the catalyst and to removing the products must be accounted for. But from among the rate processes which occur during a chemical reaction over a catalyst, only those should be chosen which affect the practicable accuracy of the simulation. During the development of the simulation of a plant, much useful information about the numerical behavior can be obtained with a reactor model which calculates the equilibrium composition. A fractional approach to equilibrium can be specified in order to obtain various conversions. The alternative extremes are either to formulate a comprehensive model from which all special cases are obtained by striking out terms, or to eliminate all unimportant rate processes before formulating the equations. The latter will be done here since this avoids carrying unnecessary detail, although the treatment is less general.

Within the bed, gradients of concentration and temperature arise by the very occurrence of reactions, by heat transfer to the reactor walls, and by variations in velocity (and thus residence time) with position. For a rigorous simulation of a reactor, one must include all transport processes which tend to alter those gradients. The resulting equations would be a set of nonlinear second-order partial differential equations which could be solved numerically but only at considerable expense in programming and computation time. A fuller discussion of the problem is provided by Petersen (1965). Here the transport processes are examined to see whether they ought to be included in the model.

Bulk transport in the fluid phase

Transport within the fluid phase is usually split into radial and axial components. One of the main mechanisms of transport is by molecular and turbulent diffusion, collectively called *dispersion.*

To limit this discussion, *radial uniformity of composition and temperature* will be assumed. This implies uniform axial velocity at any cross section, no significant radial temperature gradients resulting from heat transfer through the wall, and no variation with radius of any axial dispersion process. Under these conditions, radial variations will not arise. Radial uniformity can alternatively imply that radial transport is rapid enough to smooth out any radial variations which might otherwise exist. Thus *radial dispersion* will not be considered further.

Peterson (1965) has discussed a criterion a for deciding whether *axial dispersion* is significant for an isothermal first-order reaction in a catalyst bed. It is given as

$$a^2 = \frac{kE_A}{\bar{v}^2} = \left(\frac{k\ell}{\bar{v}}\right) \cdot \left(\frac{d_p}{\ell}\right) \cdot \left(\frac{E_A}{d_p\bar{v}}\right) \tag{8.1}$$

where k = first-order rate constant
E_A = axial eddy diffusivity
$\bar{v}$ = average velocity of the fluid
d_p = particle diameter
ℓ = bed depth

If $a^2 < 1$, axial dispersion is negligible.

For catalyst beds in turbulent flow, the last multiplicand in Eq. 8.1, the inverse Peclet number, is approximately $\frac{1}{2}$. Almost always, $\ell/d_p > 10$ and usually > 50. Finally, $k\ell/\bar{v}$ represents the ratio of residence time to reaction time and would rarely need to be as large as 10. Substitution into Eq. 8.1 yields $a^2 < 10 \cdot \frac{1}{10} \cdot \frac{1}{2} = 0.5$. Therefore, even in extreme cases of shallow beds, it is rare that a^2 would be as large as 1 for turbulent flow in a first-order reaction. Hence axial dispersion of matter will be neglected in the present model. Care should be taken however to assess the importance of axial dispersion for laminar flow where the axial Peclet number can be much smaller.

No similar criterion for axial conduction of heat is available and since heat can be transported not only by dispersion in the fluid but by conduction through the solid, a direct analogy to dispersion of matter is not valid. Axial conduction of heat will be neglected, however, with the understanding that cases may arise where it is important.

A case in which axial conduction is likely to be important is the first bed of the multibed adiabatic reactor for oxidation of sulfur dioxide. In this reactor the temperature can rise by 150°C in 18 in. At the same time, this rapid temperature rise means a very fast reaction which essentially reaches equilibrium before the end of the bed. Although axial conduction may be important in this case, it only affects the position at which equilibrium is reached within the bed and not the exit composition. Therefore, even in such an extreme case, no great error is introduced by neglecting axial conduction.

The assumption of a uniform velocity profile at any cross section is one of the most difficult to avoid because without it, one must have detailed information—almost never available—on flow patterns within the bed. The possibility of maldistribution of gas is great and requires care in designing the distributor and in packing the bed. Uniform velocity at a cross section will be assumed here.

Diffusion and heat conduction within the catalyst particle

The limitation of the rate of reaction within the catalyst particle, which is due to diffusion of reactants in and products out and to heat conduction, can be accounted for by a catalyst effectiveness, E_c. The factor E_c is defined as the ratio of the rate of reaction within the catalyst particle to the rate which would be obtained if the whole interior of the particle were at the same temperature and composition as the external surface.

The catalyst effectiveness will be assumed to be constant over an entire bed and thus it will be given by a single value, usually less than unity. Values of E_c greater than unity can arise in exothermic reactions in which the interior of the catalyst becomes hotter than its surface. This can lead to higher rates inside the catalyst than on its outer surface, despite the limitation of diffusion.

The value of the catalyst effectiveness will normally be fixed by comparison with plant operating data, although in theory it could be calculated or measured for the pellet used [Petersen (1965)]. In practice, the catalyst effectiveness can also take account of the differences in kinetics from the catalyst used in the plant to that used by workers in studying the kinetics as well as of differences in aging.

Mass and heat transfer to the particle surface

The transfer of mass and heat from the bulk gas to the surface of the catalyst particle will be assumed to limit the rates of reaction to an extent which can be accounted for by a factor multiplying the rate constant. This factor will be fixed for any one bed and will be included in the catalyst effectiveness, E_C. The absence of such a limitation amounts to operation of the reactor at a high enough flow rate that no further improvement in conversion would occur by increasing the flow rate and lengthening the bed in proportion.

8.3 Formulation of the Model Equations

The model equations are derived for a single-bed reactor at steady state in which m independent chemical reactions may occur among n components. Heat transfer to an external fluid may occur through the walls of the vessel. The assumptions to be used may be summarized as:

1. Radial uniformity of composition and temperature;
2. Flat velocity profile at any cross section;
3. No axial diffusion or heat conduction in the bed;
4. Mass and heat transfer limitations can be incorporated in an effectiveness factor;
5. Catalyst effectiveness is constant through the bed;
6. Steady state.

Assumptions 1., 2., and 3. constitute the plug-flow reactor model in that each element of gas passes through the bed as a plug, unaffected by the preceding or following elements.

Mass balance

Over a cross section of the reactor bed, shown in Fig. 8.1, the mass balance on component i gives

$$dF_i = r_i \rho_B A dz \tag{8.2}$$

where F_i = molar flow rate of i
r_i = net rate of formation of component i per unit mass of catalyst

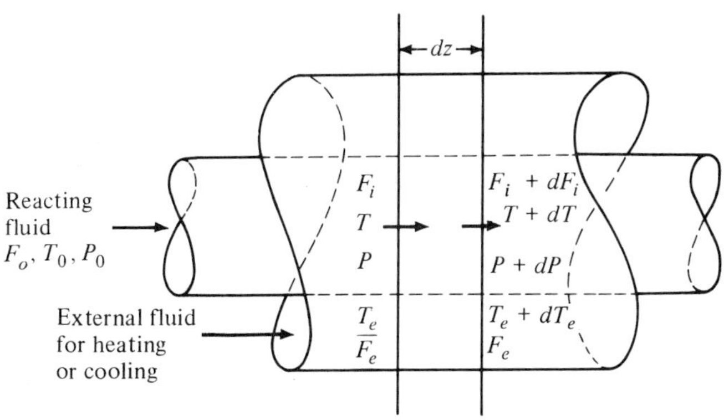

Figure 8.1. Cross section of the reactor bed.

by all reactions in which it participates. Later in Sec. 8.3, it is related to the stoichiometry of the reactions occurring.

ρ_B = bulk density of catalyst
A = area of cross section of bed
z = distance through the bed

The balance can be written across the entire cross section, having assumed radial uniformity. There is no contribution from flow by axial diffusion or by unsteady state accumulation. The inclusion in the effectiveness of mass and heat transfer limitations allows the composition and temperature at the external surface of the catalyst particle to be replaced by those in the surrounding bulk gas.

The bed has been assumed to be pseudohomogeneous in the sense that at any point within the bed the proportion of gas to solid is that of the overall bed and that a smoothly varying gas composition is defined there. The fact that properties vary discontinuously within the actual bed contradicts the preferred model, chosen for mathematical ease. For a further discussion of this, see Petersen (1965), Chapter 7.

For further use, Eq. 8.2 is written in the form

$$\frac{dF_i}{dz} = A\rho_B r_i \qquad (n \text{ equations}) \tag{8.3}$$

One such equation is needed for each of the m independent reactions occurring. Since composition changes are determined solely by the extents of reaction, all other F_i can be calculated from the stoichiometry of the reactions.

Energy balance

For a reactor, kinetic and potential energy changes and shaft work can usually be neglected. The first law of thermodynamics then gives for an element moving with the flowing fluid

$$\sum_{i=1}^{n} d(F_i\bar{H}_i) = UA_H\, dz\, (T_e - T) \tag{8.4}$$

where $\bar{H}_i$ = partial molal enthalpy of component i
U = overall heat transfer coefficient
A_H = lateral area for heat transfer per unit length
T_e = temperature of external heat transfer medium

Equation 8.4 accounts for the net change in enthalpy of the element as a result of composition and temperature changes in the flowing fluid and of heat transfer to the element through the reactor wall.

The energy balance can be expanded to give

$$\sum_{i=1}^{n} F_i d\bar{H}_i + \sum_{i=1}^{n} \bar{H}_i dF_i = UA_H\, dz\, (T_e - T) \tag{8.5}$$

which with Eq. 8.2 gives

$$\sum_{i=1}^{n} F_i d\bar{H}_i + \left(\sum_{i=1}^{n} \bar{H}_i r_i\right) A\rho_B dz = UA_H\, dz\,(T_e - T) \tag{8.6}$$

In general, the partial molal enthalpy is a function of temperature, pressure, and composition. For ideal solutions, composition has no effect on $\bar{H}_i$ and since few data are available for enthalpy variation with composition, it is usually necessary to assume that the reacting fluid behaves at least as an ideal solution. Except for gases under high pressure, enthalpy dependence on pressure is small for liquids and low pressure gases. If the pressure effect on enthalpy is important, it can be allowed for by using equations of state or reduced property correlations as described in most thermodynamics books, for example, Hougen, Watson, and Ragatz (1959). Here the enthalpy will be assumed to depend solely on temperature, and since composition effects have been neglected, the partial molal enthalpy equals that of the pure compound, H_i, so that

$$d\bar{H}_i = dH_i = C_{pi} dT \tag{8.7}$$

with $C_{pi} = f(T)$. Then Eq. 8.6 becomes

$$\left(\sum_{i=1}^{n} F_i C_{pi}\right) dT + \left(\sum_{i=1}^{n} H_i r_i\right) A\rho_B\, dz = UA_H\, dz\,(T_e - T) \tag{8.8}$$

or

$$\frac{dT}{dz} = \frac{-A\rho_B \sum_{i=1}^{n} H_i r_i + UA_H(T_e - T)}{\sum_{i=1}^{n} F_i C_{pi}} \tag{8.9}$$

By using the stoichiometry of the reactions, this equation is put in a more useful form later in Eq. 8.20.

In addition, there is an energy balance for the heat transfer medium which equates the heat lost by the reacting fluid to the heat gained by the other. Then from Fig. 8.1,

$$\pm F_e C_{pe} dT_e = UA_H dz (T - T_e) \tag{8.10}$$

where F_e = molar flow rate of heat exchange fluid
C_{pe} = molar heat capacity of exchange fluid

The plus or minus refers to cocurrent or countercurrent heat exchange, respectively. The exchange fluid temperature will rise in the direction of flow if $T > T_e$ but the flow is toward decreasing z for a countercurrent heat exchanger.

Momentum balance

For the reactor, the momentum balance simplifies to a pressure drop equation relating pressure loss to fluid friction. A suitable equation may be chosen such as the Ergun equation for laminar and turbulent flow in packed beds [Ergun (1952)]:

$$\frac{dp}{dz} = -\left(\frac{150}{\mathrm{Re}_p} + 1.75\right)\frac{(\rho_G v_o)v_o(1-\epsilon)}{144 d_p g_c \epsilon^3} \tag{8.11}$$

where p = total pressure
Re_p = particle Reynolds number = $d_p \rho_G v_o/[\mu(1-\epsilon)]$
ρ_G = gas density
v_o = gas superficial velocity
ϵ = void fraction of bed
d_p = volume-surface equivalent diameter of particle
g_c = conversion factor for force to mass units

Note that only Re_p and v_o vary with temperature, composition, and pressure but not $\rho_G v_o$ or any other quantity. In turbulent flow, the pressure drop will be insensitive to changes in Re_p. Only for gases will v_o change significantly and this change will be expressible in terms of temperature and pressure.

Kinetics and stoichiometry

For a system in which m independent reactions occur, the stoichiometry can be written as

$$0 = \sum_{i=1}^{n} \nu_{ji} A_i \quad (j = 1, 2, \ldots, m) \tag{8.12}$$

where ν_{ji} is the stoichiometric coefficient of compound i in reaction j and A_i is its molecular formula. For a product of the reaction, $\nu_{ji} > 0$; for a reactant, $\nu_{ji} < 0$; and for an inert, $\nu_{ji} = 0$. The ν_{ji} are arbitrary to a constant factor for each reaction and will be written here as the smallest set of integers. See Aris (1965) for further discussion.

The rate of reaction j will be a function of temperature, pressure, and composition as well as of catalyst properties such as age, surface area, and poisoning. R_j is defined as the rate of formation by reaction j per unit mass of catalyst per unit time of a (possibly fictitious) product with unit stoichiometric coefficient. That is,

$$R_j = R_j(T, P, \text{composition, catalyst properties}) \tag{8.13}$$

The net rate of formation of compound i by all reactions is given by

$$r_i = \sum_{j=1}^{m} \nu_{ji} R_j \tag{8.14}$$

The relation between changes in molar flow rates solely as a result of a particular reaction j is

$$\frac{(F_1 - F_{1o})_j}{\nu_{j1}} = \frac{(F_2 - F_{2o})_j}{\nu_{j2}} = \cdots = \frac{(F_i - F_{io})_j}{\nu_{ji}} = \xi_j \tag{8.15}$$

ξ_j is defined as the extent of reaction j and can be regarded as the molar flow rate of a hypothetical compound produced only by reaction j with stoichiometric coefficient equal to unity. All other molar flow rates are expressible in terms of the set of ξ_j. Thus

$$F_i - F_{io} = \sum_{j=1}^{m} \nu_{ji}\xi_j \tag{8.16}$$

Equations 8.14 and 8.16 can be substituted into Eq. 8.3 to give

$$\sum_{j=1}^{m} \nu_{ji}\frac{d\xi_j}{dz} = A\rho_B \sum_{j=1}^{m} \nu_{ji}R_j \quad (n \text{ equations}) \tag{8.17}$$

For each set of m independent equations from Eq. 8.17, the corresponding $m \times m$ submatrix of coefficients ν_{ji} has an inverse. Multiply those m equations from Eq. 8.17 by the inverse to obtain

$$\frac{d\xi_j}{dz} = A\rho_B R_j \quad (m \text{ equations}) \tag{8.18}$$

This is an alternative form of the mass balance.

The energy balance, Eq. 8.9, can also be modified by substituting for r_i from Eq. 8.14 into the term,

$$\begin{aligned}\sum_{i=1}^{n} H_i r_i &= \sum_{i=1}^{n} H_i \sum_{j=1}^{m} \nu_{ji}R_j \\ &= \sum_{j=1}^{m} R_j \sum_{i=1}^{n} \nu_{ji}H_i = \sum_{j=1}^{m} R_j \Delta H_j\end{aligned} \tag{8.19}$$

having interchanged the order of summation. Here ΔH_j is the enthalpy change for reaction j per mole of a product with unit stoichiometric coefficient. The energy balance 8.9 then becomes

$$\frac{dT}{dz} = \frac{-A\rho_B \sum_{j=1}^{m} R_j\,\Delta H_j + UA_H(T_e - T)}{\sum_{i=1}^{n} F_i C_{pi}} \tag{8.20}$$

Summary of model equations

The equations derived are collected here for reference:

$$\frac{d\xi_j}{dz} = A\rho_B R_j \quad (j = 1, 2, \ldots, m) \tag{8.18}$$

$$\frac{dT}{dz} = \frac{-A\rho_B \sum_{j=1}^{m} R_j\,\Delta H_j + UA_H(T_e - T)}{\sum_{i=1}^{n} F_i C_{pi}} \tag{8.20}$$

$$\frac{dp}{dz} = -\left(\frac{150}{\mathrm{Re}_p} + 1.75\right)\frac{(\rho_G v_o)(1-\epsilon)v_o}{144\, d_p g_c \epsilon^3} \tag{8.11}$$

$$\frac{dT_e}{dz} = \pm \frac{UA_H}{F_e C_{pe}}(T - T_e) \tag{8.10}$$

(+ for cocurrent, − for countercurrent)

$$R_j = R_j(T, p, \text{composition, catalyst}); \ (j = 1, 2, \ldots, m) \tag{8.13}$$

$$F_i - F_{io} = \sum_{j=1}^{m} \nu_{ji}\xi_j \quad (i = 1, 2, \ldots, n) \tag{8.16}$$

The boundary conditions are:

$$\begin{aligned} \text{At } z = 0, \quad & \xi_j = 0 \quad (j = 1, 2, \ldots, m) \\ & T = T_1 \\ & T_e = T_{e1} \quad \text{(cocurrent only)} \\ \text{At } z = l, \quad & p = p_2 \\ & T_e = T_{e2} \quad \text{(countercurrent only)} \end{aligned}$$

The problem then is the solution of $(m + 3)$ simultaneous nonlinear first-order ordinary differential equations. Standard numerical methods of integration of the Runge-Kutta or predictor-corrector types may be used. These are described in most books on numerical analysis, such as Fox (1962). Before proceeding to the solution the required data must be considered.

8.4 Data Required for the Model

Accuracy

The formulation of the model with its assumptions eliminates the need for some data, such as dispersion coefficients and heat capacity of the solid. Only after writing the model can the maximum list of required data be prepared.

The accuracy to which each element of the data is calculated depends upon the sensitivity of the computation to it and upon which data are least accurately known. For example, if the reactor always operates in highly turbulent flow, the Reynolds number and hence the viscosity will not greatly affect pressure drop. Therefore, an approximate and simple estimate will suffice.

Usually, the kinetics are the least reliable of all the data. Even if extensive kinetic studies have established a mechanism, the particular catalyst in use in the plant is not necessarily the same as that studied in the laboratory. In many cases, not only the kinetics but also the actual reactions occurring are unknown. This does not mean a simulation cannot be done but it does mean a greater need for plant data to fit parameters of the model.

There is little advantage in using very accurate thermodynamic data when the kinetics is uncertain. It is more important to use *consistent* thermodynamic data. For example, to assume an equilibrium constant which is independent of temperature, together with a nonzero enthalpy of reaction, is inconsistent since

$$\frac{d \ln K}{dT} = \frac{\Delta H}{RT^2} \tag{8.21}$$

If the rates of reversible reactions are written so that they are zero at equilibrium, the heat capacities, enthalpies, and equilibrium constants should be self-consistent at all temperatures so that conversions beyond those at equilibrium will not inadvertently be calculated.

Kinetic data

A rate equation is needed for each reaction and may come at one extreme from a fundamental study of the reaction mechanism, or at the other, from a statistical regression of plant or laboratory data. The lack of adequate rate equations is the most common obstacle to reactor simulation.

Only for the most common reactions, such as ammonia synthesis and the oxidation of SO_2, have extensive kinetic studies been done and even with these, disagreement persists about the nature of the true mechanism. If rate equations have been determined, one chooses the best for the simulation at hand. If no rate equation exists, one can sometimes assume a rate equation such that the reaction is elementary and the rate is zero at equilibrium, and with the rate constant dependent only on temperature according to an Arrhenius factor, $\exp(-E_j/RT)$. This for $A + B \leftrightarrows C$ one might write

$$R_j = k_j(p_A p_B - p_c/K_j)$$

with

$$k_j = k_{jo} \exp(-E_j/RT) \tag{8.22}$$

The use of models to include adsorption effects, such as those of Langmuir-Hinshelwood, is acceptable, but the insertion of more adjustable constants complicates the task of fitting the model. For further discussion of this, see Mezaki and Kittrell (1967), Boudart (1956), and Weller (1956).

For the reactor for the oxidation of sulfur dioxide, a number of kinetic studies is available [Krichevskaya (1947), Calderbank (1953), Eklund (1956), Kubota, et al. (1959), and Davidson and Thodos (1964)]. Since the particular catalyst used in the plant under study will not correspond exactly in structure, composition, or past history to those used in the kinetic studies, one is forced to choose the most reliable kinetics from those available and to adjust the magnitude of the rate constant while preserving the functional form of the rate equation. Thus in the present simulation, the kinetics of Kubota, et al. (1959) was chosen since the temperature range covered was the widest and since diffusion effects within the catalyst were corrected for. The equation for the rate of formation of SO_3 is

$$r_{SO_3}\rho_B = E_c k\,[p_{O_2} - (p_{SO_3}/K_f p_{SO_2})^2]/D^2 \equiv R_1 \rho_B \tag{8.23}$$

where E_c = effectiveness factor of the catalyst

$k = 4.88 \times 10^8 \exp(-12{,}900/T)\ \dfrac{\text{lb moles } SO_3}{(\text{atm})(\text{hr})(\text{cu ft bed})}$

p_i = partial pressure of compound i, atm

$K_a = 1.02 \times 10^{-5} \exp(7{,}025/T)$

$K_b = 4{,}090 \exp(-8{,}800/T)$

T = temperature, °K

K_f = equilibrium constant for the reaction $SO_2 + \frac{1}{2}O_2 = SO_3$

$D = (1 + K_a p_{SO_3}/p_{SO_2})$ if $T < 723$°K

$\quad = (1 + K_b p_{O_2}^{1/2}/p_{SO_2})$ if $T > 773$°K

For temperatures between 723° and 773°K, $1/D^2$ was interpolated such that it and its rate of change with temperature fitted smoothly at both extremes.

The catalyst effectiveness E_c is strictly the ratio of the rate of reaction limited by diffusion within the pellet to the rate if no diffusion limit existed. In this study, E_c was used as an adjustable parameter to fit the calculated outlet temperature of the bed to the plant measurement.

Thermodynamic data

In order to establish the heat effects of the reaction and to find the equilibrium composition, the standard enthalpy change of reactions, ΔH_j°, the ideal molar heat capacities, C_{pi}°, of all substances and the equilibrium constants, K_{fj}, in terms of fugacities are all needed. These are often available in compilations of thermodynamic data such as the American Petroleum Institute's Project 44.

As discussed in Sec. 8.3, one should strictly correct the standard enthalpy changes and heat capacities for nonideal behavior of the pure substances and of their mixtures as a function of pressure, temperature, and composition. At low pressures, all nonideality may be neglected as will be done in the present example. Then the true value of the enthalpy change ΔH_j at reaction conditions is taken equal to ΔH_j°, the standard enthalpy change for reaction j of pure reactants to form pure products, all at 1 atm and the same temperature. Similarly, the true molar heat capacities in the mixture, C_{pi}, are taken equal to those at low pressure for the pure compounds, C_{pi}°. From Kelley (1949) for the SO_2 reactor the following were used, with T in °K:

$$\Delta H_1^\circ = -24{,}097 - 0.26T + 1.69 \times 10^{-3}T^2 + 1.5 \times 10^5/T \text{ PCU/lb mole}$$
$$C_{pO_2}^\circ = 7.16 + 1.0 \times 10^{-3}T - 4.0 \times 10^4/T^2 \text{ PCU/(lb mole)(°K)}$$
$$C_{pN_2}^\circ = 6.547 + 1.389 \times 10^{-3}T - 6.9 \times 10^{-8}T^2$$
$$C_{pSO_2}^\circ = 10.38 + 2.54 \times 10^{-3}T - 1.42 \times 10^5/T^2$$
$$C_{pSO_3}^\circ = 13.70 + 6.42 \times 10^{-3}T - 3.12 \times 10^5/T^2$$

The equilibrium composition calculated at a given pressure and temperature would likewise be affected by nonideal behavior, although K_f is the true constant, dependent only on temperature. Such corrections are neglected and the equilibrium constant is given by Kelley (1949) as

$$\ln K_f = 12{,}127/T - 11.433 - 0.1309 \ln T + 8.50 \times 10^{-4}T - 3.774 \times 10^4/T^2 \tag{8.25}$$

for the reaction

$$SO_2 + \tfrac{1}{2}O_2 = SO_3$$

It can be seen that Eq. 8.21 is valid and that

$$\frac{d(\Delta H_1^\circ)}{dT} = C_{pSO_3} - \tfrac{1}{2}C_{pO_2} - C_{pSO_2}$$

as is necessary.

Heat transfer data

To define the thermal behavior of a reactor, some information concerning heat transfer is needed. This may be a detailed, rigorous calculation of local heat transfer rates, a specification of a constant heat transfer coefficient and area, or most simply the assumption of constant temperature. The local heat transfer rate to an element of fluid is

$$Q = UA_H dz(T_e - T) \tag{8.26}$$

Since the SO_2 reactor is adiabatic, the overall heat transfer coefficient U for this case was set equal to zero.

Geometric data

Under this heading are included bed diameter, bed depth, catalyst particle geometry, heat transfer area, tube arrangements, and so forth.

Plant measurements

Data routinely available from the plant under study will be essential to the fixing of the constants in the model. These data include flow rates, feed and product composition, bed temperatures, and pressures. Special runs using different data are desirable if the cost can be justified. In particular, more accurate and complete product analyses can improve the model.

8.5 Programming the Model

Choice of the method of solution

The equations of the model are summarized in Sec. 8.3. The method of solution of the simultaneous differential equations can be either the Runge-Kutta (RK) or the predictor-corrector (PC) methods [Fox (1962)].

RK methods are easier to program, but may take longer to compute. The step size for integration can readily be changed at any place during the integration. This allows more steps to be used when the changes are more rapid. The PC methods are more difficult to program, but may be more efficient. They require an auxiliary starting method, such as RK, to generate values over the first few intervals. It is more difficult to change the step size of integration during the calculation than it is for RK.

In the study described here, the standard fourth-order RK method was used for ease of programming, minimizing memory requirements, and variability of step size. It was found convenient to separate the program into four sections: input, output, integration, and evaluation of derivatives. A description of the standard RK method is available in most texts on numerical analysis.

Setting up the algorithm

The general structure of the algorithm for the reactor calculation is shown in Fig. 8.2. It can be seen from the boundary conditions in Sec. 8.3 that

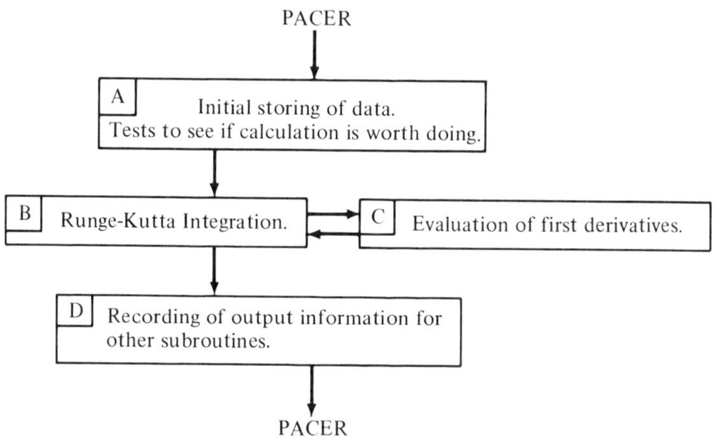

Figure 8.2. General algorithm for the reactor model.

pressure and inlet temperature for a countercurrent heat exchanger fluid are given at the exit of the reactor bed. This constitutes a split boundary problem and requires iteration to fit the conditions at the reactor exit. Thus an inlet pressure and an exit coolant fluid temperature must be assumed at $z = 0$, the calculation performed, and the calculated and required values of pressure and temperature compared at the reactor exit. The calculations are repeated through the bed until the calculated pressure and temperature at the exit are within predetermined tolerances of the given values.

In calculations with the sulfur dioxide adiabatic reactor, no difficulty was found in converging to a constant inlet pressure. However, it should not be assumed that such iterative calculations automatically converge. The programmer will have to protect against unrealistic assumed values and meaningless calculations. He will also have to incorporate efficient convergence techniques to avoid wasting time.

Detailed considerations

A number of questions need to be asked within the program in order to avoid wasted computations. For example, protection against a zero or negative pressure or too low a temperature is necessary. Further, there is no reason to perform the calculation if no reactant is provided. This can, of course, happen within a plant calculation when the reactor is encountered before nonzero values have been given for its entering stream variables.

Prior to writing the full model in accord with all the assumptions, simpler models are often conveniently written to permit computations of the whole plant early in the project. Such a model was REACT3, which calculated the equilibrium conversion of SO_2 to SO_3 and the accompanying adiabatic temperature rise. Another simple model would be one which specifies conversions of the reactions considered, given the input stream, with a protection against converting more of a reactant than was initially supplied.

An example of the algorithm for the full model of the sulfur dioxide reactor is shown in Fig. 8.3. The fitting of this subroutine, called REACT4, within the PACER framework required that input information on flow and temperature be obtained from list STRMI and pressure from list STRMO. The parameters of the bed are obtained from list EN. At the end of the computation, resulting exit flow and temperature are placed in list STRMO and inlet pressure in STRMI. The calculated pressure drop is stored in list ENC for the next time the calculation is performed.

8.6 Matching the Model to the Plant

For each bed of the reactor model, there is at least a pair of constants for each reaction assumed, namely, the preexponential factor k_{oj} and the activation energy E_j, in the rate constant. These may be known but more usually they must be adjusted to fit the available plant data. The rate constants for all reactions, given by

$$k_j = k_{jo} \exp(-E_j/RT) \tag{8.27}$$

can be adjusted to fit exit temperature and composition from the bed. If intermediate temperatures within the bed are available, a choice may be

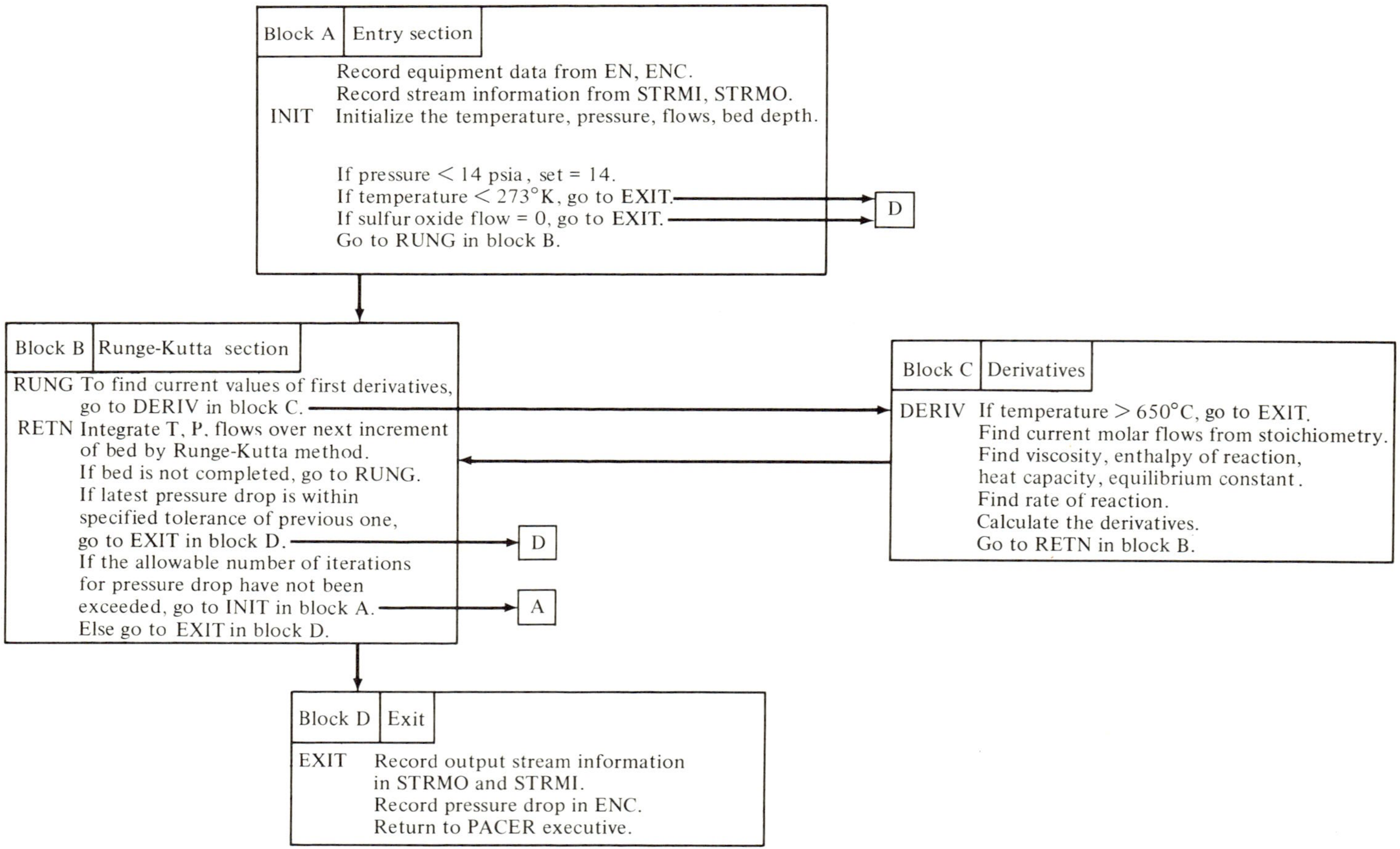

Figure 8.3. Algorithm for REACT4.

possible for a best value of E_j. Alternatively, if data on exit compositions are available for a range of inlet temperatures, E_j could be assigned a reliable value. Often the activation energy E_j would be considered constant throughout a multibed reactor and only k_{oj} would vary from bed to bed to account for different catalyst histories.

8.7 Summary

This chapter has discussed the steps to be taken in writing a model of a single-bed catalytic reactor with several reactions and with heat exchange. The fluid was assumed to be in plug flow. The various limiting transport processes, namely, bulk transport in the fluid, solid conduction, mass and heat transfer to the particle surface, and diffusion and heat conduction within the particle, were neglected.

The equations derived for the general case were applied to the reactor for the oxidation of SO_2. The selection of data, the method of solving the equations, and the algorithm used were all examined. The matching of the model to plant results for the multibed converter for oxidation of SO_2 is covered in chapter 9.

PROBLEMS

8.1. (a) For a single exothermic reaction in an adiabatic catalytic reactor, how would the temperature and concentration of reactant at a given cross section depend on radius, if the velocity is lower at the wall than at the center?

(b) Would the concentration difference from center to wall be greater or less than in (a) if the temperature at each cross section is made constant by suitable heat exchange?

8.2. Discuss Prob. 8.1 for an endothermic reaction.

8.3. Could the rate of reaction within a catalyst particle ever be greater than the rate of reaction for the particle, based on the concentration and temperature at the external particle surface?

8.4. Write an algorithm for a reactor model with countercurrent heat exchange.

8.5. (a) For a single exothermic reaction with countercurrent cooling, is it possible for the coolant temperature curve to cross that of the entering reactant temperature? See Fig. 8.4. Is there any statement which can be made about the slope of the coolant temperature with respect to bed depth where it crosses the reactant temperature?

(b) What is the physical meaning of the temperature crossover and how should the coolant flow rate be changed to prevent it?

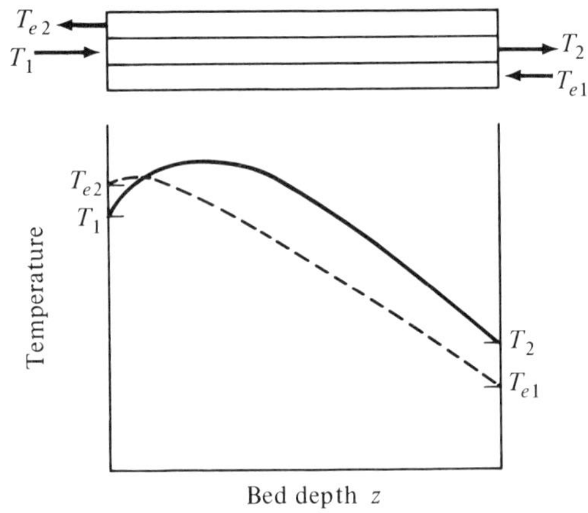

Figure 8.4. Countercurrent reactor.

(c) Show that there is a maximum exit temperature of the coolant. What would happen if a coolant exit temperature greater than this maximum were assumed and the integration of the differential equations were done beginning at the reactor entry?

8.6. In the Ergun equation (Eq. 8.11), what would the error in pressure drop be if the viscosity were in error by 10%? Consider particle Reynolds numbers of 10, 100, 500.

8.7. Most reactors using vanadium pentoxide catalyst beds for converting sulfur dioxide to sulfur trioxide have been designed assuming an adiabatic bed model in which the leaving gases reach chemical equilibrium.
(a) Create a Unit Computation based on this assumption.
(b) Compare this model with the differential model developed in Chapter 8.

8.8. Considerable work has been reported in the literature describing the cracking of ethane to ethylene. Develop a Unit Computation representing a cracking furnace, using the kinetic data of Snow and Schutt, and having the process temperature profile through the reactor specified by the parameters of the model (the latter corresponds to setting the temperature profile in the plant). See Snow, R.H. and Schutt, H.C. (1957) *Chem. Eng. Prog.* **53,** 3, 133.

9

Studies of a Multibed Reactor with Heat Exchange: Oxidation of Sulfur Dioxide

9.1 Introduction

A single-bed reactor model, as described in Chapter 8, is combined with models for the necessary heat exchangers and boilers to simulate the multibed reactor for the oxidation of sulfur dioxide. The calculated results are compared with those from an actual plant.

9.2 Description of the Reactor

The oxidation of sulfur dioxide to sulfur trioxide is carried out in a four-bed adiabatic reactor over a vanadium pentoxide catalyst, as shown in Fig. 9.1. The reaction generates heat so that its equilibrium constant decreases with temperature. Additional air is added before the second and third beds and an air inlet pipe to the fourth bed is available but not used. After the first bed and before the air addition, the reactant is cooled by passing through boiler #2. Directly following the second and third beds, the gas is cooled by two separate steam superheaters fed in parallel by saturated steam. The reactor effluent is cooled by the boiler feed water in an economizer.

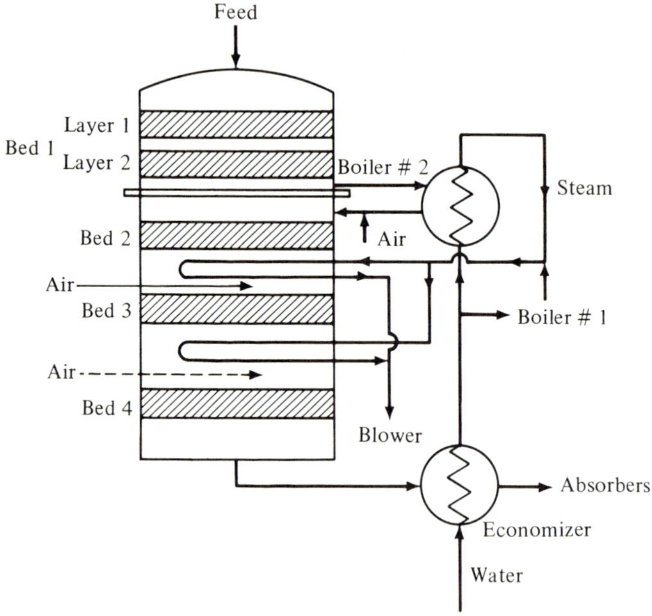

Figure 9.1. Schematic diagram of reactor showing heat exchangers and air addition.

The extensive interactions of the water-steam system with the process gas will be apparent from the flow sheet. The most important factors affecting the reactor are the total amount of steam produced by both boilers and hence the cooling capacity and heat transfer of the steam in the two superheaters. The choice of rates of supplementary air addition to the reactor affects not only the conversions and temperatures in later beds, but also the required heat exchange between beds and the amount of steam produced.

The reactor behavior can be followed on a graph of conversion of SO_2 against temperature, as in Fig. 9.2, with the equilibrium conversion shown for comparison. The slope of the adiabatic lines is, as can be shown by taking the

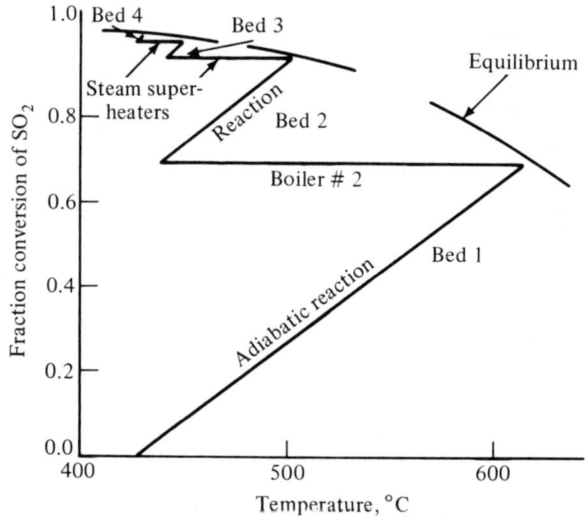

Figure 9.2. Reactor conversion behavior with temperature showing intercooling.

ratio of Eqs. 8.18 and 8.20 with $m = 1$, $U = 0$,

$$\frac{dX_{SO_3}}{dT} = -\frac{\Sigma F_i C_{pi}}{\Delta H_1} \tag{9.1}$$

where $X_{SO_3} = F_{SO_3}/S_T$, which is the fractional conversion to SO_3, and S_T is the total sulfur oxides fed in lb moles/hr. The adiabatic lines become steeper as air is added and, although they may not actually be perfectly straight, nothing is lost in considering them so. The equilibrium curve will be different after each new addition of air so that there are three separate curves required in the normal mode of operation of the reactor.

The effect of higher inlet temperature, air addition between layers of the first bed, and redistributed air to later beds can all be envisaged in terms of such a graph. An idea of the relative rates in each bed is given by the relative lengths of the adiabatic lines since bed depths are comparable. Further discussion of the optimum problem in the reactor is provided by Aris (1965).

9.3 Plant Data Available

The plant data have been presented in Tables 4.7 and 4.9.

Temperatures

There are nine thermocouples which are read routinely to determine temperatures in and out of every bed and also between the two layers of the first bed. No independent check is available of the accuracy of the temperatures. Errors could exist through radiation of heat, incomplete mixing of supplementary air streams, and variation of temperature radially across the bed. Indeed, a 12° rise in temperature is known to occur above the second bed from one side to the other, as a result of heat transfer through the blind flange separating the first two beds.

Pressures

Pressure readings are available at the same levels in the reactor as are the temperatures. Since pressure does not greatly affect the conversion, the readings are sufficiently accurate.

Compositions

The concentration of SO_2 is measured at the reactor inlet and at the outlet of the sulfur burner by an automatic analyzer. The first is checked occasionally by the Reichs analysis, namely, absorption in iodine solution. The SO_2 content of the stack gas is also measured by the Reichs test. The oxygen content of the stack gas is found by an Orsat apparatus.

Catalyst data

The depth of catalyst in each bed, the particle size, and the ages of catalyst were made available. The first two beds contain some of the original catalyst of 1948 while the third and fourth beds have been replaced within the past few years. A summary of the important catalyst data is given in Table 9.1.

TABLE 9.1
CATALYST DATA

Dimension	Bed			
	1	2	3	4
Depth, in.	18	17	15	15
Effective diameter, in.	0.242	0.242	0.242	0.202
Area of bed, ft^2	227	227	227	227

9.4 Studies of the Reactor using PACER

Summary

The reactor studies were carried out in five separate stages:

1. The effectiveness factor and void fraction of each bed were adjusted to match the exit temperature and pressure drop of that bed, using plant

data for a full capacity run. The overall heat transfer coefficient in boiler #2 was adjusted to fit the temperature into the second bed.
2. The quality of fit of these parameters was tested by comparing computed data and plant data for a half capacity run.
3. The effect of changing the catalyst and the amount of air added to bed 2 was studied.
4. In order to speed up convergence of the calculations, a Unit Computation, SETSP1, was written. It added the latest value of pressure drop across each piece of equipment to its outlet pressure to find the corresponding inlet pressure, beginning with the outlet stream of the plant. Thus, a new iteration over the plant could be started with consistent pressure in all streams.

Information flow diagram for the reactor/steam system

Figure 9.3 is the information flow diagram used to simulate the reactor/steam system within the sulfuric acid plant. The four reactor beds, represented as REACT4, are shown with the first bed split into its two layers as equipment numbered 8 and 15. The equipment lists for REACT4 for each bed and the input stream list to the first bed are shown in Table 9.2

Boilers #1 and 2 (equipment #5 and 9) were simulated initially by the simplified subroutine BOILS1, in which the heat transfer coefficients for the boilers were set at 10.0 and 10.7 respectively. The two superheaters (equipment #26 and 29) and the economizer (equipment #13) were represented by COOL7 as discussed in Chapter 7.

The MIXER6 subroutine incorporating accurate gaseous heat capacities was used for all mixers or splitters of gases, that is, all diamonds in Fig. 9.3 except equipment #34, which was MIXER3 used to handle water.

The steam drum is represented by DRUM1, which splits the steam and water flows. With the simple models for the boilers, there is no unique solution to the water flow because steam production was constant when the process side was constant. Thus with any given solution, an arbitrary amount added to water stream 70 would provide another solution. This problem was avoided by inserting a flow regulator, PUMPS1, in stream 70 to set the flow arbitrarily.

The use of SETSP1 to accelerate convergence is discussed below. The dotted line is a fictitious stream inserted to force the computer to converge on pressure over the whole main process line. Such a line is not necessary when the order of calculation is specified with the appropriate iterative paths included.

The problem as shown is a second-order recycle, since breaking two lines would allow all equipment pieces to be directly calculated. These lines are streams 12 and 70 (or 69).

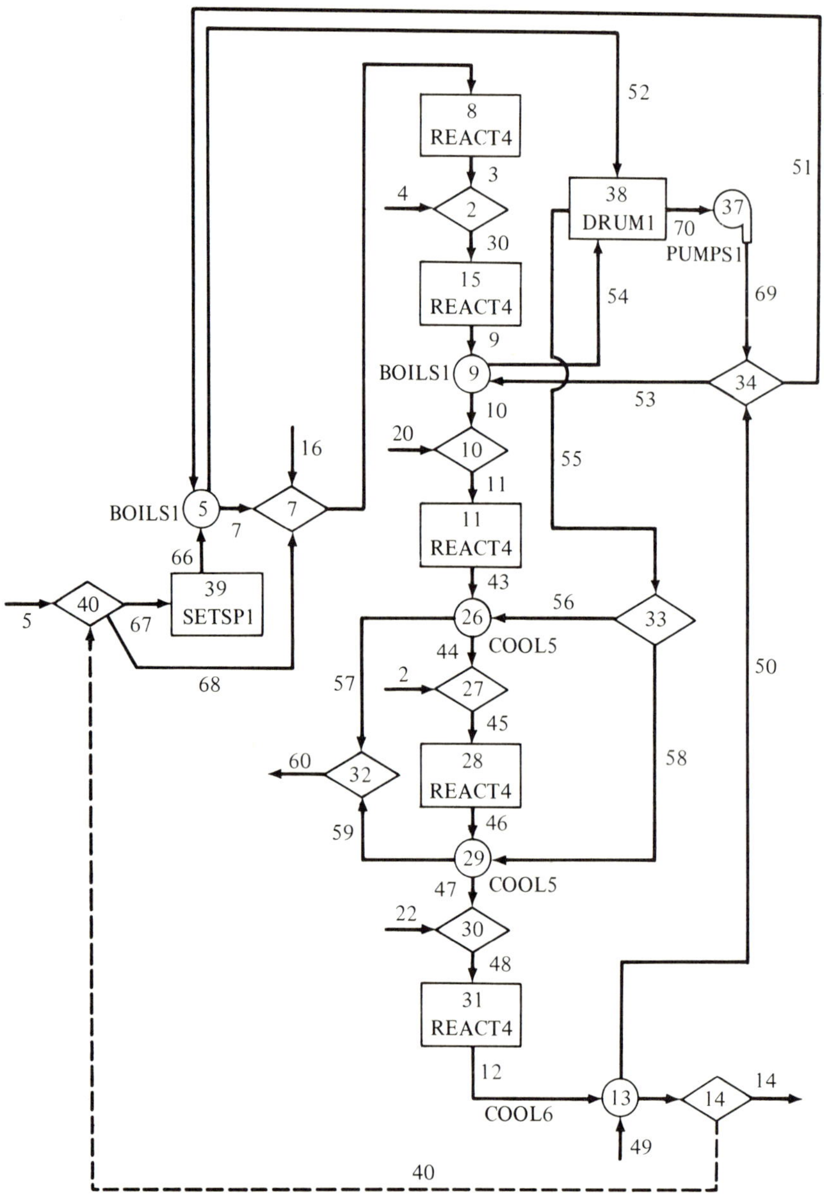

Figure 9.3. Information flow diagram for reactor-steam system.

TABLE 9.2

STREAM AND EQUIPMENT LISTS USED AS BASE CASE

Stream list into 1st bed (SN)		Equipment list for REACT4 (EN)	Values used for each bed			
			1 Layers 1 and 2	2	3	4
1. Stream number	8	1. Equipment number	8 and 15	11	28	31
2. Stream flag	0	3. Bed depth	0.75 ea	1.417	1.333	1.333
3. Total moles/hr	2926	4. Effectiveness factor	0.32	0.65	1.0	1.44
4. Temperature	425°C	5. Activation energy	12900	12900	12900	12900
5. Pressure	17.1 psia	6. Adsorption factor (1)	1.025	1.025	1.025	1.025
6. Blank	0	7. Adsorption energy (1)	−7025	−7025	−7025	−7025
7. Moles S/hr	0	8. Adsorption factor (2)	4090	4090	4090	4090
8. Moles H_2O/hr	0	9. Adsorption energy (2)	8800	8800	8800	8800
9. Moles O_2/hr	338	10. Particle diameter, ft	0.0202	0.202	0.0202	0.0169
10. Moles N_2/hr	2310	11. Bed voidage	0.310	0.318	0.345	0.350
11. Moles SO_2/hr	278	12. Area, ft^2	227	227	227	227
12. Moles SO_3/hr	0	13. No. of steps through bed	36	34	15	15

Fitting of model parameters to the plant—full capacity run

The data for the special run of March 16, 1965, at 315 tons per day at the Hamilton acid plant were matched by adjusting 1.) the effectiveness factor E_c for each catalyst bed, 2.) the heat transfer coefficient in boiler #2, 3.) the void fraction ϵ of each bed, and 4.) the split of steam flow to the two superheaters.

1. To adjust the effectiveness factors E_c, each bed was examined in turn, beginning with the first. The input stream to each bed was assigned the composition, flow rate, and temperature based on the plant data and with conversion in preceding beds. The effectiveness factor was then adjusted so that the correct temperature was computed at the outlet of that bed. A confirmation of such a fit would be possible if the intermediate gas compositions were known, but attempts to obtain such data were unsuccessful.

The reaction in the first bed approached equilibrium so closely that its outlet temperature is quite insensitive to changes in E_c. Instead, E_c was adjusted to provide the correct intermediate temperature between the two layers of the first bed. An alternative would have been to assume an adjusted depth of the first layer as inactive but with a higher E_c for the remainder of the first bed. There would be little difference between the results from these two approaches.

The values of E_c so fitted and a comparison of computed and measured temperatures are given in Table 9.3. The fact that E_c for bed 4 is greater than

TABLE 9.3
FITTED EFFECTIVENESS FACTORS

Bed	Effectiveness factor E_c	Bed void fraction ϵ	Temperature, °C		
			Assumed inlet to bed	Calculated outlet	Measured outlet
1	0.32	0.310	425	611.8	612
2	0.65	0.318	450	509.7	510
3	1.0	0.345	439	447.2	448
4	1.44	0.350	422	424.1	426

unity is justified because the fourth bed contains a newer catalyst with a bulk density 44% higher than that in other beds and because the rates were expressed per unit volume of bed. The catalyst is less effective in earlier beds because the rates of reaction are higher relative to diffusion rates and because deterioration of the catalyst is greater at higher temperatures.

2. The overall heat transfer coefficient used in the simple model BOILS1 for boiler #2 was adjusted so that the inlet temperature to bed 2 with air added was correct. The value chosen was 10.7 PCU/hr ft^2 °C compared to

11.2 calculated by the more rigorous model of the boiler, BOILS2, which is discussed in Sec. 7.8.

3. The void fraction ϵ in each bed was adjusted to reproduce the measured pressure drop through that bed at the flow conditions of the full capacity run. The values assigned to ϵ are shown in Table 9.3. The void fraction is less in the earlier beds where the catalyst is older, as would be expected from the greater compactness through use as well as from the generally higher temperatures.

4. The only information concerning how the steam was split to the superheaters after beds 2 and 3 is that the valve to the first was fully open and the valve to the second was half open. The steam generated from the boiler system was assumed to be split in various ratios between the superheaters, and the calculated steam superheat and reactor exit temperatures were compared with plant values.

The simulation of the plant at the 315 tons per day rate is compared with actual plant measurements in Table 9.4. The results agree quite well, but if

TABLE 9.4
COMPARISON OF REACTOR SIMULATION WITH PLANT RESULTS—FULL CAPACITY

Parameter		Simulation run 206	Plant measurement
% conversion of SO_2		97.59	97.55
% O_2 in stack		9.75	9.0
% SO_2 after SO_3 removal		0.230	0.248
Temperatures, °C (in/out)	Bed 1	424.2/611.4	425/612
	Bed 2	440.3/502.7	438/500
	Bed 3	443.4/450.5	439/448
	Bed 4	427.3/429.2	423/427
Temperature of superheated steam, °C		300.2	307

one tries to improve the agreement, some results are made better but others are worse. Thus any attempt to bring the outlet temperature from the fourth bed closer to the plant reading, for example, by increasing the steam flow to the first superheater, will increase the disparity between calculated and actual steam superheat. More details of the plant data for the full and half capacity base cases are found in Tables 4.7 and 4.9.

Testing the model—half capacity run

The model of the reactor/steam system was tested by simulating the same plant run at 150 tons per day—half capacity. The computation was compared with data obtained from the Hamilton plant for October 20, 1965. The results

TABLE 9.5
COMPARISON OF REACTOR SIMULATION WITH PLANT RESULTS—HALF CAPACITY

Parameter		Simulation run 2081	Plant measurement
% conversion of SO_2		97.9	98.0
% O_2 in stack		9.7	8.0
% SO_2 after SO_3 removal		0.20	0.19
Temperature, °C (in/out)	Bed 1, layer 1	405/483	405/476
	layer 2	483/594	476/603
	Bed 2	435/478	440/491
	Bed 3	432/438	434/439
	Bed 4	412/414	401/400
Temperature of superheated steam, °C		313	315

are shown in Table 9.5 with reasonable agreement between calculated and actual data.

The fact that parameters such as catalyst effectiveness and void fractions, fitted for full capacity, were adequate for half capacity gives reassurance that the model is realistic. One should note that a bypass of boiler #2 was required in the model (as in the plant) to prevent excessive cooling of the process gas from bed 1. The amount of bypass was adjusted to fit the inlet temperature to bed 2. In addition, the steam split to the superheaters was adjusted to improve the results.

The results calculated for bed 1 suggest that the low value of E_c (0.32), while satisfactory, is covering up for axial conduction and diffusion. Solution of the first bed with a higher value of E_c and with inclusion of axial diffusion would be more realistic, although the conversion and temperatures out of bed 1 are quite insensitive to E_c. The boundary conditions which should be used for axial diffusion and conduction are not obvious, in view of the blind conducting flange and side takeoff of gas at the exit of bed 1.

Study of interaction of air addition and catalyst activity

During previous computations, it was seen that when the temperature out of the second bed rose, its conversion tended to drop. This indicated that equilibrium and rate factors were in balance and that increased air addition and catalyst activity might improve the conversion in the second bed by favorably altering both factors.

Calculations were done to test this supposition for the plant operating conditions on March 16, 1965. The results are shown in Table 9.6. The general conclusions are that the use of new catalyst in bed 2 would increase

Table 9.6

Results of Reactor Simulation Conditions are as in Table 9.2 Except as Shown: Effect of Redistribution of Air and Higher Catalyst Activity in Bed 2 on Composition out of Bed 2 and on Temperatures

Simulation:	Base Case Run 125-5	Run 126	Run 127	Run 128	Run 129
Catalyst activity bed 2	0.65	1.00	0.65	1.00	1.00
Air to bed 2, moles/hr	142	142	210	210	310
Air to bed 3, moles/hr	210	210	142	142	42
Steam to superheater #1, moles/hr	1062	1062	1062	1062	885
Steam to superheater #2, moles/hr	708	708	708	708	885
Moles SO_2 in exit	7.12	6.95	7.47	6.62	6.80
Temperatures, °C					
Bed 2 in/out	438.7/502.1	438.7/503.0	430.2/489.3	430.2/494.4	418.4/478.6
Bed 3	441.0/448.8	441.8/448.9	439.6/449.7	443.9/450.5	444.6/453.0
Bed 4	426.1/428.3	426.3/428.3	427.0/429.6	427.7/429.6	429.0/431.3

the overall conversion to SO_3, raising the air addition to bed 2 would reduce the conversion, and both together would be better than either separately. This is a simple example of an optimization study which could be done with these models. More complicated examples are presented in Chapter 12.

9.5 Improvement of Convergence on Pressure—SETSP1

The input pressure to each piece of equipment was calculated from the output pressure to allow for the fact that pressure information flows backwards through a plant. Unfortunately this slowed convergence since in one pass through a line of equipment pieces, a pressure disturbance progressed backwards through only one equipment piece. To reduce computing time, subroutine SETSP1 was written to add pressure backwards from a specified stream through a specified number of Unit Computations.

A Unit Computation such as SETSP1, which operates on other units, can easily be fitted into the PACER executive. In all other aspects, it operates as any Unit Computation would. Starting from a given process gas stream, the subroutine can find the number of the unit from which it comes by searching the stream connection matrix. The process stream feeding this unit is found directly from the process matrix since process gas streams were listed first into any unit having different types of streams. The latest value of pressure drop is added to the outlet pressure, stored in the inlet stream, and the procedure continued to the next unit connected to the input of the current unit.

A second pressure addition was desired over the water/steam side of the system. But how can the stream number of a water stream into an equipment unit be found uniquely if it is sometimes the first stream and sometimes not? The solution was to place a tag number in position 6 of the input stream list. This tag would be passed forward by the Unit Computation through all the water streams but *only* those. The program then looked for the tag and added pressure from the water outlet backwards over tagged streams only. This technique is not usable when trying to pass pressure back through a heat exchanger involving reactor effluent heating up reactor feed. This is because the same tag would be in both effluent and feed streams entering the heat exchanger which prevents their being distinguished.

The reactor/stream system converged in four or five iterations with the use of SETSP1 but did not converge in ten without it. The alternative to the use of SETSP1 would be to specify all the streams over which pressure is to be added. This use of SETSP1 illustrates the way in which the rules and conventions of a particular executive program can be exploited to achieve faster or better results. Executive programs other than PACER have other rules and so require a different procedure to do the same task. It also demonstrates how a Unit Computation can be written to receive information from and pass it to any unit in the plant. The limits are the ingenuity of the programmer, the flexibility of the executive, and the computation time needed.

9.6 Summary

The reactor for the oxidation of SO_2 has been described and the available plant data have been discussed. The catalyst effectiveness and voidage were fitted in turn for each bed, beginning with the first. Temperatures and conversion of a half capacity run were found to be well predicted using the parameters fitted from the full capacity run. Thus the fitting of a reactor model to an actual plant showed that the simulation of the multibed reactor/heat exchange system could produce realistic values of exit composition and temperatures over a twofold change in capacity with one set of fitted constants. The backward flow of pressure caused very slow convergence of the reactor calculations but this could be easily overcome by the use of the Unit Computation SETSP1 for backward addition of pressure before each iteration.

PROBLEMS

9.1. What are the advantages and disadvantages of the use of pure cold oxygen instead of air as the additive between reactor beds?

9.2. Ammonia synthesis has many similarities to that of SO_3. What are they? Why is ammonia made by a high pressure process while SO_3 has almost always been made at near atmospheric pressure? Should SO_3 be formed under elevated pressure in new installations?

9.3. A recently graduated engineer has suggested that the SO_2 reactor should be built on its side because the catalyst beds could be more easily removed and replaced. What advantages and disadvantages are there in this horizontal reactor compared to the conventional vertical reactor with downward flow of gas? (See *Chem. Eng.* **74,** 152, Jan. 16, 1967).

9.4. Why has platinum been almost completely superseded as a catalyst for SO_2 oxidation by vanadium pentoxide? See Duecker and West (1959).

9.5. How accurately should the viscosity of the SO_2 reactor feed in Table 9.2 be estimated if the pressure drop is to be accurate to 5%? If the viscosity of the gas were correctly estimated at 400°C, how much error in the local pressure drop would be introduced at 600°C by assuming the viscosity proportional to $T^{0.5}$ instead of more accurate $T^{0.65}$?

Simulation of Packed Gas Absorbers and an Absorber Plant

10

10.1 Statement of the Problem

The problem of simulating a packed gas absorber can be summarized as follows: given the packing type and size, the column diameter and packed height, the inlet temperatures, flows, and compositions of each inlet liquid and gas stream, and the outlet pressure of the gas stream, calculate the outlet temperatures, the outlet flow rates of all species, and the inlet pressure.

In general, to make such a calculation account must be taken of simultaneous transfer of heat and mass by forced convection under the influence of chemical reaction in both gas and liquid phase. This is obviously a very difficult problem, particularly because of the extremely complex geometry of the gas-liquid interface. This geometry is a result of the shape of the packing element which is designed to create large interfacial areas and turbulent flow, and so to promote rapid mass and heat transfer rates. Since an exact mathematical analysis of gas absorption in packed towers is impossible, a model is at present needed which adequately approximates the transfer of heat and matter in a packed gas absorber.

The choice of a model depends on a number of considerations. These include:

1. Equilibrium model versus rate model.
   For an absorber operating near equilibrium conditions, in which the outlet gas is almost in equilibrium with the inlet liquid, a suitable model need not consider the rate processes of transfer of heat and mass. Nevertheless, a more comprehensive rate model would be suitable even if the kinetic rate constants were highly inaccurate.
2. Rigor of the rate model.
   The more rigorous a rate model becomes, the larger is the number of rate constants required. For example, a model which neglected transfer resistances in one phase would require half the number of rate constants needed by one which did not neglect them. As another example, for the absorption of two compounds from an inert gas, the use of effective binary diffusion equations, instead of ternary equations, would reduce the number of diffusion coefficients from 3 to 2.

Such simplifications are sometimes necessary because of a lack of data and are often justified by a careful examination of the transfer processes for a particular system. A rigorous model for countercurrent flow might require a complex set of differential equations with split boundary conditions for its description. The solution of these equations would involve an iterative procedure; such a model might consume too much computer time. This may be particularly serious when a complex system containing several such models is being simulated. The model used for a unit in a system can be approximate if the system as a whole is insensitive to the errors in the approximation. The sensitivity of a complex system to the variation of the parameters of such units should be examined by simulating the system with approximate models.

This chapter describes the development of a comprehensive model for the simulation of the three gas absorbers of the contact sulfuric acid plant, namely, the air drier, acid absorber, and oleum absorber. Some specifications for these absorbers are listed in Tables 4.7 and 4.9 and in Sec. 4.1. The computer subroutine for the absorber was used with other equipment subroutines to study that section of the plant which included the acid and oleum absorbers. This section, called the absorber section, contains the acid and oleum absorbers (units #3 and 9) which were modeled by Unit Computation SORB5. Its information flow diagram is shown in Fig. 10.1. The air drier, also modeled by SORB5, is in the feed preparation section.

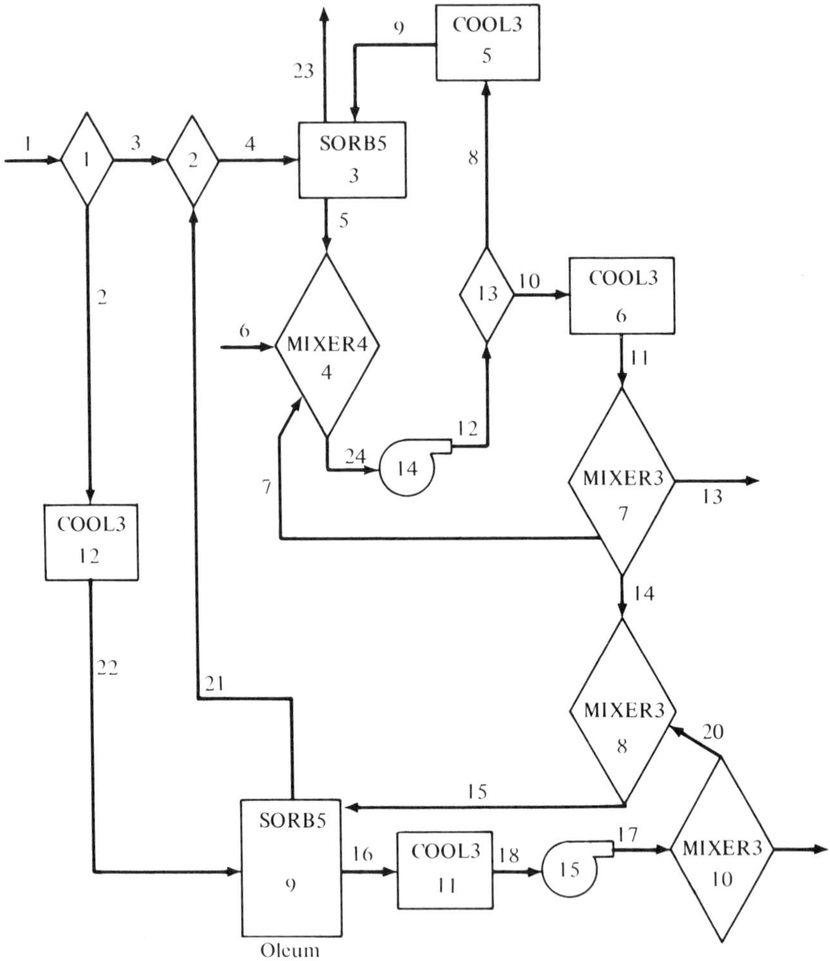

Figure 10.1. Information flow diagram of absorber plant.

10.2 Choice of the Transport Processes for the Model

Some process details

Air drier

Air saturated with water vapor enters the base of the vertical packed tower (air drier) at approximately 29°C and 18 psia. The flow rate is about 3370 lb moles/hr which is equivalent to 20 lb moles/hr ft^2 of tower cross section. Hot concentrated sulfuric acid (99.2 wt %) enters the top of the tower at 68°C at a flow rate of approximately 45 lb moles/hr ft^2. The molar flow rate of the liquid phase is calculated by adding the flow rates of both SO_3 and water. The temperature rise in the gas is about 30°C, while the decrease in the liquid is only 2°C. The water vapor in the gas phase is almost completely absorbed in the concentrated acid. It appears that a negligible amount of acid vaporizes. Further details are in Table 10.1.

Acid absorber

Nitrogen gas containing 10.7 mole % O_2 and 8.4 mole % SO_3 enters the base of the vertical packed tower (acid absorber) at about 240°C and 14.9 psia. The flow rate is approximately 20 lb moles/hr ft^2. Concentrated sulfuric acid (99.2 wt %) enters the top of the tower at 82°C and a flow rate of 100 lb moles/hr ft^2. The temperature decreases in the gas phase by 158°C and in the liquid phase it increases by 38°C. All the SO_3 is absorbed in the acid so that this absorber is operating very close to equilibrium conditions. In particular, the outlet gas temperature is almost equal to the inlet liquid temperature. It appears that the concentrated acid is virtually nonvolatile even at these elevated temperatures. Table 10.2 provides details.

Oleum absorber

The composition of the gas fed to this absorber is the same as that of the acid absorber, its inlet temperature is lower at 93°C and its flow rate is approximately 7.6 lb moles/hr ft^2. Liquid 25% oleum is fed to the top of the vertical tower at a flow rate of about 80 lb moles/hr ft^2 and an entering temperature of 40°C. The leaving gas contains appreciable SO_3 which is later removed in the acid absorber. This difficulty of removing SO_3 is due to the relatively large partial pressure of SO_3 over 25% oleum. The temperature changes in gas and liquid phases are 53°C and 6°C, respectively. For further details refer to Table 10.3.

In all three absorbers, gas and liquid with a large temperature difference flow countercurrently through random packing of highly complex geometry. The flows are turbulent and channeling may occur. For steady-state operation, time-smoothed values of pressure, temperature, and concentration will vary discontinuously with position in the towers. The mechanisms of heat and mass transfer in such a system and the difficulties of developing a comprehensive model are discussed below under separate headings.

Mass transfer considerations

Gas phase

In the gas phase, mass is transferred by molecular diffusion and forced convection. High gas velocities preclude natural convection. The rate of

Table 10.1
Comparison of Plant and Model Data

	SORB5 — Air Drier					
Stream	MODEL and PLANT Gas in	MODEL Gas out	PLANT Gas out	MODEL and PLANT Liquid in	MODEL Liquid out	PLANT Liquid out
Total lb moles/hr	3370	3330	3330	7650	7690	7690
Temp., °C	29.2	67.6	60	68	67.8	66
Pressure, psia	18.2	17.6	17.8	17.8	18.2	18.2
Sulfur lb moles/hr	0	0	0	0	0	0
Water 〃	40.4	0.27	0.13	3908	3948	3948
Oxygen 〃	690	690	690	0	0	0
Nitrogen 〃	2588	2588	2588	0	0	0
SO_2	0	0	0	0	0	0
SO_3 〃	0	0	0	3742	3742	3742

TABLE 10.2
COMPARISON OF PLANT AND MODEL DATA

SORB5—Acid Absorber						
Stream	PLANT Gas in	MODEL Gas out	PLANT Gas out	PLANT Liquid in	MODEL Liquid out	PLANT Liquid out
Total lb moles/hr	3235	2965	2965	18380	18650	18650
Temp., °C	240.0	82.0	82.0	82.0	105.9	120.0
Pressure, psia	14.9	14.8	14.7	14.7	14.9	14.9
Sulfur lb moles/hr	0	0	0	0	0	0
Water ″	0	0	0	9389	9389	9389
Oxygen ″	345	345	345	0	0	0
Nitrogen ″	2610	2610	2610	0	0	0
SO_2 ″	9.7	9.7	9.7	0	0	0
SO_3 ″	270	0.012	0	8988	9257	9257

TABLE 10.3
COMPARISON OF PLANT AND MODEL DATA

SORB5—Oleum Absorber						
Stream	PLANT Gas in	MODEL Gas out	PLANT Gas out	PLANT Liquid in	MODEL Liquid out	PLANT Liquid out
Total lb moles/hr	377.8	350.1	352.6	3980.0	4007.7	4005.2
Temp., °C	93.0	40.4	40.0	40.0	48.5	46.0
Pressure, psia	14.9	14.9	14.9	14.9	14.9	14.9
Sulfur lb moles/hr	0	0	0	0	0	0
Water ″	0	0	0	1665	1665	1665
Oxygen ″	40.4	40.4	40.4	0	0	0
Nitrogen ″	304	304	304	0	0	0
SO_2 ″	1.14	1.14	1.14	0	0	0
SO_3 ″	32.3	4.59	7.06	2315.0	2342.7	2340.2

transfer of mass is affected by the relatively large temperature gradients which exist in the gas phase and which cause properties such as molecular diffusivity, viscosity, and density to vary with position.

1. Molecular diffusion

   There are certainly more than two diffusing species in each absorber, but since N_2 is present in excess, effective binary diffusion may be assumed. The diffusion equation for the case in which trace components 2, 3, . . . , M are diffusing in nearly pure species 1 [Bird, et al. (1960)] follows:

$$\underbrace{N_i}_{\substack{\text{Total flux of} \\ \text{component } i}} = \underbrace{-c D_{ik} \nabla X_i}_{\substack{\text{contribution} \\ \text{by molecular} \\ \text{diffusion}}} + \underbrace{X_i \sum_{j=1}^{M} N_j}_{\substack{\text{contribution} \\ \text{by bulk flow}}} \tag{10.1}$$

   where N_i = flux of species i
   c = total molar concentration
   D_{ik} = binary diffusion coefficient of species i in species k
   ∇ = operator grad
   X_i = mole fraction of species i

   In the air drier, water vapor is the important diffusing species, with the possibility that small quantities of O_2 may dissolve and H_2SO_4 may evaporate. In the acid and oleum absorbers, SO_3 is the important diffusing species. Again the volatility of acid might have to be accounted for in special circumstances. For the simulation of the three gas absorbers it was considered adequate to neglect all diffusing species other than water vapor in the air drier and SO_3 in the acid and oleum absorbers.

2. Forced convection

   For two-phase flow through random packing, the hydrodynamics is so complex that the partial differential equations which describe mass balances for elements of the fluid cannot be solved. Useful insight into the important parameters and correlations can be obtained by dimensional analysis, guided by such mass balance equations. Then it can be shown that for constant physical properties, the overall Sherwood number (or overall Nusselt number for mass transfer) is a function of only two dimensionless groups, the Reynolds and Schmidt numbers. These dimensionless numbers may be defined as follows:

$$\text{Re} = \frac{V_\infty \rho L}{\mu}; \quad \text{Sc} = \frac{\mu}{D_{AB}\rho}; \quad \text{Sh} = \frac{KL}{D_{AB}} \tag{10.2}$$

   where V_∞ = characteristic velocity
   L = characteristic length
   ρ = gas density
   μ = gas viscosity
   D_{AB} = binary diffusion coefficient of species A in B
   K = mass transfer coefficient

This result, which was derived for constant physical properties, is not strictly valid for a nonisothermal gas absorber. The existence of a single V_∞ and L for a packed gas absorber is open to question. In spite of this doubt it has been successfully applied to packed gas absorbers to yield the useful generalization [Sawistowski and Smith (1963)]:

$$\text{Sh} = A\ \text{Re}^b\ \text{Sc}^c \tag{10.3}$$

Liquid phase

For the three gas absorbers (acid and oleum absorbers and air drier), the solute is highly soluble in sulfuric acid and in fact reacts with it. The resistance to transfer of solute in the liquid phase is then very much less than that in the gas phase and may be neglected. In fact the equilibrium partial pressures of SO_3 and H_2O over 99 wt % acid can be considered zero for all rate calculations. This is not so for 25% oleum where there is an appreciable partial pressure of SO_3.

Heat transfer and similarity considerations

Gas phase

Heat is transferred by forced convection in the gas phase. Natural convection and radiation can be neglected. By analogy with the Sherwood number, it may be shown that the Nusselt number for heat transfer is a function of only two significant dimensionless groups, the Reynolds number and the Prandtl number, Pr. The following expression may then be written for heat transfer:

$$\text{Nu} = A\ \text{Re}^b\ \text{Pr}^c \tag{10.4}$$

where the constants A, b, and c are the same for both heat and mass transfer. This is an application of the principle of the similarity between heat and mass transfer and will be used later when formulating the rate equations for the comprehensive absorber model.

Liquid phase

The resistance to heat transfer in the liquid phase, relative to that in the gas phase, may be neglected under most circumstances because of the very large difference in thermal conductivities.

Similarity of the transport processes in three gas absorbers

Not only is there similarity between heat and mass transfer but the transfer mechanisms for heat and mass are also essentially the same in all three absorbers. The only differences lie in the magnitude of transfer coefficients and in the boundary conditions for the diffusing species.

10.3 Formulation of the Rate Equations for the Comprehensive Absorber Model

This general model for packed gas absorbers is based on transfer equations developed by Sherwood and Pigford (1952) where assumptions and further details may be found.

Mass balance

A mass balance on a differential section of a packed absorber gives

$$-\frac{dy}{d(NTU)_M} = y - y_i \tag{10.5}$$

where y = mole fraction of solute in the gas phase
y_i = mole fraction of solute in the gas phase at the gas-liquid interface
$(NTU)_M$ = number of mass transfer units

Heat balance

Heat balances on a differential section give

$$-\frac{dt}{d(NTU)_H} = t - T \tag{10.6}$$

and

$$\frac{dT}{d(NTU)_M} = \frac{G}{LC_{PL}}\left(C_{PG}\frac{dt}{d(NTU)_M} + Q\frac{dy}{d(NTU)_M}\right) \tag{10.7}$$

where t = temperature of the gas phase
T = temperature of the liquid phase
$(NTU)_H$ = number of heat transfer units
G = gas flow rate
L = liquid flow rate
C_{PL} = heat capacity of the liquid phase
C_{PG} = heat capacity of the gas phase
Q = differential heat of solution

Application of the similarity principle

Equations 10.5, 10.6, and 10.7 have two independent variables, $(NTU)_M$ and $(NTU)_H$. The solution of this set of equations can be simplified by eliminating one independent variable. This is achieved by relating $(NTU)_M$ and $(NTU)_H$ through the principle of the similarity between heat and mass transfer by forced convection.

Assuming that Eqs. 10.3 and 10.4 apply to nonisothermal packed gas absorbers, one may show that

$$\frac{\mathrm{Sh}}{\mathrm{Nu}} = (\mathrm{Sc}/\mathrm{Pr})^c \tag{10.8}$$

and finally that

$$d(NTU)_H = \beta\, d(NTU)_M \tag{10.9}$$

where $\beta = (\mathrm{Sc}/\mathrm{Pr})^{1-c}$
Sawistowski and Smith (1963) suggest that a reasonable value of $(1 - c)$ is 0.6, and it will be used for the absorber model.

The application of the similarity principle has reduced the number of independent variables from two to one. The three ordinary differential equations for the general absorber model are:

$$-\frac{dy}{d(NTU)_M} = y - y_i \tag{10.5}$$

$$-\frac{dt}{d(NTU)_M} = \beta(t - T) \tag{10.10}$$

$$\frac{dT}{d(NTU)_M} = \frac{G}{LC_{PL}}\left(C_{PG}\frac{dt}{d(NTU)_M} + Q\frac{dy}{d(NTU)_M}\right) \tag{10.7}$$

The dependent variables are solute mole fraction y and temperature t in the bulk of the gas and temperature T in the bulk of the liquid. The independent variable $(NTU)_M$ is the number of mass transfer units. In general, such equations are nonlinear because of property variations with concentration and temperature. For the problem of simulation of a countercurrent gas absorber, the above set of equations is a boundary value problem. At $(NTU)_M = 0$ both y and t are known, but the outlet liquid temperature at $(NTU)_M = 0$ is not known. To fit the known liquid temperature at the inlet, from a sequence of outlet liquid temperatures at $N_m = 0$ that temperature must be found such that the known inlet temperature of the liquid is calculated from Eqs. 10.5, 10.7, and 10.10.

Development of a correlation for height of a transfer unit

Sawistowski and Smith (1963) have summarized empirical correlations available in the literature for the prediction of HTU_G for various operations of packed gas absorbers. The summary indicates that the available correlations are not reliable. It was therefore decided to match outlet gas and liquid temperatures predicted by the model with those from plant operating data. The number of transfer units needed to achieve this match could thus be found from the model. Since the height of the tower was known, the HTU_G could easily be calculated. An empirical correlation was developed for

HTU_G based on the correlation of Whitney and Vivian [Sawistowski and Smith (1963)] with a correction factor to match plant operating data. The correlation follows:

$$HTU_G = \text{ALPHA}\left(\frac{G^{0.3}}{L^{0.25}}\right) \tag{10.11}$$

ALPHA = 1.47 for the acid absorber, 7.50 for the oleum absorber, and 2.42 for the air drier. Because the Schmidt number does not vary appreciably with temperature for a gas, it was decided not to include such a correction in the correlation for HTU_G. ALPHA apparently depends strongly upon packing type: 7.5 for 2 in. quartz pebbles and 1.47 for 3 in. cross-partition rings.

Development of a pressure drop equation

The following equation proposed by Morris and Jackson (1953) was employed in the absorber model:

$$\Delta p = 0.0074\, N \rho V_0^2\, \ell \tag{10.12}$$

where Δp = pressure drop through packing in milliatmospheres
N = number of velocity heads lost per foot of packing, a function of the packing material and the wetting rate
ρ = gas density in lb/ft^3
V_o = average equivalent empty tower velocity ft/sec
ℓ = packing height in ft

10.4 Data Required for the Accuracy of the Model

The approach taken was to make as many reasonable approximations as possible and then to justify these by comparing plant data with the predictions of the absorber model. Poor agreement would have made it necessary to use more accurate data. Several computer experiments with the model, aimed at establishing the sensitivity to changes in certain physical properties, were most helpful.

10.5 Programming the Model

Choice of the method of solution

The equations of the model (Eqs. 10.5, 10.7, and 10.10) are summarized in Sec. 10.3. It has already been mentioned in Chapter 8 that either Runge-Kutta (RK) or the predictor-corrector methods can be used to solve simultaneous ordinary differential equations. The former method was chosen for reasons already outlined in Chapter 8.

Setting up the algorithm

The structure of the algorithm for SORB5, the model for the simulation of the gas absorbers, is shown in Fig. 10.2. For countercurrent gas absorbers the inlet temperature of the liquid stream is given at the top of the tower while inlet gas temperature and composition are given at the bottom. The use of the RK method requires that these three dependent variables be known at the same position in the tower. The procedure used was to assume an outlet liquid temperature and calculate an inlet liquid temperature. Three guesses were made such that the true value of the outlet liquid temperature was bracketed. Lagrangian interpolation was then used to estimate the true exit temperature. A final pass through the tower was then necessary to calculate the exit gas temperature and composition. Since the pressure drop in packed gas absorbers hardly affects the rate of mass transfer, the pressure drop calculation was made independently of the transfer calculations.

Detailed considerations

A number of precautions must be taken to prevent inaccuracies in computation and mathematical instabilities. For example, the step size used with the Runge-Kutta method must be sufficiently small to give the desired accuracy over a specified range of operation of the gas absorbers. An excessively small step size will waste computer time. Provision must also be made for abnormal inlet conditions to the absorbers. This occurs when a system of equipment is being simulated and all intermediate stream variables are set initially equal to zero. Division by zero can lead to severe numerical problems and must be avoided. A more subtle difficulty occurred with SORB5. When the ratio G/L becomes large, the solution of Eq. 10.7 becomes unstable and an overflow is obtained. This was avoided by introducing a protective statement in SORB5 to ensure that G/L remained at stable levels during the initial iterations of PACER.

Use of SORB5 for design purposes

The use of SORB5 for the design of a gas absorber is straightforward and does not require a trial-and-error procedure with the Runge-Kutta method. This is an initial value problem, since the inlet gas and outlet liquid temperatures and compositions are all known. For example, finding the required height for a specified separation merely requires one solution for a series of packed heights.

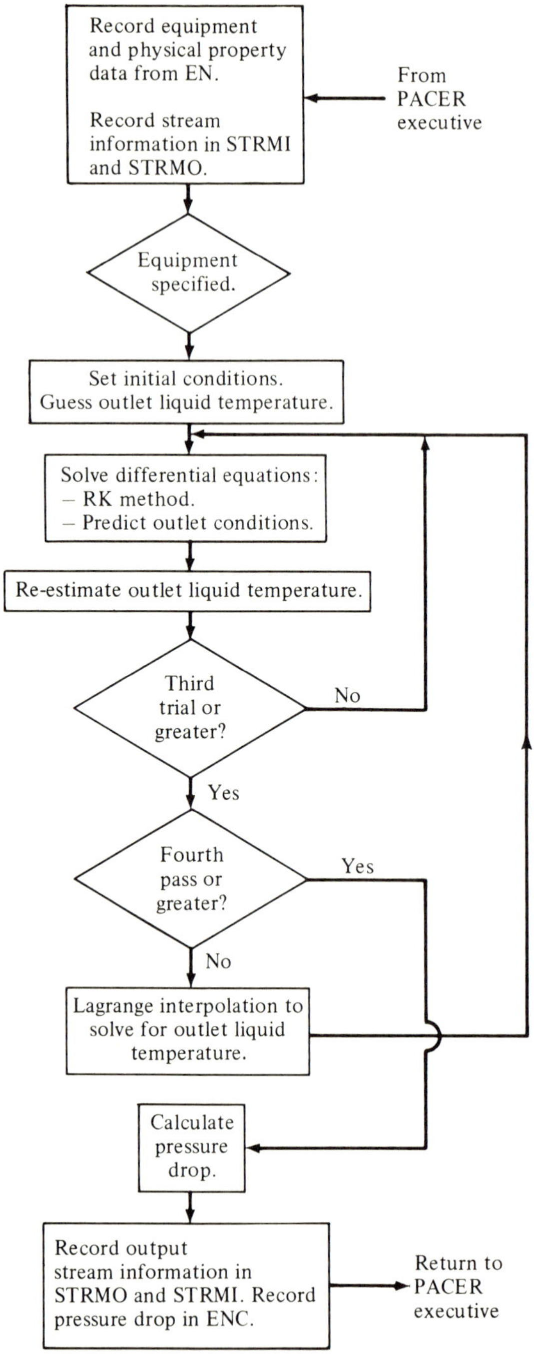

Figure 10.2. Detailed algorithm for SORB5.

10.6 Matching the Model to the Plant

Air drier

SORB5 was run using plant input conditions and the predicted outputs were compared with plant data to test the validity of the model. Unfortunately, all three gas absorbers (air drier, acid absorber, and oleum absorber) were operated under essentially equilibrium conditions. The rate expressions were therefore not really tested, and the computation was merely a heat and mass balance. The results of the test for the air drier are tabulated in Table 10.1.

Acid absorber

The results of test for the acid absorber are tabulated in Table 10.2. The exit liquid has a measured temperature of 120°C while the predicted value is 106°C. This marked difference may be due to an error in measurement of the inlet gas temperature.

Oleum absorber

The results of the test for the oleum absorber are tabulated in Table 10.3.

General conclusions

The test of the model was not entirely satisfactory because of operation of the absorbers near equilibrium conditions. The absorption rates predicted by SORB5 at higher gas throughputs might therefore be greatly in error. It would be very useful to have concentrations and temperatures at intermediate levels in the packing to test properly the rate expressions of SORB5. However, reliable measurements of this kind would probably be very difficult to obtain because of the nature of two-phase countercurrent flow in a packed tower.

10.7 Simulation of an Absorber Plant

An information flow diagram of the absorber plant is shown in Fig. 10.1. The plant consists of acid and oleum absorbers and ancillary equipment. In total there are 15 units and 24 streams. A model of such a system might be used to maximize the profit in the production of two products, 100% acid and 25% oleum. A change in market conditions would necessitate a change in plant operation, giving preference to one of the products. Such a study will not be made here. Instead the absorber plant will be simulated for only one set of operating conditions.

PUMPS1 (equipment numbers 14 and 15) are used to set the recycle flow of acid and oleum. COOL3 is simply a temperature setter. MIXER3 is used as a stream splitter. MIXER6 is used as a stream splitter and to estimate outlet gas temperature. MIXER4 is used as a splitter to estimate outlet liquid temperature and to control water flow to the plant to give a specified acid concentration.

The process matrix is shown in Table 10.4. Typical stream results are shown in Table 10.5. This case predicts after 10 iterations that stream 19 in Fig. 10.1 has 85 moles/hr of SO_3 and 55.2 moles/hr of water.

TABLE 10.4
PROCESS MATRIX FOR THE ABSORBER SECTION

Unit number	Unit Computation name	Stream numbers			
1	MIXER6	1	−2	−3	
2	MIXER6	21	3	−4	
3	SORB5	4	9	−23	−5
4	MIXER4	6	5	7	−24
5	COOL3	8	−9		
6	COOL3	10	−11		
7	MIXER3	11	−7	−13	−14
8	MIXER3	14	20	−15	
9	SORB5	22	15	−21	−16
10	MIXER3	17	−19	−20	
11	COOL3	16	−18		
12	COOL3	2	−22		
13	MIXER3	12	−8	−10	
14	PUMPS1	24	−12		
15	PUMPS1	18	−17		

The acid and oleum absorbers have very large liquid recycle flows; this generally causes an excessive number of iterations to achieve convergence. Each iteration adds but a little increment to a potentially large flow. To speed up the convergence, it is essential to have reasonable estimates of these recycle flows and to use these initially rather than to set all intermediate streams equal to zero.

The approach to convergence for one computer experiment that was made in simulating the absorber plant is shown in Table 10.6.

Thus a number of interacting units in a system may be simulated using one comprehensive model (SORB5) and several rather simple models. There is no technical reason why a much larger number of interacting gas absorbers could not be simulated in much the same manner. However, when there are large recycle flows some means of accelerating numerical convergence may be required. For a simple system it is easy to make fairly precise estimates of recycle flows, but not for a more complex system. The need for the development of general mathematical techniques for promotion of convergence is obvious.

TABLE 10.5
TYPICAL STREAM RESULTS AND SN LIST DEFINITION

Selected streams at iteration 1†				
19.0000	2.0000	140.1750	25.0000	0.0000
45.000	0.0000	55.6857	0.0000	0.0000
0.0000	84.4893	0.0000	0.0000	0.0000
13.0000	2.0000	347.8623	40.0000	0.0000
45.0000	0.0000	0.0000	0.0000	0.0000
0.0000	0.0000	0.0000	0.0000	0.0000
Selected streams at iteration 10†				
19.0000	2.0000	140.1750	25.0000	0.0000
45.0000	0.0000	55.1758	0.0000	0.0000
0.0000	84.9992	0.0000	0.0000	0.0000
13.0000	2.0000	347.5590	40.0000	0.0000
45.0000	0.0000	177.6035	0.0000	0.0000
0.0000	169.9555	0.0000	0.0000	0.0000

† Format of data follows SN list below

THE SN LIST

1 Stream number	2 Stream flag	3 Total flow, moles/hr	4 Temperature, °C	5 Pressure, psia
6 Special flag	7 Sulfur, moles/hr	8 Water, moles/hr	9 Oxygen, moles/hr	10 Nitrogen, moles/hr
11 Sulfur dioxide, moles/hr	12 Sulfur trioxide, moles/hr			

TABLE 10.6
SIMULATION OF ABSORBER PLANT: APPROACH TO CONVERGENCE OF PRODUCT OLEUM (STREAM 19)

Loop No	lb moles/hr water	lb moles/hr sulfur trioxide
10	54.83	85.35
20	55.71	84.46
30	56.32	83.86
40	56.73	83.45
50	57.01	83.17

10.8 Summary

In this chapter a general model of a packed gas absorber has been developed. The principle of similarity of heat and mass transfer was used to account for their occurring together. The boundary value problem associated with countercurrent operation was studied and a method of solving it was proposed and developed.

The comprehensive absorber model satisfactorily simulated three gas absorbers, namely, the air drier, acid absorber, and oleum absorber. It was used with several simpler models to simulate a small absorber section of the sulfuric acid plant. This focused attention on the problem of achieving convergence in a system with large recycle flows and pointed out the need for general techniques of promoting convergence.

PROBLEMS

10.1. It has been decided to increase the capacity of the contact sulfuric acid plant. Refer to the actual absorbers (air drier, acid absorber, and oleum absorber) reported in Chapter 10 and discuss factors which might limit an increase in capacity of these three units. What steps would you take to increase the capacity of the absorber plant? (See Fig. 10.1.)

10.2. Generalize the mass and heat transfer equations (Eqs. 10.5 and 10.7) to account for the absorption of two or more solutes from the gas phase.

10.3. Under what conditions of transfer in the liquid phase can the mass transfer equation (Eq. 10.5) be solved independently of the heat transfer equations (Eqs. 10.7 and 10.10)?

10.4. If natural convection were a significant process in a gas absorber, what dimensionless groups would have to be considered? Show how the similarity principle (convective heat and mass transfer) would be applied in this instance.

10.5. For the case in which the mass transfer equation (Eq. 10.5) can be solved independently of the heat transfer equations, show, by assuming that the exit gas is in thermal equilibrium with the inlet liquid, that a simple heat balance will provide the outlet liquid temperature.

10.6. When the pressure drop in a packed gas absorber is appreciable, the absorption rate will change. Write a modified set of simultaneous ordinary differential equations describing heat, mass, and momentum transfer in this instance.

10.7. The double absorption scheme has been introduced to increase the conversion of sulfur dioxide. In this scheme, a mixture of SO_3 and SO_2 is removed from the reactor after about two bed passes and sent to an absorber where the SO_3 is absorbed in sulfuric acid. The overhead gas is heated and returned to the reactor to pass through the remaining catalyst beds. The gas then goes to a final absorber. Details of such double absorption systems are given by Burkhardt (1968), Kronseder (1968), and Furkert (1969). Modify the SORB5 model where necessary and set up an information flow diagram to represent this system and simulate this double absorption system.

10.8. Mono-, di-, and triethanolamine are often used to scrub undesirable components such as CO_2 or H_2S from gas streams. A flow diagram of such a process is given in *Hydrocarbon Processing and Petroleum Refiner* **47**, 9, 220 (1968). Devise a Unit Computation to represent this absorber.

11 Studies of the Whole Plant and a Summary of the Acid Plant Case

So far in this book, single equipment units and simple recycle sections of the plant have been studied. When the simulation of the whole plant is considered, a major goal is to obtain a converged solution for a large and complex system. Once a converged solution to the simulation is attained, the questions posed in Sec. 4.3 can be answered and the technical conclusions obtained from this acid plant study can be discussed.

Finally, the behavior of PACER as an executive system for such problems can be assessed.

11.1 Complete Plant Studies

When large systems of coupled equations such as for the simulation of the whole plant are studied, three problems are encountered: ensuring that the answers from individual Unit Computations are stable and correct, that the calculations are performed as quickly as possible, and that the answers from the simulation of the complete system are correct.

In Chapter 10, the problem of stability in the individual absorber models was encountered. In Chapter 9, there was a discussion of how the system calculations could be speeded up through the use of the Unit Computation SETSP1 to assign values of the pressure throughout the whole system instead of having the pressure information transmitted back through only one Unit Computation per calculational loop.

In this section whole plant studies and the problem of ensuring that the calculations converge quickly to the correct answers are considered. Different methods of promoting convergence are introduced and the methods used in the acid plant case are described.

Defining the problem of convergence

In the present context, a converged calculation for a simulation of a complete plant requires that the absolute fractional change in prescribed variables between successive iterations be less than DELS. The method PACER uses to test for convergence has been described in Chapter 3. This method is often satisfactory but precautions are necessary for simulations with large recycle flows which converge slowly, and for simulations that have oscillating answers. This latter characteristic exists generally for nonlinear rather than for linear systems.

Slow convergence

Slow convergence has already been discussed in Chapter 2 but is restated here in Table 11.1 in terms of the tolerance TOL(n) at loop n. The tolerance for the system shown in Fig. 11.1 is the flow at loop n less the flow at loop $n - 1$ divided by the flow at loop n. This is expressed in equation form as

$$\text{TOL}(n) = \frac{|\alpha^n - \alpha^{n-1}|}{1 - \alpha^n} \qquad (11.1)$$

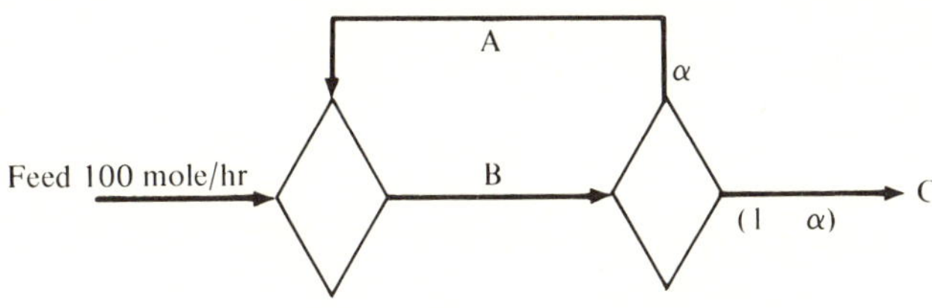

Figure 11.1. A simple recycle problem.

The loop imbalance could be defined as the difference between inputs and outputs from the loop. For this example, it is

$$100 - 100(1 - \alpha^n) = 100\alpha^n \tag{11.2}$$

From Table 11.1 it can be seen that for large recycle, a large number of iterations and very small tolerances are needed to yield the correct converged solution.

TABLE 11.1
TOLERANCE PASSED AND CORRESPONDING LOOP IMBALANCES FOR RECYCLE LOOP OF FIG. 11.1

Loop	$\alpha = 0.33$ Tolerance passed (defined by Eq. 11.1)	Imbalance (defined by Eq. 11.2)	Loop	$\alpha = 0.90$ Tolerance passed	Imbalance
2	0.079	11.1	5	0.16	59
3	0.026	3.8	10	0.06	35
4	0.012	1.2	20	0.015	12
5	0.008	0.39	40	0.0016	1.4

The slope of the line in Fig. 11.2 is proportional to the value of DELS, since DELS is the fractional change in value of a stream variable. Thus, Fig. 11.2 graphically illustrates the number of iterations required for the fractional change in the variable to become less than the specified value of DELS. Therefore, if in DELS the tolerance 0.01 were used, then the answers calcu-

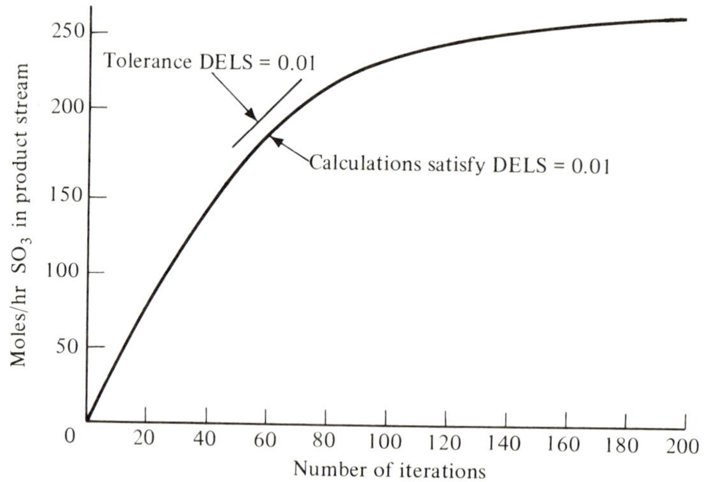

Figure 11.2. Rate of convergence of the acid plant using no convergence promotion and starting the calculations with empty lines.

lated after 60 loops would have been accepted as correct. This emphasizes the behavior illustrated in Table 11.1.

Oscillating answers

Oscillating answers such as the one shown in Fig. 11.3 can arise, for example, when there is more than one recycle stream and each is promoted independently of the other. For this situation, a long time can be spent attempting to reach convergence. Indeed, the stream test DELS is satisfied at any peak or at any valley and would therefore indicate a converged answer for that stream even for the most stringent test DELS applied.

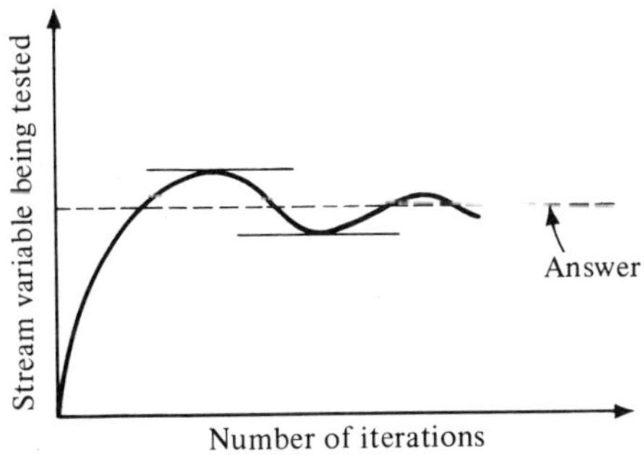

Figure 11.3. Convergence in oscillating systems.

However, the oscillating numbers could be used to advantage in finding the converged solution more quickly, since they may bracket the desired answers.

Convergence promotion

The correct answer is desired quickly. This field of research is developing rapidly and so this section concentrates on principles.

Selection of stream variables

It is not necessary or desirable to promote each variable in the SN list or in all the streams. There is no advantage to promoting stream variables in addition to those in the cut streams because the numerical values in streams other than the cut streams are destroyed in the subsequent calculations. Hence, values for the cut streams are reasonable values to promote.

Concerning the selection of the cut streams, any particular sequence of calculations implies a set of cut streams which can be found by following the sequence and recording those streams that are not already calculated. It should be realized that the set of cut parameters depends on the starting point in a given sequence of calculations. If the calculations are started at a different point in the same sequence, the set of cut streams is different. It has been asserted that the best sequence is the one that minimizes the total number of stream variables in the cut streams, i.e., the cut parameters [Rubin (1962) and Lee and Rudd (1966)]. However, there is no general proof of this assertion. For any given sequence the rate of convergence is not strongly dependent on the starting point in the sequence and therefore is not strongly dependent on the number of cut parameters. A criterion for selecting the best sequence instead of minimizing the cut parameters has not yet been formulated. Consider now methods of convergence promotion.

Use of previous stream information for convergence promotion

The use of approximate or previously computed values of stream information can speed up the calculations. For example, the data in Fig. 11.2 are calculated starting with zeroes for all stream variables but the input variables. Figure 11.4, curve B, shows the same information for the case when the cal-

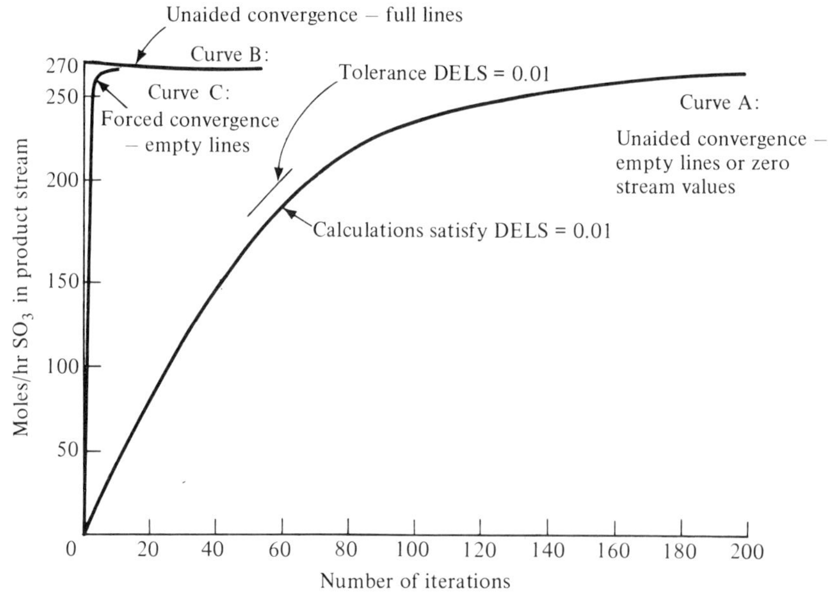

Figure 11.4. Rate of convergence of the acid plant when convergence promotion is used.

culations are started with the *best guesses* for all stream variables. Indeed, stream values calculated from the approximate simulation can be used to start the comprehensive simulation.

In addition, the executive system can be operated so that the stream and equipment matrices are not cleaned before running a new case. Alternatively, the converged solution can be punched out on cards for use in the next case.

As stated above, it is only necessary to supply starting values for the cut stream variables, since all other stream variables are functions of these.

Careful selection of the sequence of calculation

Although executive systems such as PACER find a sequence of calculation, this sequence may not quickly converge. The calculations can be speeded up by selecting particular starting points within a recycle loop or by partially converging recycle loops that exist within other loops. If good stream values are known at particular points in a recycle loop, it may be desirable to start the calculations with the Unit Computations receiving these streams.

More details of careful selection of the sequence of calculations are given by Himmelblau and Bischoff (1968), Sec. 8.2 and Rudd and Watson (1968), Chapter 3.

Examination of information flow

This suggestion combines the idea of providing reasonable values for variables with the idea of studying the sequence of calculations to see what happens to information during recycle calculations. The pressure threading already discussed in Chapter 9 in Unit Computation SETSP1 is a good example of the application of this technique. Here the calculations were speeded up because values for the pressure throughout the whole system were assigned by SETSP1 instead of having the pressure information transmitted back through only one Unit Computation per calculational loop.

Extrapolation from iteration to iteration

Present and past information can be extrapolated to yield better values for the next calculation in a recycle loop. The word "better" needs to be used with caution because the success of extrapolation methods depends upon the variables chosen to be extrapolated, the extrapolation technique, and when the extrapolation technique is applied.

For example, it is possible to select a linear extrapolation based on values for each stream variable x_j in the cut recycle stream calculated for sequential loop calculations, that is, at loops n and $n - 1$. Here the new values of the variables, $x'_{j,n}$, to be used for calculation loop $n + 1$ are

$$x'_{j,n} = x_{j,n} + k_{j,n}(x_{j,n} - x_{j,n-1}) \tag{11.3}$$

Values of k can vary with the component and with the loop number. Although values of k can be selected arbitrarily, Orbach (1969) demonstrated that an analysis of a complex linear recycle system can yield information for predicting values of k. Figure 11.5 illustrates this method.

For any component in the selected stream in the recycle loop, let

$$x_n = ax_{n-1} + b \tag{11.4}$$

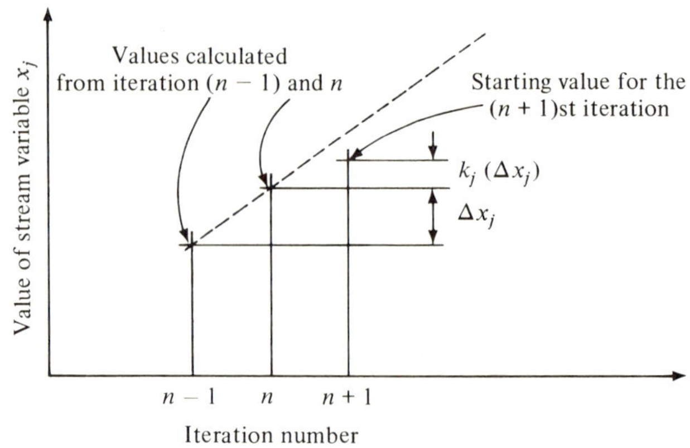

Figure. 11.5. Sample linear extrapolation of a single variable.

$$x_{n-1} = ax_{n-2} + b \tag{11.5}$$

where x_n, x_{n-1}, and x_{n-2} are three values of the variable at loop or iteration numbers n, $n-1$, and $n-2$. Values for a and b can be obtained by solving these equations simultaneously:

$$a = \frac{x_n - x_{n-1}}{x_{n-1} - x_{n-2}} \tag{11.6}$$

$$b = \frac{x_{n-1}^2 - x_n x_{n-2}}{x_{n-1} - x_{n-2}} \tag{11.7}$$

If the initial value of the variable under consideration is x_0, then

at loop 1: $x_1 = ax_0 + b$

at loop 2: $x_2 = ax_1 + b = a^2x_0 + ab + b$

at loop 3: $x_3 = ax_2 + b = a^3x_0 + a^2b + ab + b$

at loop n:

$$\begin{aligned} x_n &= a^nx_0 + a^{n-1}b + a^{n-2}b + \cdots + ab + b \\ &= a^nx_0 + (a^{n-1} + a^{n-2} + \cdots + 1)b \\ &= a^nx_0 + \left(\frac{a^n - 1}{a - 1}\right)b \end{aligned} \tag{11.8}$$

for convergence to a steady state value $a < 1$. Convergence requires that n be large enough that Eq. 11.8 reduce to

$$x_{\text{conv}} = \frac{b}{1-a} \tag{11.9}$$

Using values of a and b from Eq. 11.6 and 11.7,

$$x_{\text{conv}} = \frac{x_{n-1}^2 - x_n x_{n-2}}{2x_{n-1} - x_n - x_{n-2}} \tag{11.10}$$

Equation 11.3 can be rewritten as

$$x_{\text{conv}} = x_n + k_s(x_n - x_{n-1}) \tag{11.11}$$

where k_s is a single correction to yield the steady state converged solution after two successive iterations. For this condition, the value of k is obtained by equating Eq. 11.11 and 11.10 to give

$$k_s = \frac{x_n - x_{n-1}}{2x_{n-1} - x_n - x_{n-2}} \tag{11.12}$$

This single correction can then be applied to any component in a selected stream in the recycle loop to promote convergence.

The linear extrapolation method has been used by Wegstein (1958). It is interesting to note that techniques which assume that the errors in a variable follow a geometric progression are of the same linear form as Eq. 11.3 (see Prob. 11.2).

When there are several cut streams, the use of linear extrapolation on each cut stream separately can lead to oscillating values of the stream variables. This occurs because the changes made in two different cut streams may well be acting on the numerical calculations in opposite directions. Such problems can be avoided by assuming that the newly calculated list of all cut stream variables x_n is linearly related to the previous list x_{n-1} by

$$x_n = \mathbf{A}x_{n-1} + b \tag{11.13}$$

where $\mathbf{A}$ is a matrix of coefficients of size $m \times m$, m is the number of cut stream variables in all the cut streams, and b is a list of constants. The matrix $\mathbf{A}$ could be evaluated but this requires m iterations. Orbach (1969) has shown examples of the use of the full matrix $\mathbf{A}$.

Alternatively, Orbach examined the solution to Eq. 11.13 in terms of the eigenvalues λ_i of $\mathbf{A}$,

$$x_n = \sum_{i=1}^{m} c_i z_i \lambda_i^n + x_s \tag{11.14}$$

where the c_i are scalar constants and the z_i are eigenvectors of $\mathbf{A}$ such that $\mathbf{A}z_i = \lambda_i z_i$. The solution can be verified by substitution into Eq. 11.13. Also $x_s = (\mathbf{I} - \mathbf{A})^{-1}b$ with $\mathbf{I}$ as the identity matrix [Amundson (1966)]. If $|\lambda_i| < 1$, the limit of x_n as $n \to \infty$ is x_s, the list of cut variables at convergence.

Orbach observed that in many chemical processes as n becomes large, all of the λ_i^n but the largest in absolute magnitude, say λ_k^n, become negligibly small so that Eq. 11.14 becomes

$$x_n \doteq c_k z_k \lambda_k^n + x_s \tag{11.15}$$

To test if this equation is valid, values from successive iterations can be examined to see if they obey the relationship

$$x_n - x_{n-1} = \lambda_k(x_{n-1} - x_{n-2}) \tag{11.16}$$

for a given value of λ_k. If so, the converged solution can be obtained by solving Eq. 11.15:

$$x_s = x_{n-1} + \frac{(x_n - x_{n-1})}{(1 - \lambda_k)} \tag{11.17}$$

It may be prudent to extrapolate only part of the distance from x_{n-1} to x_s in order not to disturb the calculations unduly. One of the main advantages of this geometric extrapolation technique is that it is most effective in just those cases which are slow to converge and thus have the largest eigenvalue close to unity. The technique is most applicable when Eq. 11.15 is obeyed after only a few iterations. Orbach has also shown that this technique can give an estimate of the absolute deviation of x_n from x_s. It is then possible to specify the maximum fractional change during one iteration in order to achieve a desired absolute accuracy. In a complex distillation column calculation, Orbach was able to reduce the number of iterations by a factor of six using the geometric extrapolation technique.

So far the variables chosen to be extrapolated and the extrapolation technique have been discussed. Consider now the frequency of application of the convergence promotion extrapolation. Orbach has shown that convergence is promoted more rapidly if the successive application of the extrapolation is delayed by three or four iterations. That is, the extrapolation is applied based on data from loops n and $n - 1$ and then is applied on the data of loops $n + 4$ and $n + 3$. This method is an improvement because the act of extrapolation perturbs the calculations so that extrapolating again too soon means extrapolation on perturbed data. The perturbations die out after several iterations. More details are available from Orbach (1969).

Evans, et al. (1968) and Lederman (1968) briefly review some methods of convergence promotion. Booth and Morris (1966) discuss and give some examples of convergence promotion using three-dimensional linear extrapolation.

Overall control systems

Control systems can be used to ensure the correctness of the simulation and to speed up the calculations. Large simulations are difficult to follow. Often it is very useful, and in fact necessary, to devise an overall control algorithm. A sample algorithm is given in Fig. 11.6 for the Bayer process described in Appendix A.

Other methods

Other convergence promotion methods have been successfully used. Johnson, Aizawa, and Petryschuk (1968) and Aizawa (1967) have used the theta method for the system to converge simulations of distillation columns with a PACER executive system [see Holland (1963), Chapter 13]. Naphtali (1964) used the Newton-Raphson method.

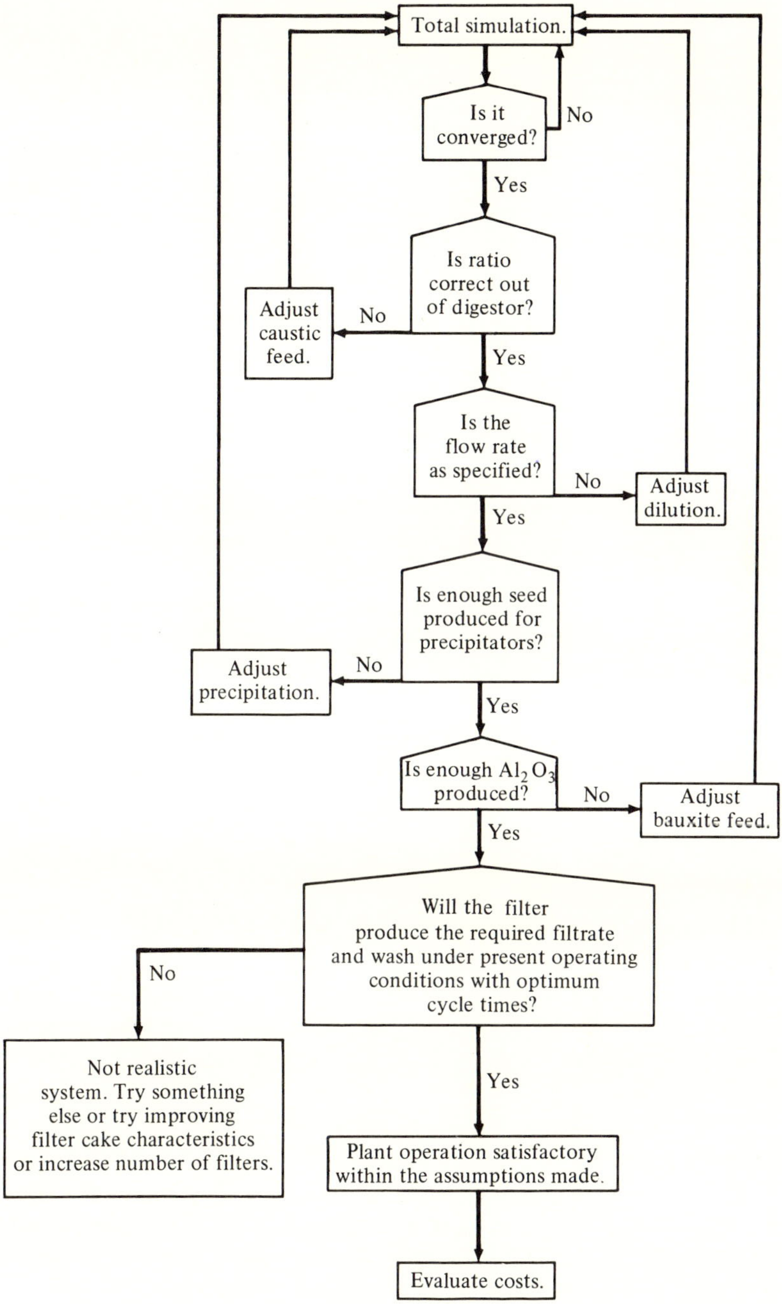

Figure 11.6. Control algorithm for the simulation to assure proper operating conditions in the plant.

Fitting convergence promotion into PACER

In general, the convergence promotion can be thought of as acting as a Unit Computation in one of the recycle streams or as a Unit Computation that works on the whole simulation externally. Whichever way it is programmed or however it is viewed, the promotion occurs in essentially the same way. Examples of these two different viewpoints are shown in Fig. 11.7(b) and (c) for the information flow diagram in Fig. 11.7(a). This represents a second-order recycle problem. Recall that the order of recycle refers to the minimum number of cut streams which will allow the problem to be solved sequential-

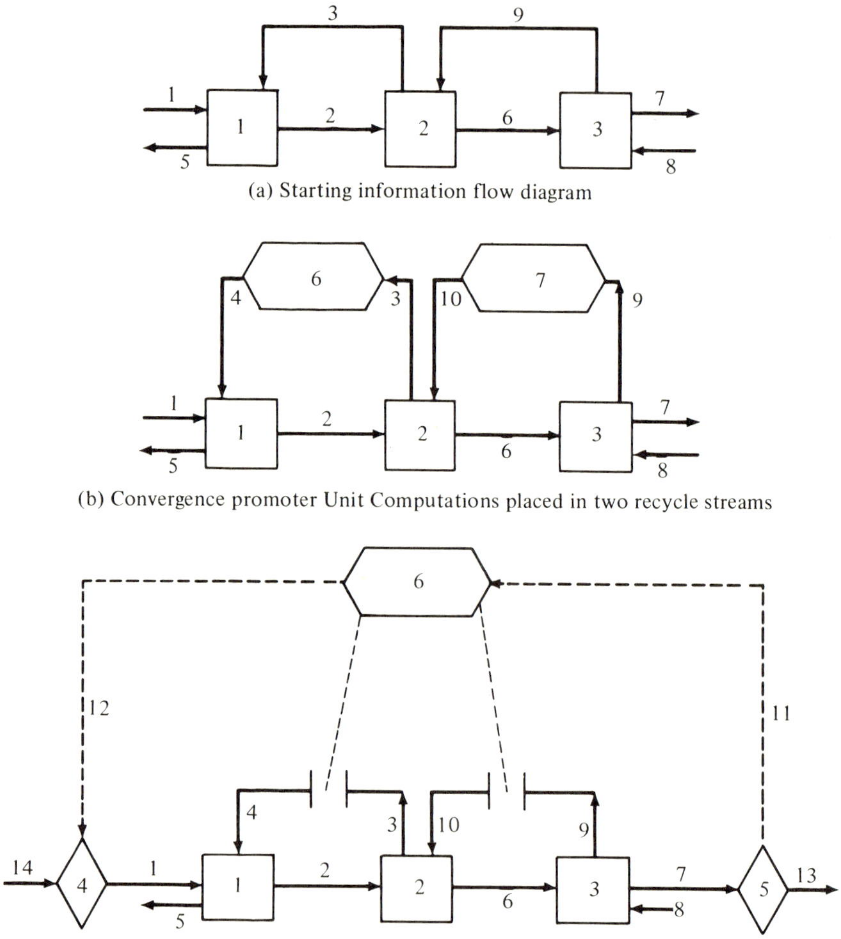

(a) Starting information flow diagram

(b) Convergence promoter Unit Computations placed in two recycle streams

(c) Convergence promoter Unit Computation placed outside the system

Figure 11.7. Fitting convergence promotion into a modular executive system.

ly. The number of streams that are promoted usually equals the order of the recycle.

Consider the operation of a convergence promotion Unit Computation CONP01, unit #6, shown in Fig. 11.7(b). Such a Unit Computation could accept the new value of stream 3 as input and calculate an improved value based on the old value from stream 4 and the new value from stream 3. The improved value would then be stored in stream 4. This type of convergence promotion acts directly on the cut stream because the Unit Computation itself cuts the stream. If two such CONP01 Unit Computations are used in the two recycle streams, it is easy to cause oscillating answers, as described in Sec. 11.1. Each of the two convergence promoters, units #6 and #7, acts in complete ignorance of the other, using only the latest change it sees in its own cut stream. What one unit does will be felt by the other on the next iteration, and each will undo or reverse the action of the other.

The single Unit Computation representing unit #6 in Fig. 11.7(c) performs a function similar to the two units #6 and #7 in Fig. 11.7(b). Here, however, instead of cutting the recycle lines directly, it manipulates the variables in whatever cut streams are selected. This approach can be more general, since each different sequence of calculations used to solve a given case has a different set of cut streams. Streams 11 and 12 are dummy streams which call in unit #6 at the appropriate time. When the Unit Computation is executed, it manipulates information in the selected cut streams. This is shown diagramatically in Fig. 11.7(c) as dotted lines that extend to imaginary cuts in recycle streams 3 and 9. Either method of using convergence promotion models within PACER is relatively easy to program.

Checking technical feasibility

When a converged answer has been obtained for a case, the technical feasibility of the case must be checked. In each Unit Computation, error and comment messages were inserted. Some are printed if individual Unit Computations were applied beyond the range of some variable; some of the messages reveal the conditions existing throughout the process. These messages should be studied to determine if the Unit Computations were correctly used and if the answers are meaningful. A sample of the output for assessing technical feasibility is shown in Table 11.2 for the simulation of the alkylation plant described in Appendix A. Each message should explain what happened, in what unit it happened, and what was done to allow the calculations to continue.

Calculating financial feasibility

Managers and engineers are interested in costs and dollar flow, and the financial attractiveness of alternatives. This financial feasibility should be

Table 11.2

Sample Technical Feasibility Printout

```
SUMMARY OF PROCESS AND CORRELATION LIMITS

THE FOLLOWING LIMITS HAVE BEEN EXCEEDED IN THE REACTOR UNIT3. CALCULATIONS CONTINUE WITH
  EXTRAPOLATED VALUE
THE CORRELATION LIMIT OF 52.5 IS EXCEEDED FOR STRMI (1, 16) WHICH IS 56.0

THE DC3 VAPOR VELOCITY IS 5.05 AND THE PRESENT LIMIT IS 13.00 CUBIC FEET PER SECOND
THE DC3 OVERHEAD CONDENSER DUTY IS 10.09 AND THE PRESENT LIMIT IS 10.00 MILLION BTU PER HOUR

THE PRODUCT SEPARATOR PRESSURE CONTROL WAS NOT REQUIRED AND THE PRESSURE REMAINED AT
  THE BASE VALUE OF 17.7 PSIA
THE FRESH MAKE-UP ACID IS 99.20 PERCENT ACID. THE ACID FEED TO THE REACTORS IS 48.0 PERCENT OF
  THE TOTAL FLOW AND IS 91.00 WEIGHT PERCENT ACID AND 1.63 WEIGHT PERCENT WATER
```

determined only after converged mass and energy balances are obtained and it is certain that the process is technically feasible. The overall relationship among these different calculations is shown in Fig. 11.8.

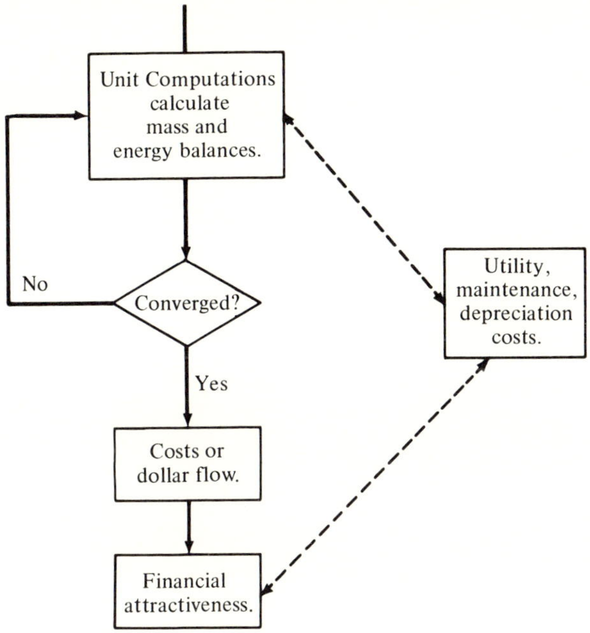

Figure 11.8. Relationship between economic and mass balance calculations.

Costs and dollar flow

Costs and dollar flow for products and by-products include the costs per unit of production and per unit of time. Table 4.1 represents a typical estimate of the cost per unit production for the single product, sulfuric acid. In this calculation, the by-products steam and sulfur dioxide were assigned values.

The costs incurred in production are summarized in Table 11.3. These costs are classified according to how they vary with production rate and how easily they can be attributed a given product. Details are given by Peters and Timmerhaus (1968), Rudd and Watson (1968), Schweyer (1955), Chilton (1960), and Bauman (1964).

In any problem, it is possible to estimate the directly attributable costs, M_D and depreciation, and/or the total cost C. Normally, for plant improvement only the directly attributable costs need to be considered. Further, the utility costs and the capital investment (and hence depreciation and maintenance) can be attributed to a given equipment unit. For example, if a pump

TABLE 11.3
CLASSIFICATION OF PRODUCT COSTS

	Classified according to variation with the rate of production	Cost item	Classified according to ability to attribute the cost to a product		
C	C_L Costs vary about linearly with production rate	Raw materials	M_D Directly attributable	M	C
		Utilities			
		Packing, containers			
		Shipping			
		Royalties			
	C_R Costs vary with production rate but are not zero at zero production rate	Operating labor			
		Direct supervision			
		Maintenance			
		Plant supplies			
		Laboratory analyses			
	C_F Costs independent of production rate	Rent	M_F Directly attributable and independent of production rate		
		Insurance			
		Taxes			
		Depreciation			
		Medical	M_1 Indirectly attributable to a product		
		Safety and protection			
		General plant overheads			
		Payroll			
		Restaurant			
		Recreation			
		Storage facilities			
		Administration	G General expenses not easily attributable to a given product		
		Sales, market research, and advertising			
		Research and development			
		Financing			

is installed, these costs for this unit can be estimated, whereas it is difficult to assign a portion of the packing, shipping, royalty, or operating labor to the pump. Normally, the utility, maintenance, and depreciation costs are calculated within each Unit Computation. This is shown as a dotted line in Fig. 11.8. The calculations can be done either when each Unit Computation is executed during the mass and energy balance calculations or after the mass and energy balances converge. To avoid calculating costs unnecessarily at each iteration, each Unit Computation can include internal switching to separate the technical from the financial calculations. When a converged mass and energy balance is attained, the internal switch is set within each Unit Computation so that the next sweep through the system calculates only the operating costs. If there is little recycle, the convenience of calculating the pertinent costs for each unit during the mass and energy balance calculations offests the additional time required to obtain a converged answer. Shaw (1969) found that, for the recycle problems and cost calculations used for the alkylation plant (described in Appendix A), calculation of the directly attributed costs concurrent with the mass and energy balance did not significantly alter convergence time.

Two different forms of costs and dollar flow can be reported. The cost information can be treated as a stream variable and dollars flowing in the streams of the system can be shown. Alternatively, it is possible to report the operating costs for each equipment unit, as well as the total operating costs. The difference in the type of information is shown in Fig. 11.9.

If the choice is to store cost information in the streams, then care must be taken to ensure that there will not be ridiculous answers because of the recycle of costs. Normally, to eliminate difficulties of recycling costs zero value is assigned to all recycle streams. This provides a convenience but does not realistically represent cost flow. However, this should not be done when part of the recycle stream is sold as a by-product as with stream 11 in Fig. 11.9 because this incorrectly implies that only the cost of unit #6 affects the cost of stream 11. Other difficulties occur. In costing the by-product stream 5, the cost of units #4 and #5 is not included if the cost split at unit #5 to stream 8 is set at zero. Yet from a technical viewpoint the cost of operating unit #3 is undoubtedly affected by the operation of units #4 and #5. Finally, in the use of streams to transmit information, there must be a selection of appropriate values for the four cost split parameters β_8, α_9, α_4, and γ_{11} for the system in Fig. 11.9. Nevertheless, the usual method is to have costs flow in the streams for overall processing plants with little recycle such as that shown in Fig. 11.10 or for serial assembly line systems such as the production of motors or automobiles.

When the process has recycle, it is usually easier to associate cost information with each equipment unit in the auxiliary equipment list ENC or in a cost matrix associated with equipment units. The costs in ENC are summed

System :

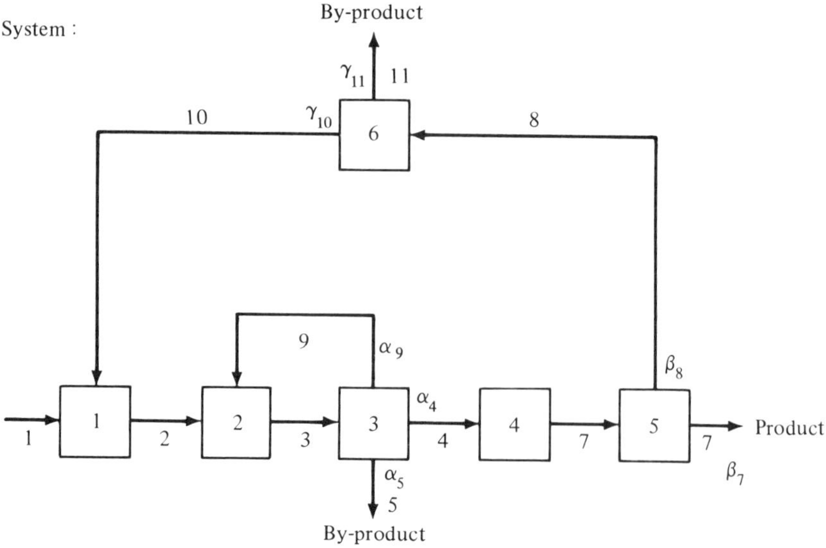

Costs associated with equipment

Equipment unit	Cost, \$/hr
1	\$ 1,000
2	2,000
3	3,000
4	4,000
5	5,000
6	6,000

Total operating cost	\$21,000/hr
Raw material cost	\$10,000/hr
Total	\$31,000/hr

Allocation factors among products 5,7, and 11: $\delta_5 = 0.306$; $\delta_{11} = 0.097$.

Stream 5 = 0.306(\$31,000) = \$ 9,500
11 = 0.097(\$31,000) = \$ 3,000
7 = \$31,000 − 12,500 = \$18,500

Costs associated with the streams

Assume $\alpha_9 = 0$; $\alpha_4 = \alpha_5 = 0.50$; $\beta_8 = 0$; $\gamma_{11} = \gamma_{10} = 0.50$.

Stream	Cost, \$/hr Intermediate	Product
1	\$10,000	
2	14,000	
3	16,000	
4	9,500	
5	—	9,500
6	13,500	
7	—	18,500
8	0	
9	0	
10	3,000	
11	—	3,000
Total		\$31,000

Figure 11.9. Dollar and cost distribution.

at the end of the evaluation of a case. The summed costs are then distributed among the products. For the example in Fig. 11.9 it is necessary to select two allocation factors, as opposed to the four factors required if costs flow in the streams, to distribute the costs among the three products.

In summary, the decisions about costs are:

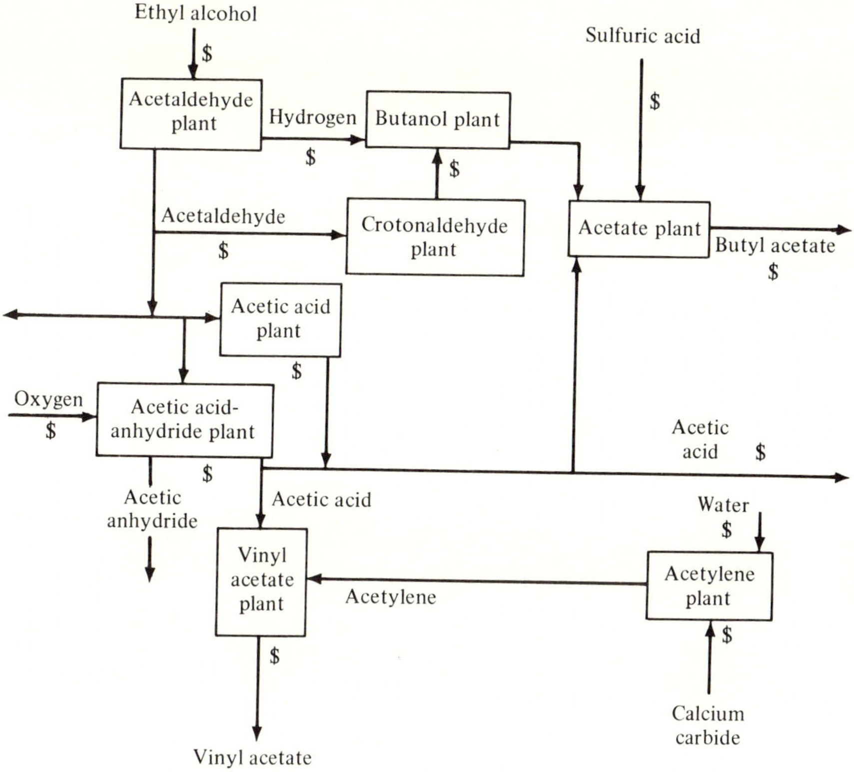

Figure 11.10. A simulation problem in which it is convenient to study dollar flow in the streams.

1. In which costs is there interest?
2. Where and when is it best to calculate the contributions?
3. Should the costs be associated with equipment units and summed or can the costs flow through the streams?

Financial attractiveness of investments

The financial attractiveness of alternative capital investments can be reported as percent return on investment, payback time, venture profit, venture worth, and discounted cash flow (DCF). Details of these methods are given by Happel (1958) and Peters and Timmerhaus (1968) and are compared by Bauman (1964a). This calculation is made after the costs or dollar flows have been calculated. James, et al. (1966) describe the use of the Kellogg Flexible Flowsheet executive program to check the financial attractiveness of processing alternatives in design. Their tables III and IV report sample economic calculations.

Fitting financial feasibility calculations into PACER

Fitting the costing and financial attractiveness models into PACER is relatively easy. If costs are not defined as stream variables, they are associated with each Unit Computation. During the execution of each Unit Computation, the operating and capital investment costs can be calculated and stored in assigned elements of the respective ENC list. Once the converged solution is available, the COST01 Unit Computation adds the costs in like elements of the ENC lists and prints out the result. A typical allocation of elements in the ENC list is shown in Table 11.4.

TABLE 11.4
SAMPLE MEANINGS FOR ENC ELEMENTS SO THAT UNIT COMPUTATIONS ARE COMPATIBLE WITH COST01

ENC element	Meaning
11	1000 lb/hr low pressure steam (50 psig)
12	1000 lb/hr medium pressure steam (200 psig)
13	1000 lb/hr high pressure steam
14	million BTU/hr fuel gas generated within company
15	million BTU/hr fuel gas purchased outside company
16	kw electricity
17	1000 U.S. gal/hr cooling water
18	1000 U.S. gal/hr process water

Normally, COST01 would sum only directly attributable costs M_D and perhaps the depreciation based on the new capital investment incurred in one case relative to the base case. If the other costs are to be included, the indirectly attributable, fixed, and general expenses are added to the product and by-product costs according to some acceptable allocation factor in another Unit Computation and need not be calculated within the process.

A Unit Computation for calculating the financial attractiveness of cases involving capital expenditure can be used after the costs have been calculated.

A sample output of a cost report showing the information of interest to the engineer is shown in Table 11.5. Here, the quantities and values or costs are reported on a monthly basis relative to a base case of operation for an alkylation process. Details of one such method as applied to the simulation of an alkylation plant are given by Shaw (1969).

Methods used in the acid plant case

For the studies of the whole plant, problems of convergence were encountered and some cost and energy information was desired.

TABLE 11.5

SAMPLE OPERATING COST PRINTOUT

OPERATING COST SUMMARY

RUN NUMBER 2
CONDITIONS

ALKYLATION SIMULATION

ENRICHED I-C4 FEED

SUMMARY OF MATERIAL FLOWS			QUANTITY PER DAY		VALUE PER DAY	
			RUN NO 2	BASE CASE	RUN NO 2	BASE CASE
FEEDS	ISOBUTANE	(BBLS)	990.	1576.	3080.	4857.
	OLEFIN	(BBLS)	3180.	3180.	9132.	9132.
	SULFURIC ACID	(BBLS)	75.	74.	648.	645.
	CAUSTIC	(100 LBS)	11.	11.	39.	39.
PRODUCTS	ALKYLATE	(BBLS)	1881.	1923.	7950.	8114.
	PROPANE	(BBLS)	723.	839.	2604.	3020.
	BUTANE	(BBLS)	1131.	1552.	3981.	5464.
UTILITIES	STEAM (50 PSIG)	(1000 LBS)	68.	84.	26.	32.
	STEAM (200 PSIG)	(1000 LBS)	327.	301.	124.	114.
	FUEL	(MM BTU)	495.	510.	140.	144.
	ELECTRICITY	(KW)	32376.	32928.	204.	207.
	COOLING WATER	(1000 GALS)	3052.	3077.	34.	34.

THE DEBUTANIZED R—3 OCTANE RATING OF THE ALKLAYTE IS 104.52

	RUN NO 2	BASE CASE	DIFFERENCE
COST FUNCTION PER DAY	13426.3	15205.4	−1779.16
PROFIT FUNCTION PER DAY	1109.1	1392.0	−282.93
COST FUNCTION PER BBL OF ALKYLATE	7.137	7.907	−0.7706
PROFIT FUNCTION PER BBL OF ALKYLATE	0.590	0.724	−0.1347

Convergence promotion

Many of the methods already described to promote convergence were used. In the plant, there is a large recycle of acid to the drier, so that unpromoted convergence is slow. To promote convergence, approximate values of the variables for the recycle streams were read in as part of the input to each case.

A study of the information flow revealed that pressure was transmitted slowly upstream against the flow of material in the reactor section. A Unit Computation SETSP1 was written to transmit pressure from the first output to the first input stream for all units within the scope indicated in its own parameter list. Furthermore, some flows had to be fixed to enforce mass balances. Thus, in the first studies of the reactor section, the flow of water to the boilers was set equal to the amount of steam generated by the two boilers. Enforcing this balance was accomplished by means of a pump Unit Computation which can fix the flow in the first output stream. (In the physical plant such a pump is necessary to enforce the balance.)

The absorber section of the plant is shown schematically in Fig. 11.11(a). Two submerged pumps are used to recycle acid to the acid absorber and drying tower and to deliver the acid product from the plant. These are located in the dilution tank in which fresh water is added to produce the acid product of desired concentration. In the first model of this system, illustrated in Fig. 11.11(b), pump and mixer Unit Computations were used to fix the flow back to the absorber and drier. The flow of dilution water was fixed equal to that encountered in the plant for the test case under study. The result was a very slow increase in the concentration of sulfur trioxide in the streams flowing in the simulation, as indicated by the unaided convergence curves of Fig. 11.2.

The method finally used to obtain rapid convergence of the acid plant, Fig. 11.4, is illustrated in Fig. 11.11(c) in which the dilution tank of the real plant has been simulated by a concentration-setting mixer. This Unit Computation calculates the amount of water that should be added to the dilution tank to maintain a desired exit concentration of sulfuric acid. A second pump Unit Computation was added to control the flow of acid leaving the whole plant. Hence in the final plant simulation, the flow and concentration of acid to the absorber were fixed and the flow of acid of a fixed concentration leaving the plant was set equal to the rate of production of new acid in the absorber. These were the ways used to promote convergence in the studies of the whole acid plant. The results were technically feasible.

Final information flow diagram and comments

Initial studies of the whole plant used fixed feeds of sulfur (stream 7), air (stream 1), and dilution water (stream 41). The pressure threading subroutine SETSP1 was not included and hence pressure was transmitted back-

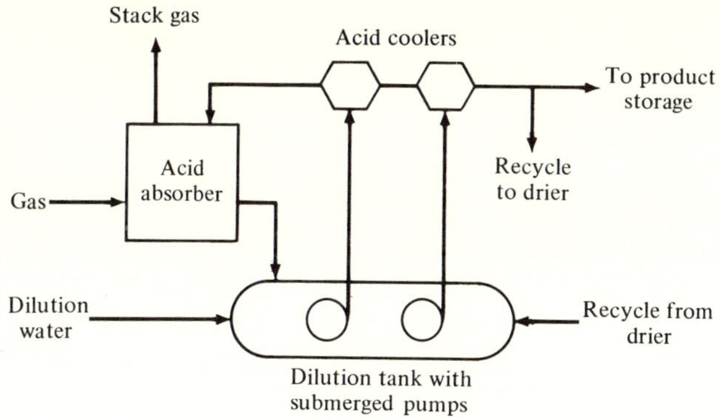

(a) Acid absorber and dilution tank system

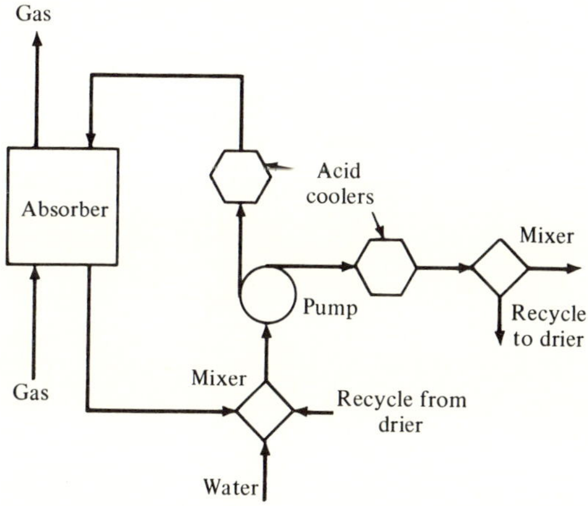

(b) Information flow diagram — slow convergence

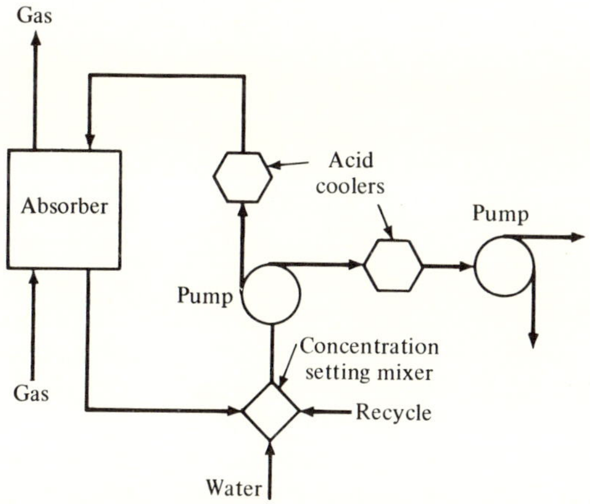

(c) Information flow diagram — rapid convergence

Figure 11.11. Improvement of overall plant convergence with a different Unit Computation.

ward through the plant at a rate of one unit per iteration. Also, the dilution tank was initially simulated by a single MIXER3 Unit Computation.

The PACER executive program caused calculations to be carried out with a normal forward feed of information, with the exception of the pressure drop in the main line as discussed above. The unaided convergence of this simulation with initially empty lines through the plant is shown in Fig. 11.4, Curve A; it can be seen that over 200 iterations were required to approach steady conditions. Even when the streams through the simulation were given approximate values, over 40 iterations were required to stabilize the calculations. In Fig. 11.4 it is seen that even when a tolerance of 0.01 (that is, one percent) has been passed, the flow of sulfur trioxide in the product stream is very far from the steady state value.

Because the backward transmission of pressure, acting against the forward flow of information, is slow during iteration, Unit Computation SETSP1 was introduced to thread the pressure back through the main line of the simulation. This reduced the number of iterations required to adjust pressures from about 20 to 4 and was particularly valuable in the simulation of the reactor section of the plant.

After observing the slow unaided convergence of the whole plant, it was noted that one would normally start the real plant by filling the dilution tank, unit #24, with concentrated sulfuric acid to promote absorption of sulfur trioxide by the liquid phase. It was similarly necessary to start with acid in the streams entering unit #24 in the simulation to bring the whole plant calculation to convergence in a short time. Otherwise the buildup of acid in the very large flows of liquid around the acid absorption column was very slow. As indicated in Fig. 11.11, the recommended simulation of the dilution tank involves a MIXER4 (unit #24) which fixes the concentration of acid flowing in streams 31, 32, 33, 36, 37, and 39, and a PUMPS1 (unit #25) which sets the flow in the streams to the absorber.

The parameter-changing Unit Computation CONT01, unit #44 (in addition to its duties in controlling water flow in the steam generating section), was used to enforce a sulfur balance over the whole plant. This Unit Computation regulates the production of the plant, the main flow from the PUMPS1, unit #28, in accordance with the sulfur trioxide absorbed in the acid production tower. It is also necessary to start the simulation with flows to the pumps; interestingly, one cannot start the plant physically or mathematically with a dry pump. With the use of the pressure threading Unit Computation SETSP1 and the material balance forcing Unit Computation CONT01, it was possible to achieve convergence of the whole plant in three or four iterations to a tolerance of 0.01%. The feed of sulfur and air to the plant, the flow of acid to the absorber and drier, and the flow of water around the steam drum must be specified, but all other lines may be initially empty. Figure 11.4, Curve C, indicates the rapidity of this forced convergence.

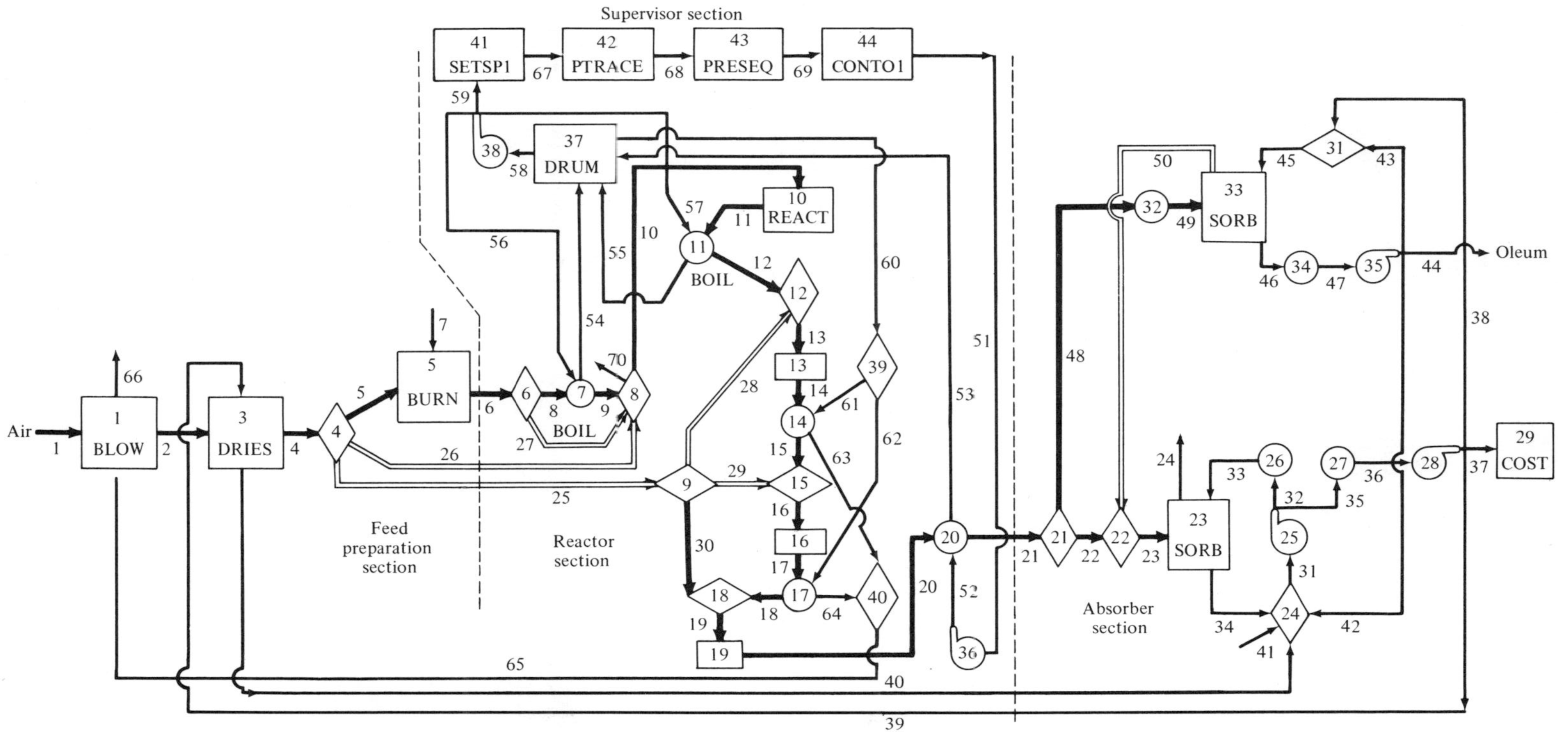

Figure 11.12. Final information flow diagram for the sulfuric acid plant shown in Fig. 4.1

Note that most of the information-gathering and parameter-varying Unit Computations have been shown together in the upper left section of Fig. 11.12. These constitute a control section of the simulation and could have been included in any recycle loop. Here PTRACE and PRESEQ also permitted pressure transmission through the steam generating system.

Financial calculations

A Unit Computation COST01 collects operating and depreciated capital costs that were calculated at each appropriate Unit Computation and stored in fixed elements of the ENC list. It also collects the total energy available from the plant.

11.2 Technical Summary and Conclusions of the Acid Plant Case

Consider the objectives of the study as outlined in Sec. 4.3. Improving the acid plant included giving engineers a better understanding of the process, the identification of sensitive variables and potential trouble areas, and the reduction in costs.

It can be seen that the product cost depends markedly on the cost of sulfur and how efficiently all the sulfur is converted to acid. Further, the addition of new catalyst to bed 2 increases conversion whereas increasing the air addition to bed 2 reduces conversion. The first bed operates so close to equilibrium that rate does not enter the evaluation of this bed.

A detailed study of the simulation of the boiler alone indicated that the boiler operated with about 15% vapor in the last tube row. This dry wall formation under these high temperatures would eventually lead to tube failures. Since the boiler limits the capacity, it is possible to increase the number of risers and add rows of tubes to the boiler to eliminate this bottleneck.

The absorbers all operated under equilibrium conditions.

The following results of the simulation suggest sensitive control conditions which might be analogous to the plant operation. The overall system control of the simulation is sensitive to flow control at the dilution tank exit and at the water feed to the boiler system. Convergence at start-up was greatly aided by filling the recycle lines with approximately correct material flows.

In the absorbers, numerical instabilities occurred when a high ratio of gas-to-liquid flows existed. This might indicate physical instabilities in the plant.

The experimental measurements which were prompted by this study provided values for the acid flows to the driers and absorbers and for the gas flow rate into the system. The former flows were previously unknown and the latter flow was experimentally much higher than was previously thought. Finally, because of its extensive testing and fundamental basis, this simulation can be used:

1. To test new ideas when new conditions arise;
2. As part of a library of Unit Computations to help in a simulation of other processes;
3. To optimize the process. (This is discussed in Chapter 12.)

The resources required for this case study were about 350 man-hours and about 6000 minutes of computer time. It is very difficult to relate this to an industrial environment in which one man may work on a simulation full-time. It is even difficult to relate to the present academic situation because of the advances in know-how since 1964 when the acid plant case was first done.

In summary, although this process is well understood, the simulation generated many ideas useful for the operation of this plant, and provided a model that can be used for future studies and operator-engineer training.

11.3 Comments on PACER and the Modular Approach

The modular approach works well because of its open-endedness and because it allows the creation of a library of Unit Computations. This library enables the engineer to start simply and then gradually increase the accuracy of the models as required.

As far as PACER is concerned, the process matrix and SN and EN lists were oriented toward the engineer. These features were easy to explain, understand, use, and modify. However, the executive had several weaknesses for the particular problem studied. The testing of convergence in every single stream is very demanding and not strictly necessary. It ought to be sufficient to test only the recycle streams for convergence. The executive could be improved markedly by the inclusion of a physical properties handling system, convergence promotion routines, and a costing system. For this problem, the original purpose of the ENC matrix as control parameter storage was not realized.

PACER finds a sequence of calculation for up to three cut streams which may not be the fastest for calculation. However, the use of convergence promotion increases the speed of calculation for any sequence, thereby overcoming the handicap of an inferior sequence. The stream used to start the sequence of calculations should be one about which there is accurate knowledge. The system used by PACER (1966) does not consider this.

In summary, the modular approach has much to offer, especially when it is necessary to start simply and add accuracy as deemed necessary and when there is a library of Unit Computations available.

11.4 Summary

Problems of obtaining correct answers quickly often occur in complex systems of information, for example, in the study of a complete plant. Examples in which these problems converge slowly and have oscillating answers are given. There was a discussion of the promotion of convergence by selecting which variables to promote and by the use of previous information, particular sequences of calculation, information flow patterns, extrapolation techniques, and control theory. The incorporation of convergence promotion into PACER was described.

The converged answers should be checked for technical and financial feasibility. Fitting costs and dollar flow and the financial attractiveness of alternative investments into PACER is relatively easy. Sample methods and a brief discussion of important considerations were reported.

Most of the discussion was in general terms. However, to illustrate how the principles could be applied, the methods used for the acid plant case were set forth.

To summarize the acid plant case, the technical results and conclusions were outlined. Finally, some comments were made about how well PACER, and modular executive systems in general, worked for this particular case study.

PROBLEMS

11.1. (a) What happens if convergence promotion techniques are applied to all the variables? Devise a simple problem to illustrate this point.
(b) How could we select the best streams and best stream variables upon which to apply convergence promotion?

11.2. A geometric extrapolation technique for convergence promotion could estimate values for x_n from the relationship $x_n - x_{n-1} = k(x_{n-1} - x_{n-2})$, having found a value of $k < 1$ on the previous iteration.
(a) Comment on this geometric technique and suggest an improvement.
(b) Prove that such a geometric extrapolation technique is the same as the linear extrapolation method given in Eq. 11.3.

11.3. For oscillating solutions mentioned in Sec. 11.1 what criteria or methods can be used to yield the correct answers rapidly?

11.4. Convergence promotion methods are used to speed up calculations. However, by inserting the convergence promotion Unit Computations into the system we increase the calculation time. What criterion could be used to help us decide when and when not to develop convergence promotion?

11.5. Variations on the linear extrapolation methods are presently being developed. The next level of rigor might be for second or third order extrapolation techniques for a single recycle stream. For example, a second order analog to Eq. 11.3 might be:

$$x'_{jn} = x_{j,n} + k_1(x_{j,n} - x_{j,n-1}) + k_2(x_{j,n} - x_{j,n-2})$$

Devise an algorithm and a working program for such a technique to be used within PACER.

11.6. Suppose we elect to study dollar flow and to allow costs to flow in the streams. Develop equations to predict the costs in the streams shown in Fig. 11.9. For the cost allocation factors, use α_4, α_5, and α_9 at unit #3; β_7 and β_8 at unit #5 and γ_{10} and γ_{11} at unit #6.

Those who have available a simulation of a sulfuric acid plant can produce quantitative answers for the following questions. Those who do not have a simulation can use these problems to test their understanding of simulation and to illustrate the type of questions that can be tested with any simulation.

11.7. Sometimes a final SO_3 cooler is installed between the economizer and the absorber. The SO_3 is on the tube side, with atmospheric air as the shell side coolant.

(a) Can the models described in the preceding chapters be used to simulate this unit? How can this change in processing be evaluated? See Simon-Carves (1958), p. 14.

(b) Under what conditions would this improve operations?

11.8. A large number of alternatives have been suggested for sulfur burning processes for the production of sulfuric acid. For example,

1. Double absorption schemes to increase conversion—Burkhardt (1968), Kronseder (1968), Furkert (1969);
2. Addition of more catalyst layers—Burkhardt (1968);
3. Ammonia scrubber to reduce effluent SO_2 concentration—Kronseder (1968);
4. Double absorption plus additional cooling of reactor feed—*Chemical Engineering* (1968)

(a) Can the models described in the preceding chapters be used to simulate these alternatives?

(b) If not, can the Unit Computations be easily modified or developed to make an evaluation possible?

(c) Under what conditions would each alternative reduce the operating cost and, where appropriate, yield a return on capital investment of more than 20%?

11.9. Sulfuric acid can be produced from gases containing sulfur dioxide. In particular, the stack gases from ore roasting plants are popular feeds. Descriptions of such processes are available from Connor (1968), Farkas and Dukes (1968), or from most contracting companies.

(a) Under what conditions can the models developed in the preceding chapters be used to simulate this process? What Unit Computations will have to have the most modifications? Why?

(b) Estimate the unit cost of producing a ton of sulfuric acid via this route, assuming that the gas is available at the stack under the existing conditions at no cost and assuming that the net cost of gas after it has been cleaned and dried ready for absorption is \$1.00 per ton of SO_2 treated.

(c) Simulate this process from details given in the literature.

11.10. In the Kachkaroff Process, sulfuric acid is produced from the reaction between sulfur dioxide, oxygen, and nitrous vitriol (80 to 85% H_2SO_4 containing 9 to 15% N_2O_3 wt/volume) [Simon-Carves (1958), p. 43]. Search the literature for details of this process. What models developed in this book can be used to simulate this process?

11.11. New sources of sulfur for the production of sulfuric acid are described by Mandelik and Pierson (1968) and Wheelock and Boylan (1968). These articles also give sample process flow diagrams. What parts of these processes can the models developed in this book satisfactorily simulate?

Optimization of the Plant Simulation

12

12.1 Introduction

After the simulation is completed, it can be used to see how to improve plant performance. The search for the best performance, according to some profit criterion and subject to constraints on possible values of the variables, is called *optimization*. Several articles and books [Wilde (1964), Wilde and Beightler (1967), Nemhauser (1966), Fan (1966), Pontryagin (1962), and Rudd and Watson (1968)] give comprehensive reviews of various aspects of optimization.

Instead of yet another review, this chapter shows how some typical methods of optimization can be applied to a simulation, taking advantage of several features of PACER. The acid plant is used as an example. Four major decisions must be made to begin the optimization. These are:

1. Choose the criterion of improved performance. It might be maximum conversion, minimum cost, or some combination of variables. In general, this criterion is called the *objective function* or *response*. The choice of criterion is normally done before the simulation is undertaken; this has been discussed in Sec. 4.3.
2. Select the variables to be changed independently during the optimization. These are the *decision variables*; examples include temperature at the inlet of a reactor or flow of coolant to a heat exchanger.

3. Specify the *constraints* which exist on the magnitudes of the process variables or on some function of them. For example, a temperature may be constrained below a maximum, or there may be a maximum total flow of air available to add to the beds of the SO_2 reactor.
4. Choose the method of optimization.

The first three decisions specify the optimization problem and are discussed next. The fourth decision will be examined in the light of optimization of the SO_2 reactor of the sulfuric acid plant.

12.2 Specification of the Optimization Problem

When choosing what to optimize, all the available knowledge about the process and its economics should be brought to bear. For example, the sulfuric acid plant could in principle be optimized as a whole but perhaps it would be too difficult because of the various recycles, or perhaps only a part of the plant offers much prospect for economic improvement.

Choice of the objective function

The objective function is almost always related to economics. Some typical choices include minimum cost per unit of production, maximum profit, or even maximum energy usage or maximum conversion, which are merely economics in disguise.

Consider the sulfur-burning sulfuric acid plant. Its economics are summarized in Table 4.1. Since the major cost is the raw material, the conversion of all the sulfur into saleable acid is important. Secondly, the efficient extraction of the heats of combustion and reaction as useful steam energy is economically necessary. Finally, the constraints on air pollution may dictate the maximum allowable concentration of sulfur dioxide that may be released to the air. If no money can be spent on capital equipment and if the capacity is limited by the existing air blower, a good choice of a realistic objective function is the maximum conversion to SO_3. (This choice neglects the effect of conversion on steam generation and superheat. It is implied then that the amount and superheat of the steam generated by the optimum case will be enough to run the turbine to supply the air required. This must be verified later, since an optimum conversion without enough air to achieve it is hardly very attractive!)

Choice of the decision variables

The choice of which decision variables should be used depends partly on how long a single case takes to compute but primarily on what variables can, in fact, be varied.

For the surfuric acid plant, and specifically the optimization of the SO_2 reactor, the possible decision variables could include air flow rates to each bed, the inlet temperature and gas concentration to the first bed, temperatures at the inlet to every bed, and the distribution of a given amount of catalyst among the beds. These are given roughly in the order of decreasing ease of alteration.

The choice of any of these means that only the reactor, boiler #2, and the superheaters must be computed during the optimization. The final optimum choice should be checked to see that the steam superheat produced is indeed enough to run the air blower. The illustrative examples in this chapter will use temperatures at the inlet of one or several beds as the decision variables. More detailed results using other decision variables for this system have been reported by Chartrand (1967).

Choice of the constraints

To obtain a realistic optimum, constraints must usually be placed on the process variables through limits on linear or nonlinear functions of those variables. For example, in the SO_2 reactor, vanadium pentoxide catalysts deteriorate above 620°C so that most plants limit the temperature out of the first bed to 600° or at most 610°–612°C. This choice of the temperature constraint is a balance between faster conversion at higher temperatures but shortened catalyst life. Duecker and West (1959) discuss this problem in more detail. The optimization problem in this chapter is done both with and without an upper temperature constraint of 600°C to assess the effect of the constraint.

Because the blower is being operated at its limit, a fixed total amount of air is available for burning sulfur and for addition before each bed. A constraint should then be placed on the total air available. If the problem of optimum catalyst redistribution were considered, a constraint might be placed on the total catalyst available.

Summary of the optimization problem

The examples to be considered are:

1. Objective function—maximum conversion to SO_3 at the reactor exit;
2. Decision variables—inlet temperature to the first bed (single variable) or to all beds (several variables);
3. Constraints—$T < 600°C$ at all points; maximum amount of air available is 3370 lb mole/hr; inlet composition and catalyst distribution are fixed.

Many other problems could be specified but the essential factors will be illustrated by the ones chosen. The technique of optimization to be used must now be chosen.

12.3 Choice of the Optimization Technique

Optimization can be achieved by theoretical means or by efficient search among values of the decision variables. Theoretical techniques, such as Pontryagin's maximum principle, are beyond the scope of this chapter and indeed are difficult to apply to complex problems, especially those with constraints.

This chapter deals with search techniques and their adaptation to the PACER framework. An example of a single-variable search, golden section, is treated first. Then a multivariable search, Hooke and Jeeves's direct search, is discussed. Finally, dynamic programming is examined as a technique for breaking a large staged problem into a series of smaller problems.

Single decision variable—golden section search

Consider the problem of maximizing the conversion of SO_2 by varying the single decision variable, the temperature to the first bed of a given four-bed reactor. It is assumed that the feed rate, composition, amount of catalyst in each bed, and the air distribution to the beds are fixed. The constraint of a maximum temperature of 600°C is applied. This is a realistic problem for the acid plant example because the setting on the bypass of boiler #1 controls the inlet temperature to the reactor bed.

A reasonable optimization technique to use for this problem is the method of golden section. This method is almost as efficient as the Fibonacci method but is more convenient [Wilde (1964)]. It prescribes the sequence of trial values of the decision variable (inlet temperature) at which the objective function is to be calculated so as to make maximum use of the information from previous calculations. The sequential values of the trials are located within the remaining range of the decision variable such that the length ratio of the total range to the greater part equals that of the greater part to the lesser. If the range is unity and the length of the greater part is α,

$$\frac{1}{\alpha} = \frac{\alpha}{1 - \alpha}$$

and

$$\alpha = \frac{\sqrt{5} - 1}{2} = 0.618034 \tag{12.1}$$

The application of this technique requires that the range be specified. For the SO_2 reactor example, a reasonable range for the inlet temperature is between 400° and 450°C. For convenience this can be represented as having unit length, as in Fig. 12.1. Utilization of this technique for the four-bed reactor means that for each trial temperature T_i, the conversion of SO_2 from the fourth bed is computed. If there is only one maximum of the objec-

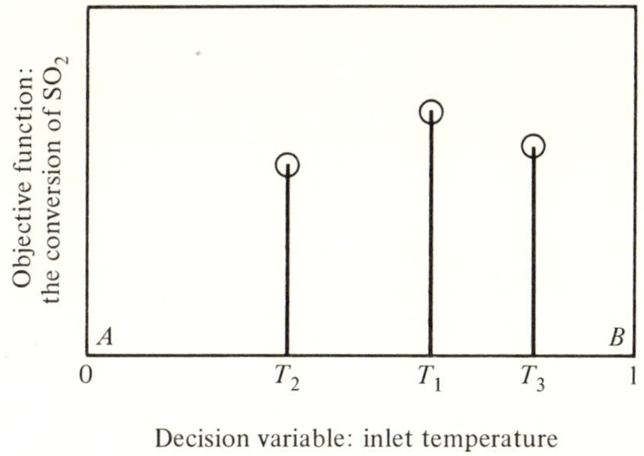

Figure 12.1. Conversion of SO_2 from the fourth bed as a function of inlet temperature to that bed.

tive function within the temperature range, that maximum cannot lie to the outside of the lower of any pair of responses. Two trials, T_1 and T_2, are then necessary before any part of the temperature range can be eliminated. The region AT_2 is eliminated, which removes $(1 - \alpha)$ or 38.2% of the current length, AB.

The next trial, T_3, must be so placed that it and the remaining point T_1 are symmetric in T_2B since the location of the maximum is not known beforehand. T_1 can be shown to cut T_2B in golden section. In general, the remaining point cuts the remaining length in golden section. The new point T_3 (or T_{i+1} in general) also cuts the remaining length in golden section. The region T_3B is now eliminated and a new point T_4 will be placed in T_2T_3.

The search continues until the remaining length has been narrowed to within a preset tolerance. Since a known fraction of the remaining length $(1 - \alpha)$ is eliminated at each step, this is equivalent to fixing the number of steps to be done. It must be emphasized that golden section depends on having only one optimum within the range of interest. This is called *unimodality*.

Fitting golden section into PACER

The Unit Computation OPTIM1 was written to optimize with a single decision variable by golden section search. OPTIM1 differs from an equipment model in two respects. First, it can operate on streams not connected to it and on other equipment parameters. It finds the decision variable and the objective function somewhere in the SN or EN matrix from data read into its own EN list. For example, the stream number of the inlet to the first bed and the position in that SN list (element 4) of the inlet temperature are supplied in the EN list of OPTIM1. Similarly, the stream number of the outlet from the fourth bed and the position (element 12) of the SO_3 flow rate in that

SN list are also given. If the objective function is not already found anywhere in the SN or EN matrices, it may be calculated and placed there by another special Unit Computation.

The second difference between OPTIM1 and a typical Unit Computation is that it causes the entire list of equipment to be recalculated as long as it has set a new value of the decision variable. The computation is forced to be recalculated by setting an integer KBACK2 = 1 (for a complete printout after each step) or 2 (for no printout). KBACK2 is set to zero when the optimization is complete. PACER tests KBACK2 whenever it finishes an entire list of calculations. If KBACK2 is positive, the list is recalculated, but if it is zero, a new problem is begun.

A constraint such as $T < 600°C$ may be handled in this case by interposing a special Unit Computation between the process and OPTIM1 which examines the temperature out of the first bed. If it exceeds 600°, the reciprocal temperature is stored for the objective function. Since the SO_3 flow $\gg 1$, this forces the optimization towards the temperature constraint. Otherwise the actual SO_3 flow is stored as the objective function. Since this method of

TABLE 12.1

OPTIMUM INLET TEMPERATURE OF THE FOUR-BED REACTOR BY GOLDEN SECTION

Range:	400°–450°C
Feed composition:	278 lb moles/hr SO_2
	338 lb moles/hr O_2
	2310 lb moles/hr N_2
Objective:	Maximum flow of SO_3 out of bed 4
Constraint:	$T \leq 600°C$
Assumption:	Unimodal

Trial no.	Trial temp. Low T, °C	High T, °C	Calculated molar flow rate at new trial temperature FSO_3, lb moles/hr	Range of temperatures rejected
1		430.90	0.00162*	—
2	419.10		268.99	$T > 430.9$
3	411.80	419.10	211.81	$T < 411.8$
4	419.10	423.61	0.00163*	$T > 423.6$
5	416.31	419.10	266.66	$T < 416.3$
6	419.10	420.82	0.00165*	$T > 420.8$
7	418.03	419.10	268.64	$T < 418.0$
8	419.10	419.76	0.00166*	$T > 419.8$
9	418.69	419.10	268.89	$T < 418.7$

Remaining range: $418.69 < T < 419.76$
$FSO_3 \geq 268.99$

*Temperature constraint exceeded, $FSO_3 = 1/T_{max}$

handling the constraint introduces a discontinuity in the objective function, it should not be used without great caution.

An example of the use of OPTIM1 is shown in Table 12.1. Note that the best temperature in nine trials was found on the second trial. The computation time for this problem was 5 minutes and 18 seconds on the IBM 7040, exclusive of loading time. If the constraint on temperature is removed, the optimum inlet temperature is 442.3° with a resultant molar flow rate of SO_3 (FSO_3) of 270.91 lb moles/hr.

Several decision variables—Hooke and Jeeves's search

Of the many search techniques for optimizing with several decision variables, the direct search method of Hooke and Jeeves (HJ) will be used as an example [Wilde (1964) and Hooke and Jeeves (1961)]. Basically the HJ pattern consists of sequentially choosing a new level for each decision variable X_i in turn. If the response at the new level is better, it is kept. If not, it is rejected and the previous level of that variable is retained. After the decision variables have been examined and if any improvement occurred, the line joining the last points of the latest two search patterns is extended and a new starting point is established. If any move after a pattern search is the same as the previous move, the length of the move is increased by a factor in order to accelerate the search.

The method is illustrated for two variables in Fig. 12.2, where a point * means an improvement and a point 0 means none. The numbers show the sequence of examining the points. When no step in any direction improves the latest point, one is either at an optimum or the step size is too large. The

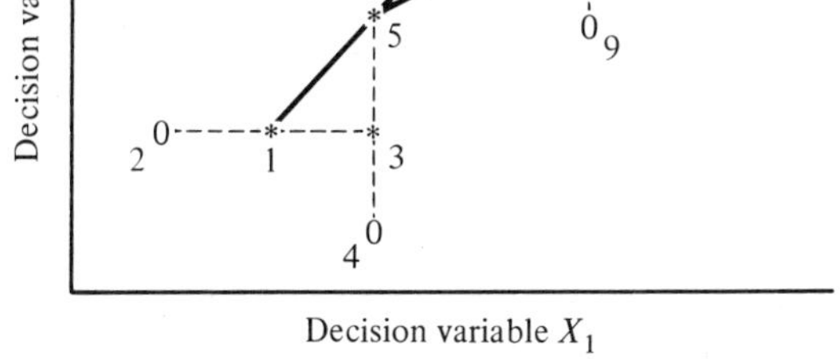

Figure 12.2. Hooke and Jeeves's direct search for two decision variables.

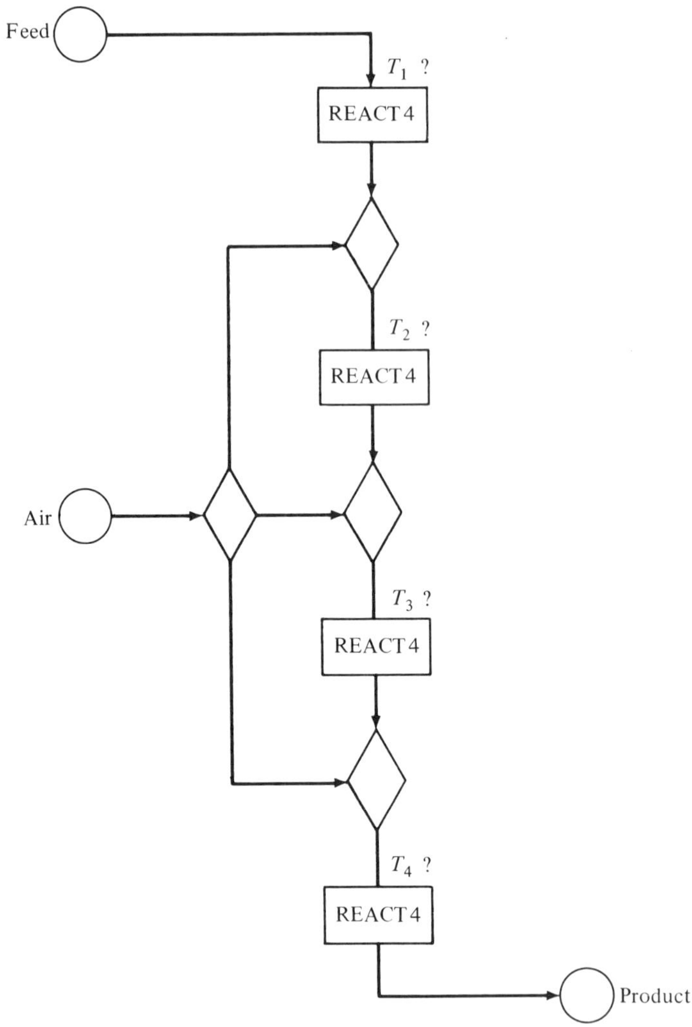

Figure 12.3(a). Information flow for the selection of the optimum inlet temperatures to the reactor beds.

step size is usually reduced by half and the search is resumed. The search will only stop when no further improvement occurs and when the step size has been reduced to the preset minimum. Further details of the method can be found in Wilde (1964) and Hooke and Jeeves (1961).

An example of a four-variable problem is the choice of the inlet temperatures to the four beds of the SO_2 reactor which will optimize the conversion of SO_2. This implies that the inlet temperatures are freely variable. Although the actual heat exchangers in the acid plant do not permit complete freedom of choice of the inlet temperatures to the beds, the optimum sequence of temperatures found can be compared with the actual sequence in the plant. Thus one would find how far the plant is from optimum temperature conditions. The information flow is shown in Fig. 12.3. This problem will also be treated later in this chapter by dynamic programming.

Fitting Hooke and Jeeves's search into PACER

The use of HJ within PACER has the same requirements as the golden section method. A subroutine, say OPTIM2, is written for the HJ method. The location in the SN or EN matrix of each decision variable and of the objective function is supplied to OPTIM2 in its own EN list. For example, the stream numbers for the inlet to each bed and the position in the SN list of the temperature (element 4) are supplied as decision variables to the EN list of OPTIM2. In addition, the stream number out of the fourth bed and the position in the SN list of the SO_3 flow rate (12) are also supplied as the objective function to OPTIM2.

The calculation is performed and OPTIM2 changes the values of the decision variables according to the HJ method, stores the new values in the right places in the SN or EN matrix, and causes the process to be recalculated. As before, a control number, KBACK2, is used to indicate whether the optimization has been completed (= 0) or whether the calculation sequence should be repeated (= 1 for printout, = 2 for none). The results of an optimization by the method of Hooke and Jeeves of the temperatures to the beds of the SO_2 reactor are shown in Table 12.2.

TABLE 12.2

OPTIMIZATION OF INLET TEMPERATURES BY HOOKE AND JEEVES'S DIRECT SEARCH [CHARTRAND (1967)]

Temperature:	$\leq$ 600°C
Feed composition:	278 lb moles SO_2/hr
	338 lb moles O_2/hr
	2310 lb moles N_2/hr
Air addition before bed 2:	142 lb moles/hr
Air addition before bed 3:	210 lb moles/hr

Bed no.	Outlet conversion $X(I)$ lb moles/hr SO_3	Inlet temperature $T^*(I)$, °C	Outlet temperature, °C
1	184.1	418.5 (425)†	598.8 (612)†
2	259.6	432.3 (438)	501.1 (500)
3	269.8	453.5 (439)	462.2 (448)
4	271.9 (271.30)†	440.8 (423)	442.5 (427)

†Values in parentheses are calculated for plant conditions (base case).

Reduction of a multivariable staged problem—dynamic programming

Instead of optimizing the entire reactor at once with several decision variables, the beds can be done one at a time, starting from the last, each bed having only a few variables. Use is made of the principle of optimality of Bellman (1957); this will be explained using the four-bed reactor as an example and then it will be stated in its general form.

Suppose the problem is the same as for the search of Hooke and Jeeves, namely, to choose the inlet temperature to each bed so as to maximize the final conversion to SO_3, given a fixed feed rate and composition into the reactor. It is again assumed that the temperatures into the beds are freely variable even though this is not true in the real plant. The temperatures will be constrained below 600°C, although the existence of a constraint does not change the basic method.

The fourth bed is examined first. The actual conversion between the third and fourth beds in the optimum case is not yet known. But whatever it may be, the temperature into the fourth bed should be chosen to maximize the performance of that bed. This is based simply on the fact that without recycle, whatever is done in the fourth bed cannot alter what has already been done. When the optimum is found and if the fourth bed does not operate optimally with its own feed, the whole reactor is not optimal because an improvement in the fourth bed would be possible.

In general, whatever the conversion into a particular bed may be, in order for the reactor as a whole to be optimal, the temperature into that bed must be chosen so as to optimize total conversion of *all* the remaining beds. This is Bellman's principle of optimality and forms the basis of dynamic programming. Bellman (1957) expressed his principle of optimality as follows: "An optimal policy has the property that whatever the initial state and initial decisions are, the remaining decisions must constitute an optimal policy with respect to the state resulting from the first decision." Thus the second, third, and fourth beds must be optimal with respect to the outlet stream from the first bed. This in turn implies that the third and fourth beds must be optimal with respect to the outlet stream from the second bed.

The one difficulty is that until the optimum has been found, the actual conversion into any bed will not be known. Thus a table of optimum inlet temperature versus inlet conversion must be generated for the fourth bed, so that the actual range of conversions into the fourth bed will be covered by the table.

Once the table has been generated for the fourth bed, no further calculations are necessary for it. When the third bed is considered, another table of optimum inlet temperature versus conversion into that bed is generated. Since there is a single decision variable, golden section can be used to search for the optimum inlet temperature for each level of conversion. With each candidate temperature for a given conversion, the behavior of the third bed is calculated and an outlet conversion from it is found. This is then the inlet conversion to the fourth bed and is entered in the table for the fourth bed to find the optimum inlet temperature and corresponding outlet conversion of the fourth bed. The optimum temperature into the third bed for each conversion is entered in the table for that bed. After completing the third bed, the second bed is examined and another table is generated for it. The tables already available for the third and fourth beds are used to find the optimum temperatures into those beds. A general algorithm for this problem is shown in Fig. 12.3(b) and the numerical results for one optimization are given in the next section.

Dynamic programming has been applied to the sulfuric acid plant by Lowry and Pike (1967) by lumping various units together to eliminate most of the recycle. The actual results obtained were not comparable to the acid plant under study because the temperature in its first bed was allowed to reach 727°C and because the actual water flows to the two boilers are not

```
Note:    Names INIT, START, GOLD, FINAL are statement labels to permit
         branching
                                   BEGIN;
                                   I = 5;
INIT:                              I = I - 1;
                                   J = I;
START:   CHOOSE NEXT LEVEL OF X(I - 1), CONVERSION OUT OF
         BED(I - 1)[OPTIM3];
GOLD:    CHOOSE NEXT TRIAL T(I), INLET TEMPERATURE TO BED(I)
         BY GOLDEN SECTION [OPTIM1];
         CALCULATE BED(I) TO FIND X(I);
         IF J < 4, USE TABLE(J + 1) TO FIND OPTIMUM TEMPERA-
         TURE, T*(J + 1), AND X(J + 1) FROM X(J); J = J + 1;
         REPEAT UNLESS J = 4;
         IF OPTIMUM TEMPERATURE HAS BEEN FOUND, ENTER T*(I),
         OPTIMUM TEMPERATURE FOR X(I - 1), AND X(I) IN
         TABLE(I) VERSUS X(I - 1);
         ELSE GO TO GOLD;
         IF ALL LEVELS OF X(I - 1) HAVE NOT BEEN DONE, GO TO
         START;
         ELSE CONTINUE;
         IF I > 1, GO TO INIT; ELSE CONTINUE;
FINAL:   FOR J = 1 STEP 1 UNTIL 4, TAKE CONVERSION X(J - 1),
         FIND T*(J), X(J) FROM TABLE (J), NOTING THAT X(0) IS
         GIVEN; PRINT RESULTS;
         IF MORE VALUES OF X(0) ARE GIVEN GO TO FINAL, ELSE
         STOP.
         END.
```

Figure 12.3 (b). Algorithm for dynamic programming.

separately controllable, as assumed by Lowry and Pike. The reactor system itself has been studied using dynamic programming by Lee and Aris (1963) for the problem of optimal addition of cold feed before each bed.

Dynamic programming is not a useful method when the number of state variables exceeds two or three. The computer memory required to store all the tables of numbers becomes quickly excessive. Also, because Bellman's optimality principle assumes that there is no recycle of information, recycle processes need special treatment for handling by dynamic programming [Nemhauser (1966)].

Fitting dynamic programming into PACER

To apply dynamic programming within PACER two new subroutines are needed as well as the existing OPTIM1 for golden section. The first is a subroutine TABLO1 which takes as its input the output conversion from BED(I). It produces the optimum temperatures and conversions for the subsequent beds from the arrays TABLE(J) for J = (I + 1) to 4.

The second subroutine, say OPTIM3, controls the calculations in bed I by deciding (a) whether a better inlet temperature T(I) to bed (I) is still being sought for a particular conversion, X(I − 1), out of bed (I − 1), (b) whether a new value of conversion X(I − 1) should be assumed, or (c) whether the calculations for the current bed (I) are finished with. The first two duties of OPTIM3 demand the same access to the SN or EN matrices that OPTIM1 and OPTIM2 had. The third duty, however, requires an additional ability, to reduce the number of the bed being worked on by 1 (statement INIT in Fig. 12.3(b)). At the same time, this will oblige TABLO1 to process one more TABLE(J).

If all the beds have been dealt with, OPTIM3 must reset the equipment calculation order in NELIST so that the whole reactor can be calculated forward with the given feed condition(s) (statement FINAL in Fig. 12.3(b)). The information flow diagram will thus be changed during the computation from that shown in Fig. 12.4(a) to that in Fig. 12.4(b).

The results of performing an optimization of the bed inlet temperatures by dynamic programming are shown in Table 12.4. The four tables were generated in turn from right to left. TABLE(1) only has one entry because only one initial conversion was specified. The inlet conversion X forms the leftmost column and is common to all four TABLE(I). Each TABLE(I) contains the optimum inlet temperature T*(I) to bed I and the corresponding outlet conversion X(I) against the inlet conversion X(I − 1) to that bed in the leftmost column. The arrows in Table 12.4 show how the exit conversion X(I) from bed I is found in TABLE(I) and is used in turn as the entry to TABLE(I + 1) for the next bed (I + 1).

Thus the conversion X(1) out of bed 1 and into bed 2 is 182.42 lb moles/hr from TABLE(1). This value is used in TABLE(2) as shown by the arrows to

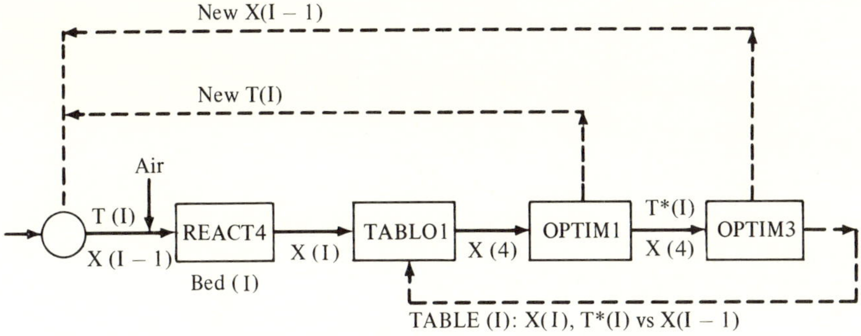

Figure 12.4(a). Initial information flow diagram.

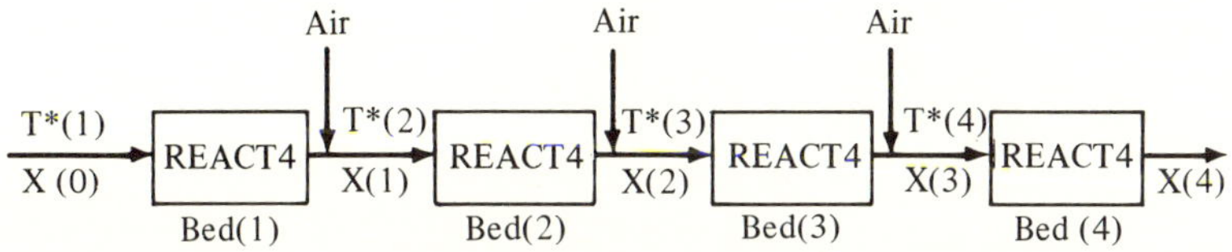

Figure 12.4(b). Final information flow diagram after OPTIM3 is completed.

interpolate for T*(2) and X(2). X(2) is 259.5 lb moles/hr and again is used in TABLE(3) to interpolate for T*(3) and X(3). With X(3) = 269.7 lb moles/hr, TABLE(4) is used for T*(4) and X(4). The summary of the optimal conditions is given in Table 12.3.

TABLE 12.3
OPTIMUM CONDITIONS BY DYNAMIC PROGRAMMING

Bed	Inlet conversion, lb moles/hr	Outlet conversion, lb moles/hr	Inlet temperature, °C
1	0.0	182.42	418.0
2	182.42	259.5	431.6
3	259.5	269.7	453.2
4	269.7	271.77	440.4

By comparing Tables 12.2 and 12.3 it is seen that Hooke and Jeeves's method and dynamic programming lead to the same results to within numerical accuracy when applied to the same problem.

12.4 Summary

Three methods of optimization have been outlined and applied to the SO_2 reactor. The fitting of each method within PACER has been discussed.

Golden section search was used as an example of a single-variable technique and the optimum inlet temperature to the first bed was sought in order

TABLE 12.4
DYNAMIC PROGRAMMING OF THE SO_2 REACTOR
[CHARTRAND (1967)]

Catalyst temperature: <600°C
Feed composition: 278 lb moles/hr SO_2
338 lb moles/hr O_2
2310 lb moles/hr N_2
Air addition before bed 2: 142 lb moles/hr
Air addition before bed 3: 210 lb moles/hr

X, lb mole/hr SO_3	TABLE(1) for Bed 1		TABLE(2) for Bed 2		TABLE(3) for Bed 3		TABLE(4) for Bed 4	
	T*(1)	X(1)	T*(2)	X(2)	T*(3)	X(3)	T*(4)	X(4)
0 →	418.0	182.42						
172			429.2	257.65				
174			429.2	257.97				
176			431.6	258.33				
178			431.6	258.69				
180			431.6	259.03				
182			431.6	259.36				
182.4			→ 431.6	259.5				
184			431.6	259.68				
186			433.6	260.02				
188			433.6	260.36				
190			433.6	260.68				

250	449.7	268.95		
252	450.6	269.09		
254	451.9	269.24		
256	451.9	269.39	447.8	269.99
258	452.7	269.56	447.8	270.13
259.5 →	453.2	269.7		
260	453.4	269.74	448.6	270.28
262	453.2	269.94	448.8	270.44
264	452.3	270.18	449.2	270.64
266	452.3	270.49	447.5	270.89
268	449.4	270.90	444.3	271.29
269.7 →		→	440.4	271.77
270			439.7	271.87
272			432.2	272.78
274			417.9	274.18

T*(I) = optimum inlet temperature to bed I for X(I − 1)
X(I) = moles SO_3/hr out of bed I

to maximize the conversion to SO_3 out of the fourth bed. A subroutine OPTIM1 for performing golden section needed to know where to find the decision variable and objective function in the SN or EN matrices.

Hooke and Jeeves's direct search was an example of a multivariable technique. The best set of inlet temperatures was sought to give the maximum conversion from the fourth bed. A subroutine OPTIM2 would similarly control the search and would need to locate decision variables and objective function in the SN or EN matrices.

Finally, dynamic programming was discussed as a technique for breaking a multivariable staged problem into several smaller problems. The subroutine OPTIM3 required the additional ability to alter the sequence of calculations as found in NELIST.

It has generally been found that Hooke and Jeeves's method is faster for a single problem of the type illustrated here than was dynamic programming. However, dynamic programming generates a whole set of optimum solutions for the four-bed problem, as is seen in Table 12.4. With very little extra computation, many more optimum temperature sequences could be obtained, each for a different intitial conversion to the first bed. Thus when several optima are wanted, dynamic programming becomes more and more attractive since each new optimum requires the same time by Hooke and Jeeves's method.

PROBLEMS

12.1. (a) Suppose that you are given a set of plant data for the SO_2 reactor, together with all the requisite Unit Computations for the reactor beds and heat exchangers. Write the algorithm for an optimizing Unit Computation which would choose, by Hooke and Jeeves's method, values of effectiveness factors, boiler bypasses and split of steam to the superheater so as to minimize the (weighted) sum of squares of errors between calculated and plant values of, say, conversion, steam superheat, and bed temperatures. Try to keep the algorithm general enough so that it could apply to different types and numbers of decision and state variables and to other plants.

(b) Are any other Unit Computations needed or convenient to supply? Consider placing constraints on the admissible values of the decision variables, and how they would be imposed. The constraints ought to be freely altered without affecting the optimizer. Similarly, the quantity to be minimized should be capable of being altered easily, (for example, to a weighted sum of absolute errors).

12.2. (a) Within what fraction of the original range of a single decision variable could you locate a maximum by golden section with ten separate trial calculations?

(b) Show that the following two statements are equivalent:

(i) A function $f(x)$ has a unimodal maximum on $x_1 \leq x \leq x_2$.

(ii) For any x_a, x_b such that $x_1 \leq x_a < x_b \leq x_2$, there is a maximum of $f(x)$ in the interval $x_a < x \leq x_2$ if $f(x_a) < f(x_b)$.

12.3. Pick a multivariable search technique not discussed in this chapter and show how it could be adapted to the framework of an executive program of the PACER type. What differences would there be in comparison with the Hooke and Jeeves method? See Wilde and Beightler (1967) for reference to various methods.

The Strategy of Simulation–A Summary 13

It has been shown how the digital computer may be used to simulate a sulfuric acid plant. With this relatively complex process, it was demonstrated how PACER functions, what information is required, and how Unit Computations are formulated. The details of the input/output information were also indicated. This example demonstrated a strategy for handling complex problems. It was then shown how this simulation could be used to evaluate the technical feasibility of changes in plant operating conditions or process layout and, finally, how an optimization could be incorporated to adjust operating conditions for maximum conversion. Each process simulation has its own characteristic problems, however, and much experience is required to achieve efficient and effective simulations with minimum cost and effort.

In this chapter a general strategy for simulating any process is suggested. This strategy is discussed in the PACER framework, but is sufficiently general to be applicable to all executive systems. The overall steps in this strategy, outlined in Fig. 13.1, are:

1. Define the problem. Decide if a simulation should be done; decide on questions to be answered; select the overall accuracy required to answer the questions.
2. Learn the process. Identify areas sensitive to cost. Collect mass and energy balance data, operating and control procedures. From the

process flow diagram, devise an information flow diagram to satisfy an approximate mass balance. The units that comprise the information flow diagram represent functional operations. Approximate models, for example, SEPA01, MIXER1, are written if a library of such models is not already available. These approximate models must relate to the actual process operation.

3. Develop the base case and analyze for sensitivity. Establish a base case for a single set of operating conditions. Identify those units and parameters where minor perturbations about the base case operating conditions and equipment parameters make major changes in those variables which are important for the objectives of the simulation.
4. Decide about simulation and resources required to answer the questions. Revise objectives subject to the availability of manpower, time and money, and the accuracy possible in sensitive units. Plan data collection, research, or plant measurements as necessary. Decide how general, as opposed to specific, the models are to be and select the method of handling physical properties. Based on these decisions, prepare more comprehensive Unit Computations and a new information flow diagram, representing more closely the actual operation of the process units. Develop the necessary schemes for overall control and convergence promotion to ensure that all system constraints are satisfied and that alternative cases have comparably accurate balances of material and energy.
5. Develop the Unit Computations.
6. Perform studies on parts of the plant and then on the whole plant.

Each step in this strategy is discussed in turn and illustrated by examples. The reader will see as he follows the steps that there is a strong interaction among them. For example, there is strong interaction between whether a simulation is attempted and the questions to be answered. An executive-aided simulation may be chosen, but this decision may later have to be reversed because the questions posed may not require an executive-aided simulation. Thus these two decisions are strongly interrelated.

13.1 Define the Problem and Learn the Process

First it must be decided whether to simulate. If the answer is affirmative, then it is necessary to identify the questions to be answered, study the process, specify a stream list, collect data on the plant and physical properties, and construct a simple information flow diagram.

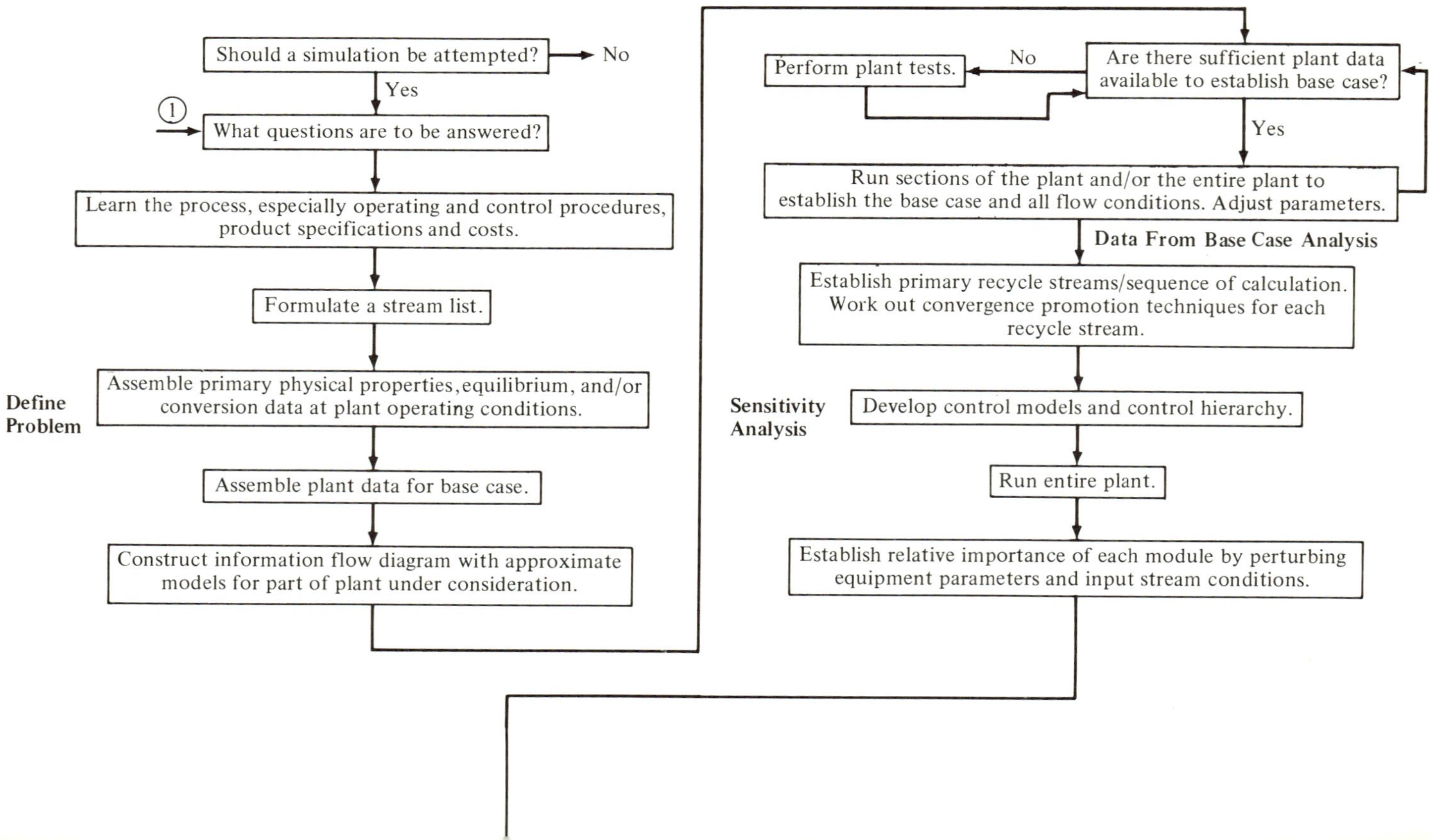
Should a simulation be attempted?
No
Yes
1
What questions are to be answered?
Learn the process, especially operating and control procedures, product specifications and costs.
Formulate a stream list.
Define Problem
Assemble primary physical properties, equilibrium, and/or conversion data at plant operating conditions.
Assemble plant data for base case.
Construct information flow diagram with approximate models for part of plant under consideration.
Are there sufficient plant data available to establish base case?
No
Perform plant tests.
Yes
Run sections of the plant and/or the entire plant to establish the base case and all flow conditions. Adjust parameters.
Data From Base Case Analysis
Establish primary recycle streams/sequence of calculation. Work out convergence promotion techniques for each recycle stream.
Sensitivity Analysis
Develop control models and control hierarchy.
Run entire plant.
Establish relative importance of each module by perturbing equipment parameters and input stream conditions.

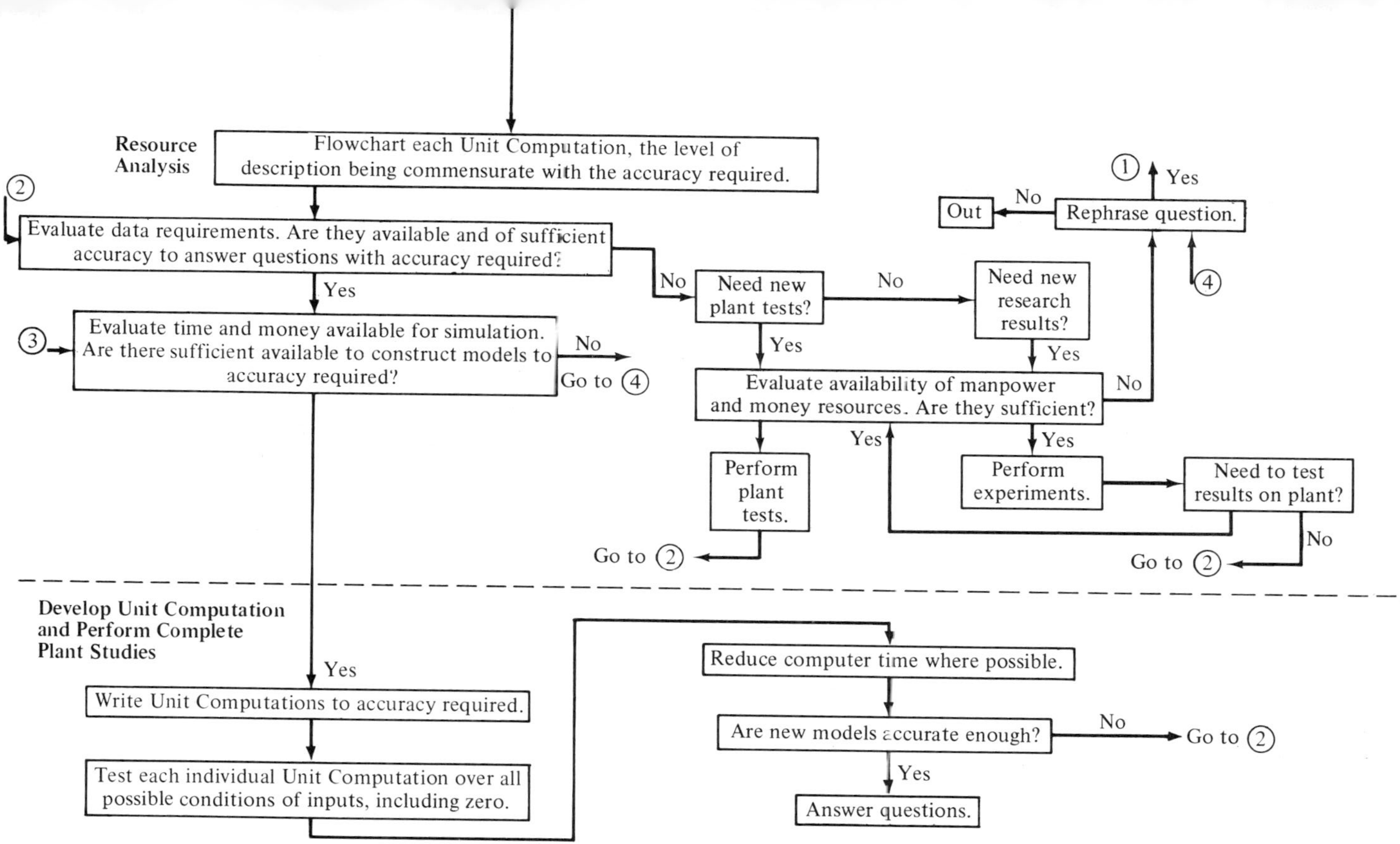

Figure 13.1. Flowchart of simulation procedure.

Is a computer simulation necessary?

A process simulation should usually be attempted:

1. When the process is complicated by recycle;
2. When the interaction among equipment units is important;
3. As a design technique for a new process that requires evaluation under a variety of conditions [Fair and Bolles (1968) or James, et al. (1966)];
4. When a large set of numbers must be manipulated and the accuracy of the bookkeeping becomes a serious problem.

An engineer, whether he is concerned with evaluating a new process design or the performance of an existing process, must balance two factors in deciding whether or not to simulate a process. One is the complexity required of the individual Unit Computation to answer the questions. The other is the number of repeated calculations of the entire process which are required to evaluate not only each case, but all cases to be studied. The greater the number of repeated calculations, the more attractive the process is for simulation. But the more complex the individual Unit Computations, the less possible it is to simulate the *entire* process. Questions about an individual unit can be answered by a particular complex model for that unit alone while questions about the process as a whole require a more simplified Unit Computation. Nevertheless, if a plant does have units requiring very complex calculations, rigorous simulations of these can be used to supply simple models for use within the complete plant simulation.

The economics of the process and the relative return to be achieved from the simulation are major criteria. A preliminary evaluation of the needs of the simulation as outlined in Table 13.1 should be attempted before proceeding. Rudd and Watson's (1968) admonition "don't use complex and involved techniques to solve problems that might be solved by simple analysis" is worth repeating because many rush into ambitious simulation projects when simple hand calculations would suffice. In other words, the problem must be evaluated in the light of the questions to be answered.

TABLE 13.1
A SUMMARY OF THE NEEDS IN CREATING A SIMULATION

1. Realistic, worthwhile, and well-defined objectives
2. Restraint so that simple models are used first to help establish accuracy level
3. Fundamental and/or corroborating plant data
4. Access to computer
5. Possibility of describing equipment behavior
6. Availability of data
7. Computational and programming experience
8. Time, money, and people

Questions posed for the simulation

The simulation must reflect its final use. Some uses are indicated in Table 13.2. In general, the questions to be answered by the simulation will be related to technical feasibility, quality of material, and/or economic optimization. The definition of the objectives is a very important first step in the development of a simulation since these questions determine many of the later decisions, particularly those relating to stream variables and the sophistication and accuracy of the models.

The questions about technical feasibility arise from many sources: proposed changes in operating conditions, process equipment layout, equipment type, equipment size, and feed purity. Whatever the question, the Unit Computations must be formulated to respond correctly to the changes that are to be tested.

TABLE 13.2
SUMMARY OF USES OF A COMPLETE SIMULATION

1. Predict effect of changing conditions, physical layout, and capacity
2. Rapidly complete mass and energy balances as an aid to design and for monthly production studies of an existing plant
3. Optimize operation rapidly and safely
4. Detect and eliminate production bottlenecks
5. Provide knowledge in depth about complete system behavior
6. Improve control and evaluate new or better control schemes in preparation for on-line computer control
7. Facilitate cost calculations, budget control, and planning of operations
8. Train operators and engineers

For example, in the sulfuric acid plant simulation some questions which were considered were: What is the optimal use of air? What is the best split of steam to the superheaters? Can the operation of the heat exchangers, absorber, and driers be changed to reduce the loss of sulfur? Was the catalyst distributed optimally? Note that questions concerning steam production were not included in the optimization studies of Chapter 12. Questions such as the one relating to burnout of boiler tubes should not be included in a plant simulation. To answer that question, the boiler should be modeled independently and the inlet gas temperature calculated for any given gas flow rate for a given maximum tube-wall temperature. Once this sophisticated model has been devised and the questions answered, this information can provide constraints for any optimization, or the model may be used in the evaluation of any proposed process change.

Usually, questions relating to product quality require more sophisticated Unit Computations since the engineer is interested in small changes in materials of low concentration. Here it is important that these "impurities" can be traced accurately through all equipment, so that information must be available at the desired level of accuracy to achieve this.

On the other hand, optimization programs may or may not require a high accuracy. If, for example, the problem is to obtain the best operation of an entire petrochemical complex composed of many interacting plants, such as ammonia, nitric acid, and urea, or as shown in Fig. 11.10, then dollar flows among the various units determine the objective function. This may need relatively crude estimates of utility and material (plant and stored inventory) costs to answer the questions posed. Control models for such economic constraints as minimum production rate would have to be incorporated in this simulation.

Other typical questions will be illustrated when specific decisions at various stages in the development of the simulation are dealt with. The important point is to define clearly the major questions which the simulation is to answer before the project begins.

Learn the process

Before attempting any simulation, it is necessary to learn the process, its operation, the desired control, and the actual operating and control procedures performed by the operators under all conditions. In the initial versions of the simulation, these latter procedures will probably have to be built in to the simulation.

> For example, in the alkylation plant described in Appendix A, the depropanizing column accepts feed from the product separator. This column separates isobutane and inert light hydrocarbons (mostly propane) from the system. The operators must not let the propane concentration in the bottoms become greater than 5%. The column is operated with maximum boil-up rate as determined by the capacity of the overhead condensers. Thus the maximum purity of overhead product is obtained.
>
> The authors have had considerable experience in simulating nonideal multicomponent distillation columns [Petryschuk and Johnson (1966, 1965)], employing either a modification of the Ball (1961) and Burman, et al. (1966) block relaxation method, or the theta method which has been advocated by Holland (1963). In each simulation, the number of stages, input feed, product-to-feed ratio, and reflux ratio have to be specified. A comparison of these simulations of a standard distillation column with the actual alkylation plant operation revealed that the mathematical

model had to be modified to properly reflect the actual operation and operating procedures employed in the plant.

It is emphasized that the overall simulation must react to perturbations as the real plant and its operators do. In addition to learning the technical details, the economics of the process should be studied to identify areas of economic importance.

Formulate a stream list

Experience indicates that almost invariably, particularly when reactions occur, the material flows should be expressed as molar flows per unit time. In isolated cases, for example, in simulating grinding, screening, or filtering operations in a pulp mill, mass flows per unit time were more useful.

More important decisions bear on what information must be transmitted from unit to unit. The points of particular interest are enumerated here.

1. The information must relate to the questions that have been posed for the simulation. Sometimes it is useful to anticipate future needs of the simulation but, in general, the list should not be too long and detailed. In many instances, components can be grouped together and an average molecular weight may be assigned so that material balancing is possible. As a general rule, those components comprising the recycle streams are most important because otherwise they would not be recycled at all.

   For example, in the alkylation plant described in Appendix A, the isobutane recycle from the two distillation towers is important in the economics of this process. Since material heavier than C_5 leaves the system as bottoms from the flashing system and the distillation column, these components should be grouped together and assigned a fictitious molecular weight to maintain material balance.

2. The list must include those components which are not specified in the objectives of the simulation but interact strongly with the main components.

   An example is in the Bayer process for extracting alumina from bauxite. Here caustic usage affects the cost of alumina. Caustic loss arises from inadequate washing of the red mud and hydrate precipitate, and also from the precipitation of silica as a sodalite-type solid material containing a relatively large amount of caustic. Therefore silica must be included but titanium or vanadium, for example, are not required in any economic study because these compounds are removed from the system into the mud lakes.

3. The modification of the information in input streams can be described with the same accuracy in all units in the simulation. The state of knowledge of the basic phenomena occurring in a particular equipment unit may be inadequate to describe the action of that unit to the required accuracy.

For example, in the Bayer process the production of alumina of the required quality specifications depends upon the particle size distribution of the seed and product from the precipitators. Because of the recycle of seed, these particle characteristics must be predictable from the settling and washing tanks. If they are not, little is to be gained by considering models for the precipitators which do predict particle size distribution; simpler models may be just as useful.

In summary, molar flow per unit time is a reasonable unit of measurement and the stream variables selected must relate to the questions being answered, the main components in the recycle streams, and the accuracy possible in the Unit Computations.

Assemble primary physical properties: chemical equilibrium and yield, conversion, and/or rate data

The requirements again depend on the final use of the simulation. Obviously if a heat balance is not required, neither are specific heats. In general, however, heat capacity, density, and basic equilibrium concentration data will be required, and these properties can be assumed constant for the preliminary simulation. Based on judgment and experience with the plant, those units of particular importance to the simulation should be easily identifiable.

Experience of the authors indicates that chemical reactions are not usually described adequately; moreover the fluid mechanics in the reactors may not fit the usual plug-flow or back-mix models. However, it is useful to use reasonably simple descriptions of the reaction mechanisms and to obtain with these the rate constants from the plant conversion data through some search technique. Again it is important to account for the main components so that the reactor will respond realistically to changes in operating conditions.

As an example, in the Bayer process it is known that the dissolution of the gibbsite alumina ($Al_2O_3 \cdot 3H_2O$) is very fast and that the residence time in the digester reactors is required to allow the reprecipitation of the dissolved silica. A complete description of the alumina dissolution kinetics will not be required; the desilication reaction, however, will be most important because it requires two moles of caustic for every mole produced and represents an important impurity in the final product. Therefore, a description of this reaction mechanism and rate is important, and data should be sought early.

Assemble plant data for the base case

Since almost all correlations in chemical engineering contain at least one adjustable parameter, it is imperative that adequate plant data be available to set values on such parameters. The completeness and accuracy of the information should be evaluated early in the light of later requirements. In particular, the major components in the stream list must be calculable with sufficient accuracy from the plant data. The accuracy of the data will have to be critically examined since these data will be assumed to be exact later on when evaluating the adjustable parameters. Many times this requires that data be checked by various means.

For example, in the sulfuric acid plant, air flow was measured, a good value for sulfur usage was available, and the concentration of SO_2 in the gas in the entrance to the reactors was measured.

These data are not independent and should be consistent in a material balance calculation. Reevaluation of measurement accuracy ought to be made whenever inconsistencies are uncovered. If the data cannot be evaluated by independent measurements, the placement and geometry of the measuring devices and their calibrations must be checked.

For example, during the simulation of an ethylene cracking furnace [Petryschuk (1967) and Johnson, Aizawa, and Petryschuk (1968)], a sophisticated model was formulated for the combustion process for calculating the radiation and convective heat transfer to the tubes on the furnace walls. Adjustments were made on the model to account for inaccuracies in the assumed combustion and mixing processes to make the calculated outlet temperature agree with that which was measured. Later it became clear that the measured temperature was very much lower than that actually occurring. An examination of the thermocouple indicated that it was subjected to severe radiation and conduction errors amounting to over 200°F. A new element, redesigned to minimize these effects, provided the correct information which allowed a reasonable adjustment of the model. In this case, judgment and evaluation had suggested an error only after considerable time and money had been expended; a priori evaluation might have made this measurement suspect from the outset and simple plant tests might have substantiated these suspicions.

Since the flows in most chemical processes change continually from perturbations in feed or environmental conditions, a complete set of data should be obtained either after the plant has operated under close supervision and control for a time sufficient to achieve steady state or, alternatively, from an average over a number of days of normal operation. Experience of the authors suggests that the former is preferred since sufficient measurements cannot usually be taken during normal operation because of limited manpower. In all cases the data must be consistent as indicated by material and heat balance.

When a primary measurement is not available and considerable disruption to plant operation would result if a device were installed, a stream which is important to the simulation can be indirectly calculated by material or heat balances or indirectly measured. For example, the acid flows to the drier and absorber were measured by radioactive tracer. Alternatively, the expected value on the basis of pump or blower characteristics may have to be used until a more exact measurement is available. In any event, measurements, analyses, or estimates of these must be available if a simulation is to be achieved.

Construct and work with a simple information flow diagram

The first information flow diagram should include simple or approximate Unit Computations to:

1. Show whether more rigorous models are needed and which ones are most important. This further pinpoints requirements for better operating, physical property, or fundamental data, or for a better understanding of the phenomena taking place in the various equipment units.
2. Identify problems of control, recycle, and interaction and develop convergence promotion schemes.
3. Assess the SN list and objectives, and answer some questions early, perhaps even to suggest new plant tests.

In principle, simple Unit Computations are the simplest representation of fundamental behavior. Yet, the models must also reflect the method of plant operation. Care should be taken in selecting approximate models, such as given in Table 4.3, since indiscriminate choices may introduce system instabilities and convergence promotion problems.

> For example, consider the difference between SEPA01 and SEPA03 and the approximate representation of filtration with simple models such as in Table 4.3. SEPA01 splits the solids and the liquid to the filtrate and cake according to respective elements in the EN list. SEPA03 specifies a split of solids and the concentra-

tion of solids in the underflow. The remaining filtrate leaves in the overflow. SEPA03 therefore represents the operation of a filter press better than SEPA01. Perhaps some of the convergence promotion difficulties which were experienced in approximate simulations resulted because wide use was made of the SEPA01 model without scrutinizing the differences in control, operation, and fundamental principles between separation equipment units.

Once a set of sufficiently diversified approximate Unit Computations has been developed, approximate simulations are relatively easy to patch together. Nevertheless, universally applicable Unit Computations may not be worth writing for some operations such as reaction. Instead, models unique to the problem can easily be developed. Above all, the approximate Unit Computations should be simple; the common temptation to write overly sophisticated models should be avoided.

Now that the merits of using simple models have been examined, consider some implications. First, the correlations within the Unit Computations must be continuous to obtain a converged solution. Some acceptable and unacceptable correlations within a Unit Computation are given in Figure 13.2. Secondly, approximate models are often linear. Such models in complex

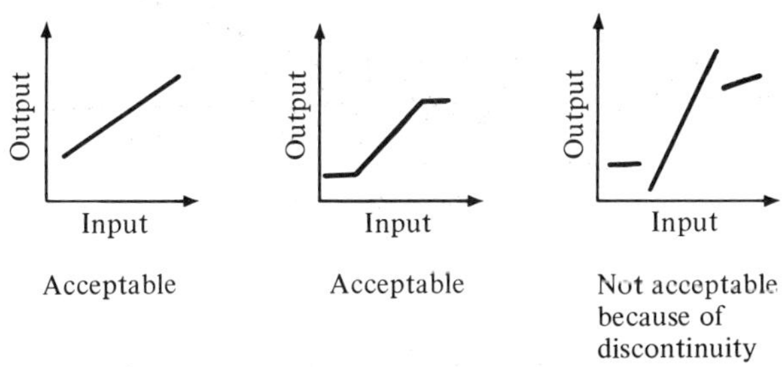

Figure 13.2. Acceptable and unacceptable linear models.

recycle problems may converge much more slowly than if the Unit Computations are nonlinear because the linear approximations may not contain as many realistic constraints as would naturally occur in the operation of the plant. When this condition arises, minor deviations from reality snowball through the plant and give rise to nonconvergence.

For example, a buildup of components in approximate simulations of the alkylation and Bayer processes was encountered. Instead of converging, the concentrations of some components continued to build up from iteration to iteration while those for other components decreased to zero and negative values. In other cases, once a converged solution was obtained, a step increase imposed on one component in a closed recycle stream yielded another converged solution. That is, if the linear correlation in Fig. 13.3 is given as $y = \mathbf{A}x$ then the solution is $x = \mathbf{A}^{-1}y$. For a system within a recycle set, the matrix of correlating coefficients **A** needs to be inverted to obtain the solution for the variable x. If the matrix is difficult to invert or the errors in the coefficients are large, then, as shown in Fig. 13.3, a broad range of values of x satisfy the equation. Such a problem occurred in the section of the acid plant shown in Fig. 13.4. Here, a fixed quantity of steam was incorrectly predicted to be produced regardless of the water circulation rate sent through the two boilers. This occurred because the operation of the boiler was linearized and thus the steam rate was made independent of water circulation flow rate. Many converged solutions resulted because insufficient technical information was included in the model. This same type of difficulty occurred in the alkylation and Bayer processes.

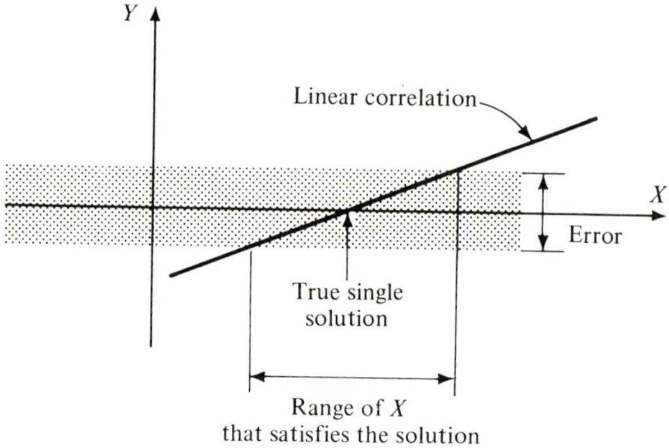

Figure 13.3. Errors in coefficients resulting in multiple solutions.

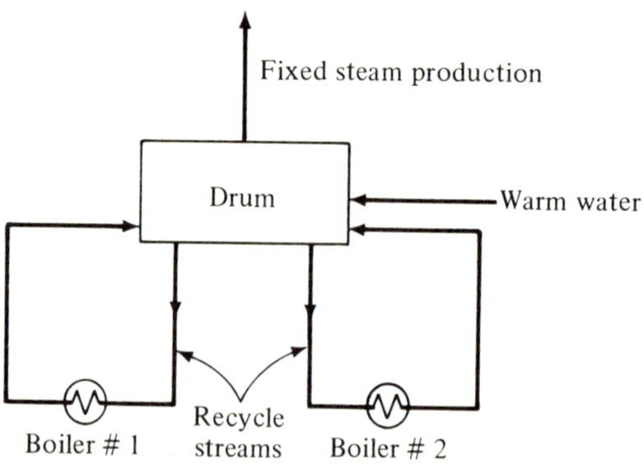

Figure 13.4. Boiler system for the sulfuric acid plant.

Hence, convergence problems can occur in recycle systems for approximate simulations comprised of linear Unit Computations because of inaccuracies in the correlating coefficients or because the linear approximation is too simple. Once a set of approximate Unit Computations has been developed, the approximate information flow diagram can be constructed. Attempts should be made to anticipate the final appearance of the flow diagram so that approximate models can be easily replaced with sophisticated ones as necessary.

For example, a set of information flow diagrams for the alkylation plant is given in Fig. 13.5. In the first version, Fig. 13.5(a), only twelve models are used with the simplified models of Table 4.3: MIXER, SEP, REACT, and temperature and pressure setters SETST1 and SETSP1. Such a simple approximation parallels the process, anticipates future developments, and minimizes the recycle interaction. In the next level of development, Fig. 13.5(b), the com-

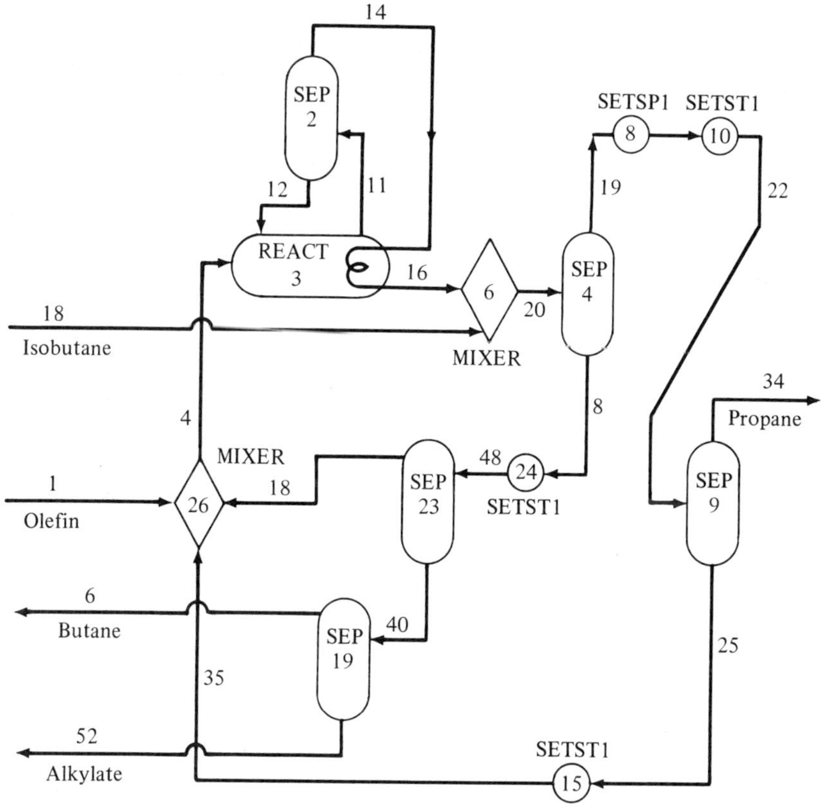

Figure 13.5(a). Information flow diagram for the alkylation plant.

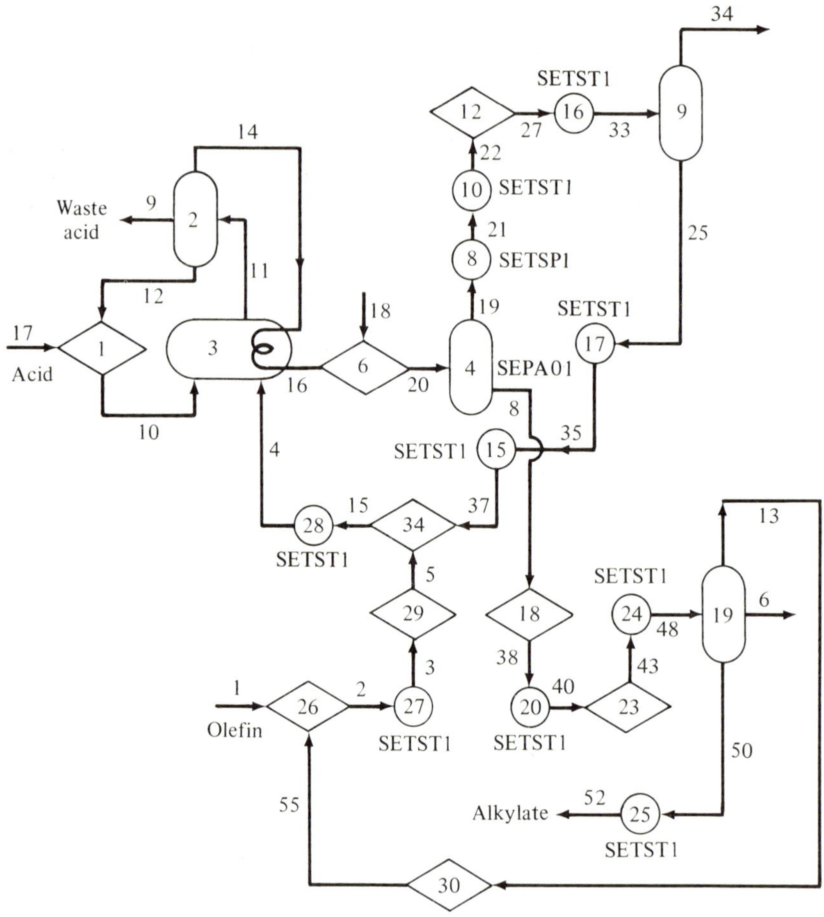

Figure 13.5(**b**). Information flow diagram for the alkylation plant.

plexity of the recycle has not been changed; however, simple models have been introduced for almost all the units which will be studied. Unit #1 allows consideration of acid consumption. The compressor model will eventually replace unit #8, a pressure setter; heat exchangers will eventually replace the SETST1 Unit Computations.

At the next level of accuracy, Fig. 13.5(c), the simple SEPA01 and SETST1 of units #4 and #15 are replaced by a more accurate Unit Computation representing the product separator. Note that recycle is thus introduced in stream 35. Furthermore, two SETST1 units #20 and #27 are replaced by a single heat ex-

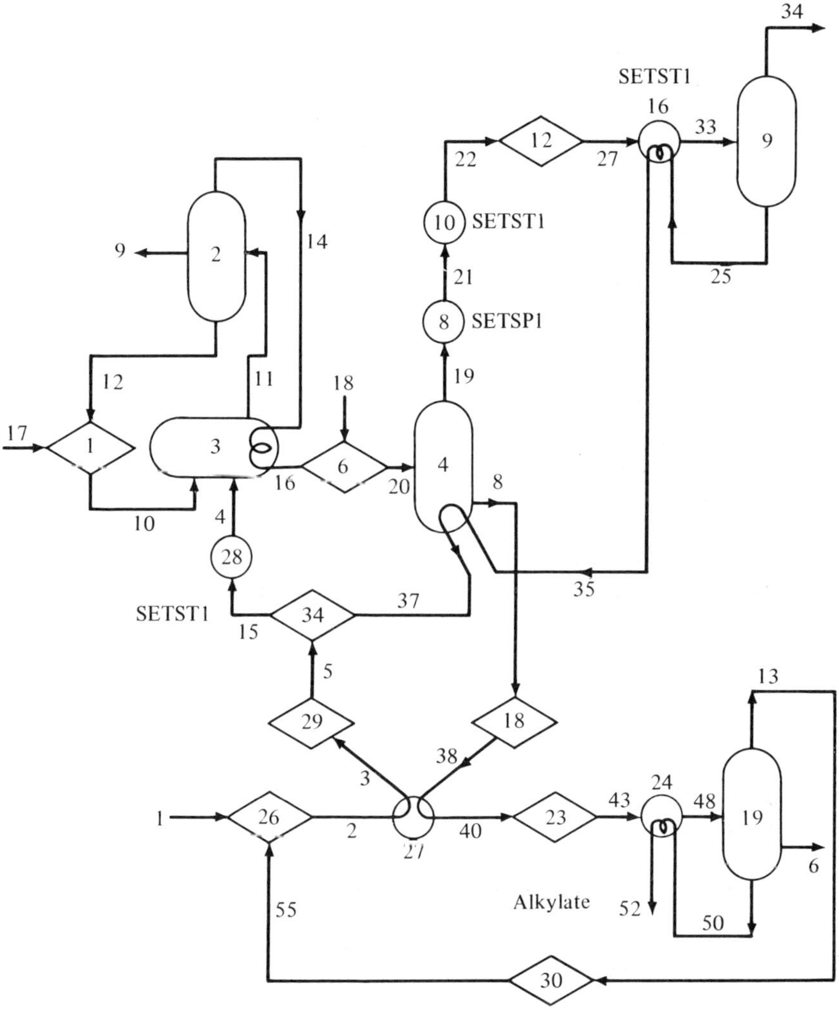

Figure 13.5(c). Information flow diagram for the alkylation plant.

changer, unit #27. This introduces more recycle. Note the two other new recycle interactions in units #16 and #24. In Fig. 13.5(d) there is a further level of accuracy. The compressor and its control system (Prob. 6.6) units #7 and #8, the washing system units #12, #13, #21, and #22, and the coalescers units #14, #23, and #30 are introduced. A control model CONTR3, unit #53 is also used.

In this four-stage development, simple models were used at the outset and these were easily replaced as required.

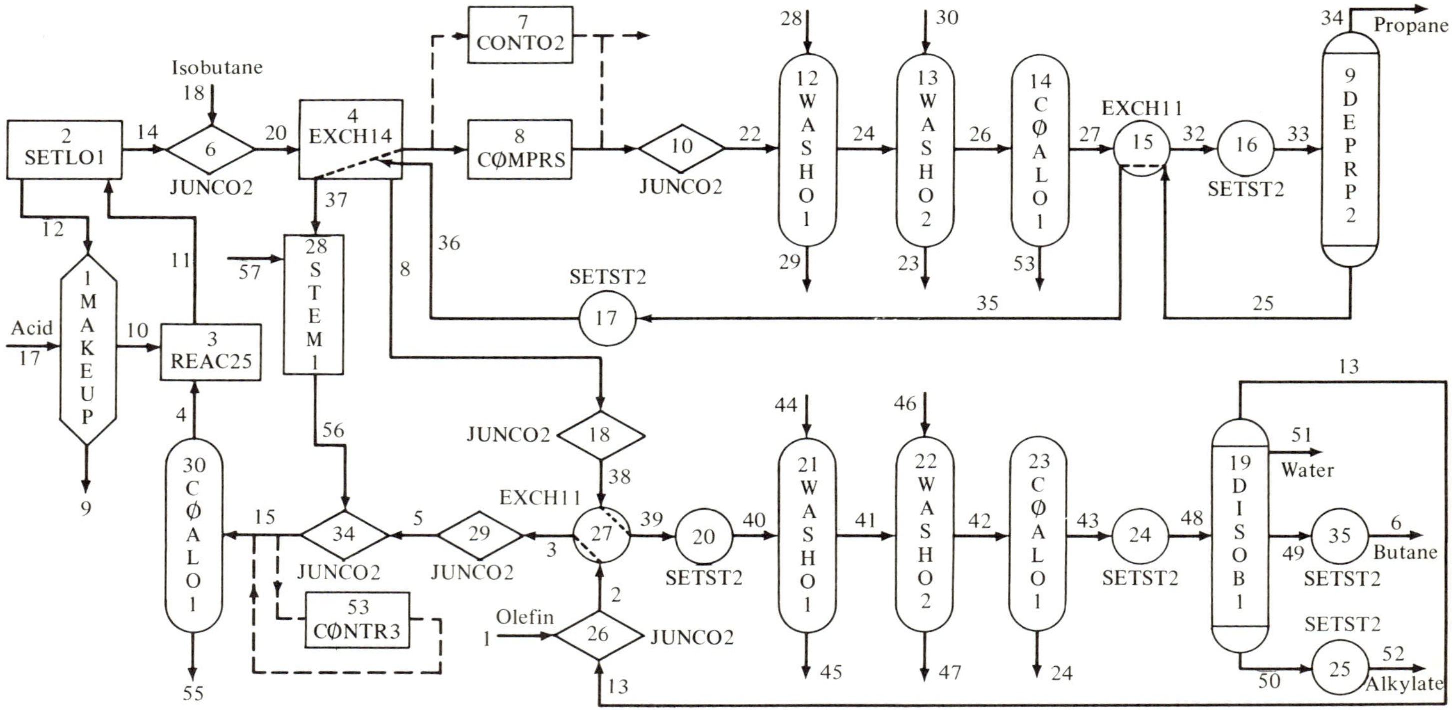

Figure 13.5(d). Information flow diagram for the alkylation plant.

13.2 Develop the Base Case and the Sensitivity Analysis

First, approximate or simple models are used to simulate a balance on material to within an expected acuracy for a single set of consistent operating data. This is called a base case simulation. Then the behavior of the system at other than base case conditions is studied. With this approximate simulation a sensitivity analysis for a range of conditions consistent with the overall objectives is done.

To establish the base case, a set of equipment parameters for the Unit Computations, such as fraction splits, concentration, conversions, and exchange efficiencies, is determined to yield the flow as measured for plant conditions corresponding to the base case.

This quickly identifies the accuracy and consistency of the plant data available, the data needed, the tests to be done, and the streams and stream information which are important to establish the base case. It is necessary to appreciate the difference between what should and what can be measured.

> For example, in the simulation of a pulp and paper process, questions arose as to what variables were important and could be measured concerning the suitability for printing, strength, and texture of paper. The only variables and data that could experimentally be measured and related to actual shape of the wood fibers were the burst, bulk, and tear of pulp handsheets. Forgacs (1963) showed that fiber length distribution and the specific hydrodynamic surface area correlate well with these strength properties. These results dictated that only questions about the strength of the paper could be asked and only a base case considering these variables could be established. Additional research is necessary to find relationships between measurable pulp characteristics and paper qualities other than strength before additional meaningful questions can be asked.

While a base case is being established, control models and a control hierachy are developed to ensure realistic answers, to identify primary recycle streams and a sequence of calculation, and to establish convergence promotion techniques. A sample control hierachy for the Bayer process was given in Fig. 11.6. Such a control scheme ensures that the simulation will operate so that computed cases can be compared to plant data under similar operating conditions and that the simulation is controlled according to known plant practice. The models themselves ensure a material balance. Once the base case is established, the equipment parameters and stream variables are varied away from the values needed to establish the base case. By systematically perturbing the system away from base case operations, the sensitivity of the whole simulation to the individual Unit Computations is determined. This establishes the relative importance of each Unit Computation and variable. This work with the approximate simulation also helps identify problems in control and convergence promotion.

13.3 Resource Analysis and Simulation Decisions Required to Answer the Questions

The approximate simulation with linear models has limitations. Nonlinearities in the real system and their interactions can make significant responses that may not be identified in the approximate linear models. The interaction of heat and mass transfer will not be accounted for in the simple models, and the assumptions made in the simple models may limit the reliability of the answers.

Those Unit Computations have been identified whose accuracy needs to be increased if questions about behavior away from base case conditions are to be answered. Before proceeding, it should be decided whether the information, the time, and the money to continue are available.

Since the models must accurately predict behavior away from the base conditions, data are needed at more than one set of operating conditions. The models must correctly respond to the range of conditions to be studied in the simulation and accurately predict how the key components in the stream list are affected.

For some equipment, the formulations can be based on fundamentals and the need for extensive data is not crucial. Nevertheless, for other equipment there may be insufficient knowledge and more data may be required to develop reliable models. This is particularly true for reactors because of the uncertainty of catalyst activity, mechanisms, and fluid mechanics. Meaningful results can often be generated from simplified mechanisms. Nevertheless, plant data are needed to determine parameters, rate constants, and activation energies and to discriminate among the various mechanisms. Recent work on discrimination techniques in determining parameters is reviewed by Lapidus (1968).

In general, the need for plant data is great. Sometimes previous records of the operation of the plant supply the necessary data but inevitably key measurements are either not made or not recorded. Obtaining data from the present operation of the plant presents problems, mainly because operators are reluctant to change conditions while still maintaining production rate and product quality. Carefully designed statistical experiments minimize the number of experiments and make interpretation relatively easy. The variance in the measurements should be determined through judicious replication of data to indicate the accuracy achieved in the results.

For example, the reactor of the alkylation plant was difficult to model. It was necessary to predict the effect of temperature, pressure, residence time, feed flow rate and composition, and acid concentration on conversion, acid consumption, and octane rating of the product. At first, an analytical model was attempted. A back mix, or completely stirred, reactor was assumed and a set of possible reaction equations was devised to explain the products that were believed to exist in the crude alkylate stream. Linear programming was then used to select a smallest set of reactions and their conversions which gave the product distribution for the single set of data obtained for the base case.

These results could not be used for studies of variations in the operation of the unit, so that the model was inadequate. Further modeling was postponed until a three-month plant study of reactor behavior was complete. Since these data indicated that components were involved which were not in any kinetic mechanism models, further attempts at a kinetic model were postponed. Instead, a linear regression model relating output compositions to input compositions was developed and used in the subsequent simulations. Details are available [Johnson, et al. (1968) and Freel (1969)]. This example illustrates the difficulty in obtaining reliable data for reactors, the interplay between simulation and data collection, and the modification of modeling policy because of the availability of data.

Consider the pulp and paper system of screens, grinders, and hydrocyclones. The physical phenomena in these operations are complex and preclude accurate mathematical description based on present fundamental laws. The practical alternative is to isolate the units from the existing system in the mill and to operate these units under a variety of conditions according to an experimental design. The responses are correlated with input stream variables and equipment parameters using multivariable regression analysis.

In theory, this approach appears easy. In practice, however, the development of significant regression equations was tedious because the variances are large and many experiments were needed. Indeed, based on the resources available, the simulation of a screening-refining-cycloning system in this pulp and paper mill was discontinued.

The importance of the data cannot be overestimated.

In the preceding examples, it was seen that the simulation of the pulp and paper system was stopped because the significant variables were not identified, sufficient information was not available, and extensive tests to develop correlations of the equipment were needed that overreached the resources available. On the other hand, consider the simulation of polymer reactor systems. With the arrival of the Gel Permeation Chromatograph, detailed knowledge of the characteristics of homopolymerization products became available. Indeed, quantitative kinetic relationships for the polymerization process, including high conversions of monomer, could now be obtained. Hence, the development of accurate reactor models permitted reliable simulation of reactor systems. See, for example, Hui and Hamielec (1969). Of course, this degree of sophistication would be used only if the questions to be answered demanded such rigor.

The data requirements can be summarized as follows:

1. Are plant data available and of sufficient accuracy to answer the overall simulation questions with the required accuracy?
2. Is the understanding of the phenomena sufficient?
3. Are rate data available?
4. Are physical property data available?

Consider the time required to obtain data, prepare the models, and study the system and its convergence and control problems. From the preceding analysis of data, it is possible to estimate the time required for data collection.

The effort required to develop the necessary level of accuracy of Unit Computations can be estimated by flowcharting or devising their algorithms.

At this stage the evaluation requires strong interaction between the research and the plant. For a design problem, the interaction must be extended to include the design team. Regardless of how difficult it may be, the time and money requirements should be estimated for the overall objectives.

The questions to be answered can be reformulated or some restrictions can be relaxed to make the project pay off; otherwise the project should be scrapped. The evaluation is probably the most difficult in the sequence shown in Fig. 13.1 until know-how and experience develop.

13.4 Prepare Unit Computations

Unit Computations based on fundamentals are constructed as simply as possible. General Unit Computations are usually written. They must be compatible with whatever physical property handling scheme is used by the system. Because of the principle of gradually increasing accuracy, a variety of Unit Computations are usually developed for each functional operation. Regardless of the accuracy of the model, the Unit Computation must always work for the range of variables indicated in the specifications. The Unit Computation should test input stream information to ascertain if the values of the variables correspond to an allowable range for the unit. If these values are outside the range, an appropriate error message should be printed to indicate what was wrong, in what unit, and what was done to allow the calculations to continue.

The Unit Computation should still execute, that is, pass information on to the next unit. However, it is generally poor policy to keep the calculations going by resetting a variable that went beyond a constraint to the base case value. A better policy is to set the variable at the constraint value. Adequate debugging printout within each unit must be allowed. Mass and energy must balance over each Unit Computation. Comment cards for each Unit Computation should give adequate and careful documentation, SN and EN list meanings, specific physical properties and other parameters, and programmer credits and dates. This documentation is important for those who later use and maintain the program. The steps in devising a Unit Computation are summarized in Table 13.3.

TABLE 13.3
STEPS IN THE FORMULATION OF MODELS

1. Decide on the type of model to be used and the assumptions to be made.
2. Formulate equations using the assumptions.
3. Obtain data required to solve the equations.
4. Choose a method for solving equations; prepare the program.
5. Match the model to reality and test it.

13.5 Answer Questions by Simulating All or Part of the Plant

The principle of starting simply and gradually increasing the complexity is worthy of application here. In other words, an approximate simulation of the complete plant and more accurate models for individual units are both available. First, sections of the plant are simulated; there must be overall mass and energy balances over each section. The control and convergence promotion schemes can then be worked out.

Convergence promotion inevitably presents the major problem. General principles were given in Sec. 11.1. Here are some examples of how they have been applied to processes other than the acid plant.

1. A pair of distillation columns, Fig. 13.6, separates the products of the dehydrogenation of ethyl benzene to styrene at the Polymer Corporation, Sarnia, Ontario. The columns illustrated form part of a larger distillation train but are of particular interest because of the large recycle stream from the top of the second column to the bottom of the first. When the two columns were modeled,

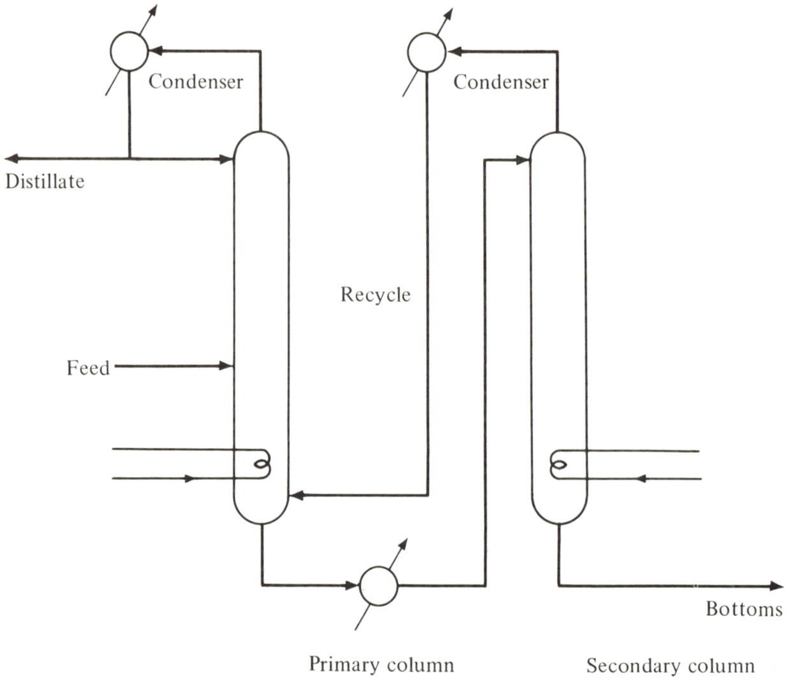

Figure 13.6. Double column assembly.

with only the theta method of convergence promotion for the distillation column Unit Computations, the convergence of the system was slow. However, a Newton-Raphson convergence promoter for forcing overall mass balances was applied to the recycle stream and satisfactorily improved convergence, as is shown in Fig. 13.7. Details are given by Aizawa (1966).

2. A light ends recovery unit illustrated in Fig. 13.8 processes feed stocks from nearby refineries at the Polymer Corporation, Sarnia, Ontario, to produce ethylene and butane. The block relaxation technique has been used to converge the multicomponent distillation columns. Two recycle loops include an inner loop of an absorber stream from the bottom of the depropanizer to the top of the demethanizer, and an outer recycle loop of ethane from the bottom of the ethylene-ethane column, through the dehydrogenation reactor, to the feed of the demethanizer. The former recycle flow is large while the latter stream has a relatively small flow. A satisfactory convergence promotion scheme for this whole plant is partial convergence of the inner loop with a system theta method (that is, a coarse value of DELS for the convergence tolerance was used). This was coupled with sequential iteration in

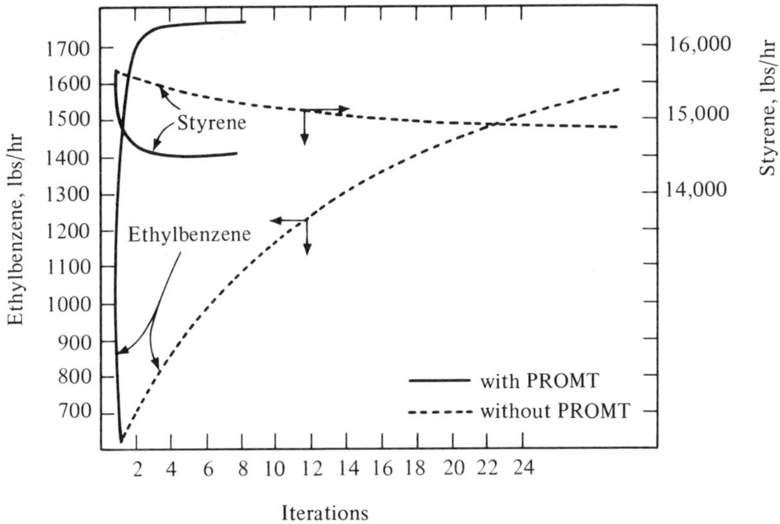

Figure 13.7. Convergence of styrene distillation assembly with and without promotion.

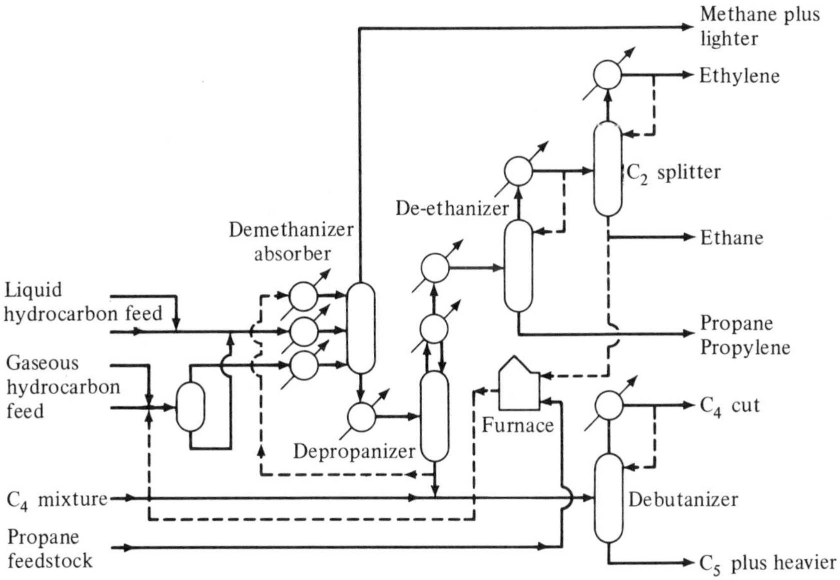

Figure 13.8. Schematic diagram of the light ends recovery unit showing the main recycle flows.

the outer loop, of which the inner system forms a part. In such a scheme, it is important that the internal column concentrations at the end of each pass through a column be stored for the initiation of the next pass through the column. When the whole system approaches convergence, the number of iterations required within the columns can then be reduced. Details are given by Petryschuk (1967).

3. For the simulation of the alkylation plant (described in Appendix A), the most challenging part of the study was finding a sequence of calculation and of convergence promotion. The process is essentially first order with respect to the flows of the components, because if all components in stream 11 [Fig. 13.5(c)] were known exactly as well as the equipment parameters, the component flows everywhere in the process could be obtained by a direct calculation. However, if one examines the heat exchange loops (for example, around the product separator and depropanizer), it is apparent that the recycle has a considerably higher order. Nevertheless it has been found possible to suppress the heat recycle and to promote convergence on stream 4 [Fig. 13.5(c)], the input stream to the reactor unit. The convergence promotion technique used is the "dominant eigenvalue" procedure devised by Orbach (1969). Figure 13.9 shows the flows

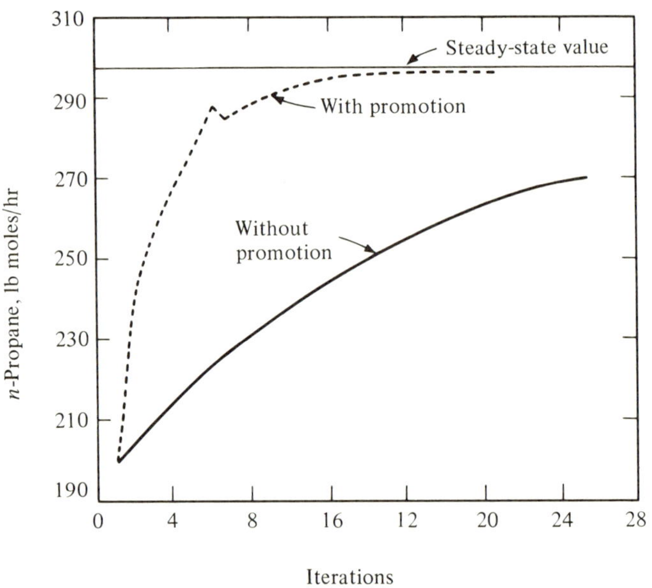

Figure 13.9. Effect of convergence promotion on the alkylation plant.

of propane in stream 4 with and without convergence promotion. It should be emphasized that this convergence promotion technique is based on a linearization of the system, a condition that can be assumed only when the simulation is near the converged solution. If good starting values in the selected stream are available, this technique is attractive.

Once a simulation of the whole plant that converges rapidly to the correct answer is in hand, the simulation can be used imaginatively to answer the original questions. Simulation is not an end in itself; it is a means to an end. Through the use of these techniques, engineers can rapidly evaluate their innovations and ideas.

13.6 Summary

The strategy of simulating processes was developed step by step in Chapters 4 through 12 and illustrated mainly by one process. In this chapter there was a summary of the steps in the strategy and further examples from different processes to illustrate its general application. The uses of and requirements for simulations were given.

PROBLEMS

13.1. In this chapter we summarized a strategy for solving complex problems, in particular, the problem of simulating an existing process. Modify the algorithm shown in Fig. 13.1 so that the strategy can be applied to the design of a *new* process.

13.2. In this book we have emphasized the usefulness of starting simply and gradually adding accuracy where needed.
(a) Wei (1966) describes the "Principle of the Optimum Sloppiness." Are these two approaches the same?
(b) Under what conditions should these concepts be applied to engineering problems that take 2 to 40 hours of hand calculations?

13.3. In simulating parts of the plant, we obtain rapid convergence. When we move from simulating the subsystem to simulating the whole plant, we find the complete system does not converge. Instead, we have a buildup of all components in the recycle streams. Suggest possible causes of the lack of convergence and indicate how the problem can be corrected.

13.4. The accuracy and decisions made about the simulation depend on the use. Table 13.2 summarizes some uses for a simulation. This book has stressed the strategy for preparing a simulation and not of using a simulation. Consider briefly methods of identifying means of improving processes. Table 13.4 is a checklist summary of sources of improvement.
(a) Which sources of plant improvement listed in Table 13.4 could be ac-

Table 13.4
Checklist for the Identification of Sources of Process Improvement

Main area	Subdivision	Questions and comments
Raw materials	Yield	Can we improve the yield by new catalyst, configurations, or operating conditions? Why?
	Recycle	Does it overload any system? How could we eliminate buildup of contaminants?
Exit streams	Specifications on products	Can we change product specifications by changing concentration, phase, form? Are the specifications too rigid, realistic? What do the customers really want? What is the incremental cost of adding each specification?
	Effluents and vents	Is there waste thermal energy, mechanical energy, concentration? What is the amount and condition of drains, vents, samples, blowdown, purge, bleed, blow out, residues, tar, ash, catalyst, washings, disposable solids, condensate, or flash steam?
	Safety and control system	Details of the flare, blowdown, and sampling systems.
	Start-up and shutdown	Volumes involved and disposal.
Time	Maintenance and downtime	When to replace? Maintenance costs, maintenance policy? Main sources of maintenance and spare equipment policy?
	Cycle time	What is it for reactors, filter, batch distillation?
	Trouble-shooting problems causing downtime losses	What are the probable trouble areas? Anticipate corrective measures.
Utilities	Thermal energy	Is high-pressure steam used to produce energy and low-pressure steam for heating? Identify all sources of heat; how can they be utilized? Condensate return? Flash steam used? Establish levels of worth for utilities and study if we carelessly mix different levels without obtaining the net worth.
	Electricity	How can the power factor be improved?
	Inert gas, water, steam, refrigeration, compressed air.	Can the material or the energy be utilized?
Labor		Good relations, cooperative with plant tests, good housekeeping? Safety conscious?
Operating conditions		What is the effect of changes in temperature, pressure, concentration, flow rate, bleed-off rate, recycle rate, catalyst charge size, rpm?

TABLE 13.4 (CONT.)

Main area	Subdivision	Questions and comments
Capacity	Increased (bottlenecks)	How many major bottlenecks exist? Ease in removing? Identify Unit Operations that have a large fundamental uncertainty as opposed to those with a small uncertainty.
	Reduced	Are there any operating problems when plant capacity is reduced?
Minor processing alternatives		New piping, packing, material of construction, equipment, additional coolers?

curately studied from the simulation of the sulfuric acid plant described in this book?

(b) Which sources for plant improvement would be difficult to evaluate from any simulation based on present knowledge and executive system capabilities?

(c) Apply this checklist to processes selected from *Hydrocarbon Processing and Petroleum Refiner* (1967), Faith, et al. (1966), Shreve (1967), and *Industrial and Engineering Chemistry* (1950). For example, we might ask "How we can improve ammonia plants?" See *Chem. Eng.* (1967), Miller (1967), Dingman, et al. (1966), *Hydrocarbon Processing and Petroleum Refiner* (1967), and Gary (1967).

13.5. For different processes, different steps in the algorithm shown in Table 13.1 are critical. For example, in the pulp and paper plant simulation, the selection of the variables for the SN list and the resources required to collect correlating data for all pieces of equipment were critical. For the processes studied in Prob. 13.4(c), identify the steps in the algorithm that may cause the main problems.

13.6. Reactors are usually difficult to model. As a follow up to Prob. 13.5, the main constraint in some simulations is modeling the reactor. For which of the following processes is this likely to be true:

(a) Acetic anhydride from acetic acid;
(b) Acetic acid from butane;
(c) Adipic acid from cyclohexane;
(d) Vinyl chloride from acetylene, HCl, and $HgCl_2$.

The article by Grumer (1967) may be of assistance in this problem.

On the Horizon and Beyond 14

A strategy for handling large, complex problems has been described. The simulation of the steady-state behavior of a chemical plant illustrated this strategy and demonstrated how modern computers can best be used to improve the performance of existing plants. Today, simulation is leading to increased profits and to a greater knowledge of processes.

This strategy, computers, and information handling ability can be effectively used for a variety of applications. The extent and speed with which these come to fruition depend upon the development of computing hardware, upon the successful solution of problems common to many engineering activities, and upon the speed with which the required fundamental understanding of the phenomena in question is gained.

14.1 Variety of Applications of Computer-aided Executive Systems

Executive-aided analysis helps decision making in many problems. The design of new processes, simulation of unsteady-state behavior, and control of systems and business and economic models are areas in which the combination of an executive system for information handling and a high-speed computer will benefit engineers and managers.

Design

Comparison of design with simulation

One might suggest that design refers to ideas, plans, and engineering drawings for plants and equipment *to be built* whereas simulation is an analysis of an *existing* plant. This definition is too restrictive since simulation techniques are also used to design. Instead, design is defined here as the sizing of the device to achieve a required change from input to output stream variables; in simulation the output stream variables are calculated, given the inputs and the behavior of the device. In other words, design is a boundary value problem and simulation is the analogous initial value problem. This is summarized in Fig. 14.1(a).

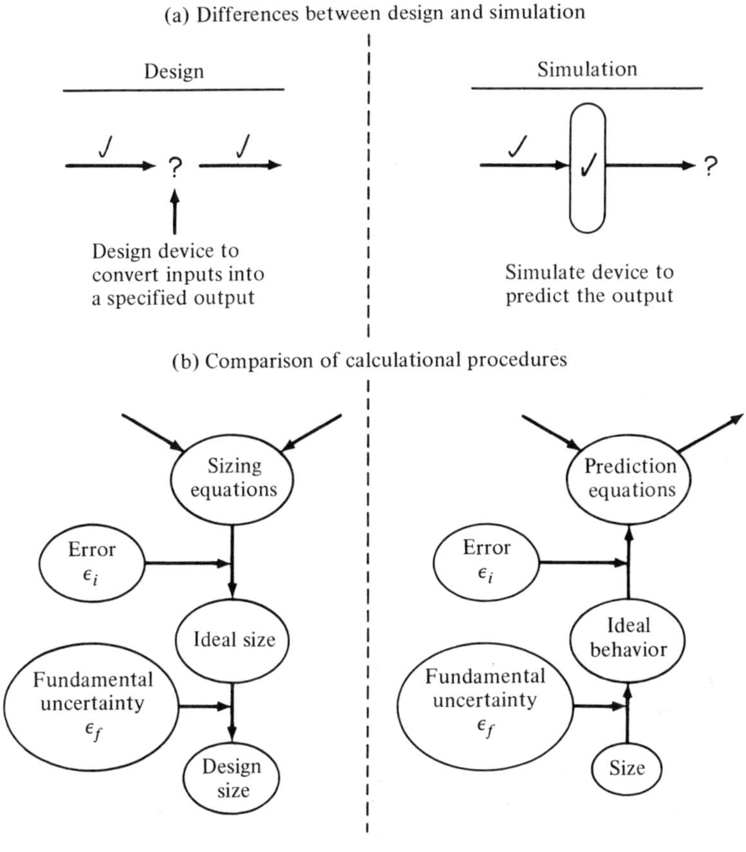

Figure 14.1.

In design, questions about the equipment are answered; in simulation, about the streams. Figure 14.1(b) shows how this distinction affects the calculations. Design starts with values of the input and output stream variables, and use is made of design equations to predict the equipment size. Associated with this calculation are the errors inherent in the design equations, ϵ_i. For example, heat transfer correlations in terms of the Nusselt, Reynolds, and Prandtl groups can be in error by $\pm 10\%$. However, because the only design equations available may contain assumptions not fulfilled by the problem at hand, a fundamental uncertainty error ϵ_f may be used by the designer to ensure that the design works under nonideal conditions. The fundamental uncertainty error ϵ_f is largest for behavior in the mixed regime where several mechanisms are important and also in particulate systems. For more details of this, see Johnstone and Thring (1957), Kline (1965), and Rudd and Watson (1968), Chapter 12.

An example of the procedure and the errors involved is the design of a primary settler for a waste treatment plant. From batch settling data for the flocculating solids, both detention times and overflow rates can be determined for a given desired percentage removal of solids and a given depth. These experimentally determined design criteria (whose accuracy ϵ_i may range from $\pm 5\%$ to $\pm 50\%$, depending on the material beging tested and upon the experimental technique) are used to obtain the idealized size, in Fig. 14.1(b). For the actual plant installation, with such nonidealities as turbulence, scour, rake action, and short-circuiting, this fundamental uncertainty ϵ_f is accounted for by decreasing the design overflow rate by a factor of 1.25 to 1.75 and by increasing the detention time by a factor of 1.50 to 2.00. Thus the design size includes two errors, ϵ_i of $\pm 5\%$ to 50% for the data from idealized experimentation and ϵ_f of 25% to 100% to adapt ideal conditions to real situations. In this example, ϵ_f is large because of a lack of fundamental understanding. A description of this design procedure is given by Rich (1961), p. 91 ff.

The prediction equations can be used with less confidence for simulation than for design if the fundamental uncertainty and correlation errors, ϵ_f and ϵ_i, respectively, are large.

For example, the standard design data for primary settlers in waste water treatment are correlated in terms of the removal and the overflow rate. These commonly cited correlations for generalized application are shown in Fig. 14.2. Hence, design is possible using either the generalized information of Fig. 14.2 that includes both ϵ_i and ϵ_f, or the batch data analysis described previously, in which the designer uses individual values for ϵ_i and ϵ_f.

For simulation of waste water treatment, the overflow rate is calculated from the size of the existing settler and from the flow rates, say 1000 U.S. gal per ft^2 day. The removal rate cannot be predicted accurately for this settler

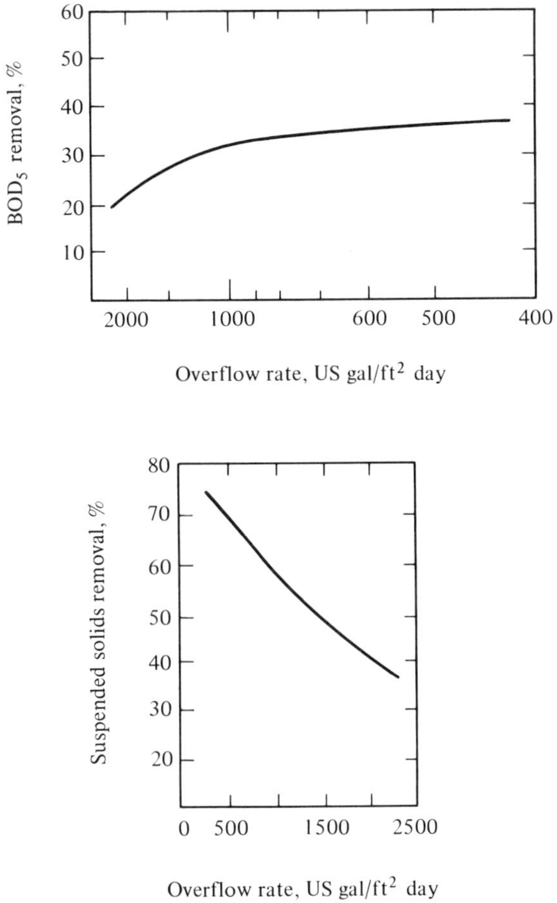

Figure 14.2. Generalized design removal rate for primary settlers [Ten State Standards (1960)].

from the data of Fig. 14.2 because of the two uncertainties ϵ_i and ϵ_f. Indeed, operating data for waste treatment plants, superimposed on Fig. 14.2 and shown in Fig. 14.3, illustrate just how large ϵ_i and ϵ_f are for settlers. For an accurate simulation, data for the performance of the actual settler are needed.

In summary, regardless of the values of ϵ_i and ϵ_f, a plant can be designed and an executive-aided program may help. However, when ϵ_i and ϵ_f are large, generalized simulations cannot be done; data specific to the operation of the process are needed.

The similarities between the strategy in design and simulation were em-

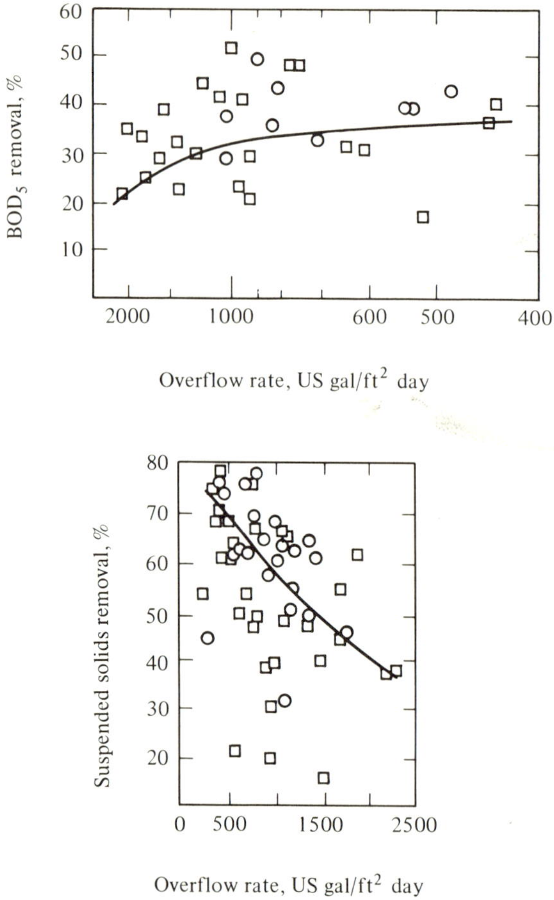

Figure 14.3. Removal rates for installed primary settlers [Water Pollution Control Federation (1967), p. 90].

phasized in Figs. 4.3 and 4.4. Figure 14.4 presents another view of the strategy for design. The blocks of interest are those which represent the calculation of the preliminary mass and energy balance and the size of the equipment. During the former, values for the streams are calculated whereas during the latter the equipment is sized. Thus within the structure of the design is a step during which there is simulation. Indeed, even the stage of sizing the equipment can employ simulation techniques [Fair and Bolles (1968)]. Hence, although simulation and design are different, they are closely interrelated. It is necessary to recognize these differences and limitations, and to capitalize on the advantages and interrelationships wherever possible.

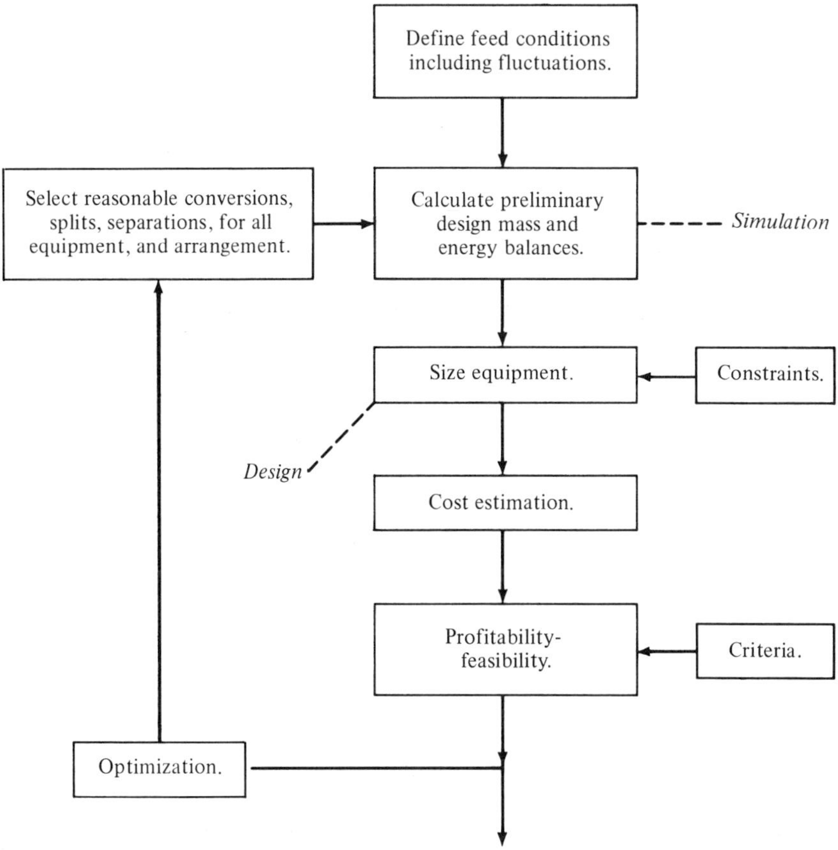

Figure 14.4. Overall design strategy.

Fitting design into PACER

Design can be readily fitted into modular executive systems like PACER. Indeed, the main difference from the simulation mode of operation may be that the answers from the Unit Computation are stored in the EN list rather than in STRMO, and that every unit to be designed adds a recycle of information to the system. The overall strategy is similar to that shown in Fig. 13.1 insofar as an approximate simulation and base case are developed from technically feasible values for the adjustable parameters in the approximate Unit Computations. The next steps are devising design Unit Computations, calculating the total costs and criterion of financial attractiveness, and optimizing.

For example, the models for the simulation of the sulfur-burning sulfuric acid plant described in this book were very easily altered for the design of a metallurgical off-gas plant such as the one described in Prob. 11.9. The reactor

Unit Computation was changed to allow the user to specify a leaving gas temperature. Thus for given input gas conditions and a specified catalyst effectiveness, the depth of bed required to achieve the given leaving temperature was computed. The output gas compositions were obtained simultaneously. With this modified reactor model, the amount of catalyst required for different numbers of beds and different exit temperatures could be investigated quickly, using approximate heat exchanger models. It was also easy to explore the effect of the total pressure in the reactor on the amount of catalyst required to achieve desired conversions. Details are available [Crowe, et al. (1966)]. The absorber model was modified to compute the depth of packing required to achieve a specified leaving gas composition. When the main units of the process had been sized, either using Unit Computations modified from a simulation, or in conventional hand calculations (for example, for pumps), the whole new plant was assembled using the original simulation models and computer runs made to observe interactions. Capital and operating costs were then calculated. Thus design can easily be fitted into PACER.

Relationship among design, simulation, and research

In Chapter 13 the interplay between simulation, plant operations and tests, and research for more fundamental data was clear. In the previous sections, the interaction between design and simulation was illustrated. Indeed, contractors now supply working simulations with plants they design and construct. Furthermore, the financial advantage in the use of simulations during the design phase is illustrated in Fig. 14.5. When a process is being brought on line, there is an induction time between attempted start-up and

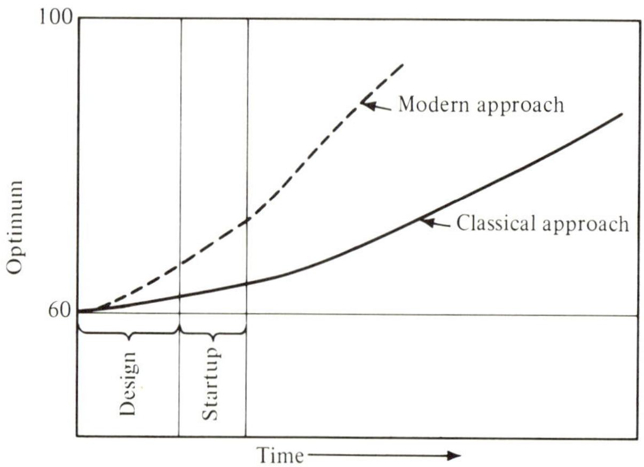

Figure 14.5. Rate of learning with classical approach and modern approach (simulation).

optimal operation. During this time, designers remove any unforeseen difficulties and operators learn the process and its control. Armed with a simulation, a rapid method for evaluating alternative conditions, both designers and operators learn the process faster than during the classical start-up procedure in which learning is more by trial-and-error. Further details on how learning curves can help improve operations are given by Hirschmann (1964) and Andress (1960).

In summary, simulation activity represents an all-important link between the three traditional activities of operation, design, and research as shown in Fig. 14.6. This important link should also exist between people and departments within any organization.

RESEARCH ⇄ SIMULATION ⇄ DESIGN

⇅

OPERATION

Figure 14.6. Relationships among engineering activities.

Unsteady-state problems

A variety of problems can be classified as unsteady. The short-term transient behavior of equipment is important for control around a fixed operating condition. In addition, unsteady-state scheduling problems occur in:

1. Operation of batch vessels in line with continuous processes;
2. The start-up and shutdown of processes;
3. Testing reliability, safety, and loss prevention in the event of a failure of equipment, utilities, or feedstocks;
4. The receipt, distribution, and inventory policy of discrete quantities of resources;
5. Plant operation during changeover from one product to another (for example, for multiproduct plants, or for maintaining product specifications when the feed changes);
6. Adjustment of the process to new operating conditions.

A typical scheduling problem occurs in some polymerization systems. Reactant is fed to parallel trains of polymerization reactors. While the reactant product mixture passes along a train from stage to stage, samples are analyzed to tell the process operator what conditions should be set downstream. The results may indicate that a reactor should be brought into or removed from the train. Most companies have depended upon experienced operators to correctly interpret the results and act accordingly, but this can result in a variable product. With a simulation of the process, the operator could supply plant observations and receive accurate and reliable instructions for the downstream setting.

Unsteady problems arise in individual equipment units or in entire systems. Both control and scheduling problems for individual equipment units have been widely studied [Williams (1969)]. However, less work has been done on the control and scheduling of complex systems. There are two extremes in behavior:

1. Scheduling problems dominated by limited inventory storage. Description of the unsteady state of individual units is avoided by assuming that they operate in the steady state. This implies a perfect control system within each unit to assure constant product quality and flow.
2. Unsteady-state behavior of individual units and the impact of upsets on the process as a whole. Here, the unsteady-state description of the units dominates.

Systems in which inventory dominates

When the inventory storage limits action and the units are assumed to operate at steady state, such executive systems as GPSS, DYNAMO, and SIMSCRIPT are applicable. These are described by Meier, Newell, and Pazer (1969), Forrester (1961), and Markowitz, et al. (1964). Some chemical engineering examples of executive systems and typical problems, summarized by Rudd and Watson (1968), Chapter 15, are:

1. The simulation of a fire water distribution system [Wright (1962)];
2. A Monte Carlo simulation of the arrival of tankers [Dagnall (1965)];
3. A Monte Carlo blending of oils problem [Tayyabkhan and Richardson (1965)];
4. Planning steam supply systems [Jackson, Klomparens, and Westbrook (1965)].

Programs like GPSS and DYNAMO are particularly suitable for study in business systems in which probability factors and queues are important, such as in sales prediction and the flow of orders to plants. Some units with various functions have been provided, but it is difficult for the user to create a new unit to represent, say, a complex chemical transformation required for a business problem. Apparently, these programs presently require very large computers for realistic problems.

Systems in which technical units dominate

Executive programs have been developed for handling control and scheduling problems for large systems in which the technical description of the processing units dominates. For example, two DYNSYS programs, one simultaneous and the other sequential, have been developed by Johnson and co-workers (1970), based on the modular approach. These programs have been applied to the dynamic simulation of a butadiene extraction unit, a chlorine compressor plant, and a bauxite digester system.

Digital simulators of analog computers, such as MIMIC [Sansom and Petersen (1967), Northcott (1967), and Franks (1967)] and CSMP [Gordon (1969)] offer one approach to the solution of large sets of coupled differential equations which arise in dynamic simulations. Hybrid computers, which combine features of both analog and digital computers, offer another approach to the same problem. Undoubtedly the simulation of large dynamic systems will progress dramatically in the years ahead.

Examples have been reported in which specific scheduling problems have been tackled when the technical units predominated [Reilly and Greene (1966) and Johri (1968)] but so far, flexible executive programs have not been developed.

Such programs may be applied to simulate the start-up of a newly designed plant, to eliminate the uncovered bottlenecks, and to redesign for a rapid and safe start-up of the plant when it is built.

Business analysis

Computer-based business analysis has been rather extensively developed. However, the emphasis has not yet been on chemical processes and their intimate interaction with the technical analysis, such as described in this book, and with the financial decisions. The blending of these two aspects will undoubtedly be very fruitful. In the future, one can imagine the simulation of a complete factory, such as shown in Fig. 11.10, in which each Unit Computation represents a technical simulation of a plant. Included in the simulation would be inventory, maintenance, and budgetary control. Uncertainty and such queuing problems presently handled by GPSS would be built into the system to account for changes in future events, variability of feeds, and constraints.

14.2 Computer Hardware Developments

The rate of increase in the use of executive-aided computer analysis depends partly on improvements in computers. The sulfuric acid plant and other simulations described in Chapter 13 were studied on a medium-sized digital computer operated in a batch mode. Computers with faster execution times and larger core storage will broaden the complexity of the problems that can be economically and conveniently analyzed. The development of computers with remote terminals will encourage the widespread use of the techniques described herein, since once an executive system is operating on a given computer and data are assembled for existing or proposed plants, the user needs adequate access to his computer to encourage him to use the model. It is now possible for an engineer to develop Unit Computations and evaluate alternative cases for moderately complicated processes using a teletype beside his desk or even from the control room of his operating plant.

For example, a file-oriented, interactive computation system called GEMCS has been developed by Johnson and Toong (1968) in cooperation with Canadian General Electric. The system is modular and can handle about 100 streams (each containing 30 components) and 50 Unit Computations, each with 85 equipment parameters. Developed initially for the GE 265 computer, GEMCS has been adapted to several computers. The executive was designed so that whole systems could be easily studied using approximate Unit Computations and the approximate models could easily be replaced by more comprehensive ones so the latter could be studied and applied within the whole process system. The input of data is *conversational*; the program asks the user for stream and Unit Computation information. Previous stream and EN parameter information can be stored on the disc files to simplify evaluation of many alternatives. The system produces short, management-oriented reports of the results of a simulation. Details of this GEMCS executive program are given by Johnson (1970).

In summary, computer hardware developments will have a marked impact on the acceptability, widespread use, and complementary software developments of executive-aided analysis.

14.3 Solution of Problems Shared in Many Applications

The successful solution of a number of software problems will greatly increase the applicability of executive-aided computer analysis. In particular, the handling of physical properties is required by most Unit Computations. The handling of uncertainty or risk and the use of statistical correlations also require attention.

The common data problem

Chapter 5 discussed the principles of developing methods to handle physical properties. Properties of pure components and of mixtures are required by, and yet common to, most Unit Computations. Because the data are common, it is often advantageous to read and calculate the properties in a common *property-handling* block that is independent of both the Unit Computations and the executive. Such an arrangement makes it easy to change values of the properties and to write Unit Computations that are independent of the compounds. The development of a property-handling system is outlined here because it is an attractive addition to executive-aided analysis.

The common data system for physical properties developed for the MACSIM executive system is modular, accepts correlations or tabular data, allows a range of accuracy to be reported, and permits intermediate storage of calculated properties. The physical properties are identified by codes; three subroutines, READIN, PROCAL, and PROMIX, read in the property data, calculate a pure component property for a specified temperature and pressure, and calculate a mixture property, respectively. Sample identification codes for physical properties are given in Table 14.1. Two

TABLE 14.1
TABLE OF ALKYLATION PLANT PHYSICAL PROPERTIES

Physical property identification code	Physical property	Units
1	Molecular weight	lbs/lb mole
2	Critical pressure	psia
3	Critical temperature	° Rankine
4	Critical compressibility	
5	Heat capacity ratio	
6	Cox chart ratio	
7	Heat capacity of liquids	BTU/lb mole °F
8	Heat capacity of gases	BTU/lb mole °F
9	K ratio (18 psia)	
10	(130 psia)	
11	(275 psia)	
12	Thermal conductivity	BTU/ft^2 hr (°F/ft)
13	Density of liquid phase	lb mole/ft^3
14	Viscosity of liquid phase	lb mole/ft hr
15	Viscosity of gas phase	lb mole/ft hr
16	Vapor pressure	psia

different code numbers may refer to the same physical property with different calculational accuracies. For each code, the property is calculated for all components in the stream list. READIN is treated as a dummy equipment unit which is calculated only once before any units in the simulation. It stores the data in a matrix PROP (NOPP, J, K), where NOPP is the physical property code, J is the component number, and K denotes the Kth constant in the correlation. PROCAL calculates the property of the pure component, given the code NOPP, the independent correlating variable X, such as temperature or pressure, and stores the result in the matrix [PPSC (NOPP, J), J = 1, NOCOMP]. The method of calculation is specified by a code NTPC such as those shown in Table 14.2. The form of the regression equation should naturally reflect the fundamental equations as has been discussed in Chapter 5. PROMIX calculates the mixture property given NOPP, X, the component flows XMOLE (J), and the type of calculation to be used, M. The answers are stored in PPMX (NOPP).

TABLE 14.2
TYPES OF COMPONENT PHYSICAL PROPERTY CALCULATION

Calculation types	Designation, NTPC
Use existing values in properties vectors	0
Use table look-up to obtain new properties vectors	
Linear interpolation	1
Lagrangian interpolation	2
.	.
.	.
.	.
Parameter equations:	
$Y = A_0$	11
$Y = A_0 + A_1X$	12
$Y = A_0 + A_1X + A_2X^2$	13
$Y = A_0 + A_1X + A_2X^2 + A_3X^3$	14
$Y = A_0 + A_1X + A_2X^2 + A_3X^3 + A_4X^4$	15
.	.
.	.
.	.
(Other types are to be added)	

This problem is not limited to physical properties. A common data system could be utilized when any set of data is used by most Unit Computations. For example, this might be used for a safety analysis, with the hazardous properties of the materials being in the common data system. Alternatively, the unit costs could be in the common data system for management analyses of systems such as is shown Fig. 11.10. Some common data systems have been developed, but improvements in storage capacity, generality, accuracy, and time required are needed.

Other developments and needs

Some useful developments are handling uncertainty and risk, calculating reliability, developing statistical correlations, use of consistent nomenclature, training of personnel, and collection of data.

Reliability, uncertainty, and risk

Executive-aided analysis can be used to incorporate uncertainties and risk about the future and about fundamental data, variation in inputs, and the reliability of systems. Rudd and Watson (1968) introduce many of these in their Chapters 11 to 14.

For example, error distributions could be assigned to many of the parameters and Monte Carlo experiments could be performed with parameter values within these error limits to ascertain their effect on plant operation and final product quality.

Statistical correlations

Techniques are needed for automatically producing *correlation* Unit Computations based on rigorous Unit Computations. For example, the sophisticated multicomponent distillation model requires a large amount of computing time and should be replaced in most simulation applications by a simple correlation model. The latter permits the outputs to be calculated for a restricted range of inputs and operating variables using a correlation based on the repeated application of the complete model. With correlation models, the operation of a complete plant under a given condition should be evaluated in less than thirty seconds.

Consistent nomenclature

Consistent nomenclature is needed. Many users have changed nomenclature, sometimes through necessity, sometimes not. This has happened both in the development of executive systems and of individual Unit Computations. Because of the effort required in a simulation, consistent nomenclature will speed up executive-aided analysis.

Personnel

There is at present a shortage of people with experience in executive-aided simulation. This is partly because real plant problems are so difficult to analyze, and partly because idealized mathematical exercises do not really give the required experience.

This problem rests partly with education. The authors encourage others to use real plant problems in undergraduate and graduate programs. The main constraint is that the questions asked of the simulation or the design be attainable by the student-staff-industrial team available. Nevertheless, within such realistic constraints, valuable experience can be gained which will be vital for future applications. Some experience with the use of such programs in undergraduate education is available [Crowe, et al. (1969)].

Industries can also profit by establishing groups for computer-aided process analysis. As was suggested in Sec. 13.3, the evaluation of the resources required to profit from executive-aided simulation is probably the most difficult step in Fig. 13.1 until know-how and experience have been gained by engineers.

Data

There is often a lack of accurate data. New processes are being designed with the data requirements for simulation and process improvement in mind. Nevertheless, most present plants still lack the data necessary to yield meaningful results. Although management personnel are for the most part reluctant to put money into instrumentation over and above that which is required to control the plant and make it run, many are realizing that such instrumentation can result in a long-term return on investment.

14.4 Summary

On the horizon and beyond are applications and extensions of executive-aided analysis of great significance to the engineer and to industry. Design, plant scheduling, analysis of unsteady-state processing, and simulation of business activities—these are just some of the applications.

The speed with which these are realized depends partly on computer hardware developments but mainly on the evolution of more powerful executive systems and the availability of data and people experienced in simulation.

PROBLEMS

14.1. For the physical property system developed for MACSIM, the NOPP data had to be supplied through the ENC list. If a meaning for the code could be agreed upon by all users, the need for the code could be eliminated.

(a) List all the properties that you think are needed for any problem.

(b) Which of these properties do you consider to be fundamental and which derived?

(c) Order the list so that calculations are the easiest. Is this of any significance?

(d) Select Standard International Units for dimensions. For which properties does this cause major difficulties?

(e) If for any property we want a range of accuracies as was described in the MACSIM system in this chapter, how can we easily modify or adjust the coding system?

14.2. Write algorithms and working programs for the subroutines READIN, PROCAL, and PROMIX.

14.3. The physical property system developed by Motard, et al. in CHESS is different in philosophy and mechanics from the method illustrated in this chapter.

(a) Compare the two.

(b) Choose the overall philosophy and mechanics you think are ideal.

14.4. In this book the equipment units were represented in the information flow diagram by nodes and the streams by lines connecting the nodes, to indicate the relationship among the nodes. However, a diagram in which the streams were represented by nodes and the equipment units by lines could show the the same information. Most Critical Path Methods use the directed lines to represent the job or equipment [Moder and Phillips(1964)].

(a) Compare the relative merits of using equipment units as the nodes against using streams as nodes. Use a simple recycle scheme such as that shown in Fig. 2.8 for your comparison.

(b) Prepare a simple CPM program based on job-on-the node. Compare this with the traditional job-on-the-arrow CPM that is probably already available to your computing center.

14.5. Sometimes all the Unit Computations in a section or even in the whole process are linear or to a first approximation can be assumed linear [Nagiev (1964, 1957)]. For linear systems, it is easier to solve the equations simultaneously rather than sequentially as PACER does. Develop an executive system to handle a complete linear system. For example, we might devise a program DREAD and DPRINT to read in and print out data, DSORT to establish the matrix of coefficients **A** and the vector of constant known terms, and MINVSE to execute the matrix inversion.

Appendix A: Brief Descriptions of Processes

Throughout the book reference was made to problems of simulating processes other than the sulfuric acid plant to emphasize and illustrate that the strategy of simulation is quite general. For convenience, brief descriptions of these processes are included here.

A.1 An Alkylation Plant

Alkylates are used to improve the octane rating of gasoline for internal combustion engines. They are produced from the reaction of propylene and butylenes with an excess of isobutane in the presence of a liquid concentrated sulfuric acid catalyst. The usual olefin and isobutane feed streams contain propane and butane which pass through the reactor untouched and are removed as by-products. Thus the plant has three product streams: alkylate, propane, and butane. The process has three sections: feed preparation, reaction, and product/by-product purification. These are shown in Fig. A.1.

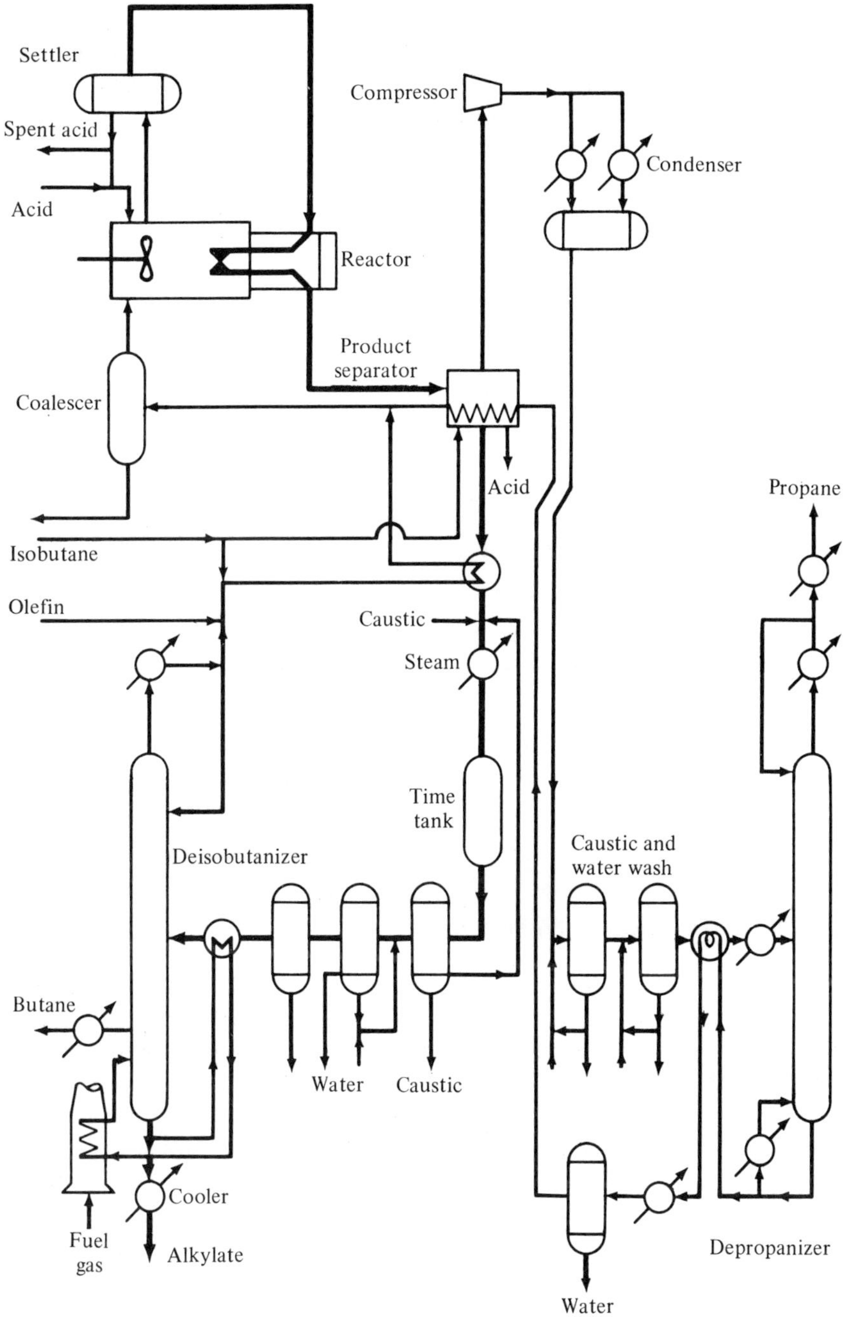

Figure A.1. Flow diagram of an alkylation plant.

Feed preparation section

Recycle streams from the product/by-product purification containing unreacted isobutane are combined with fresh isobutane and olefin as feed to the reactor. Water is removed from the feed because it dilutes the sulfuric acid and reduces catalyst activity.

Reaction section

Two typical reactions are

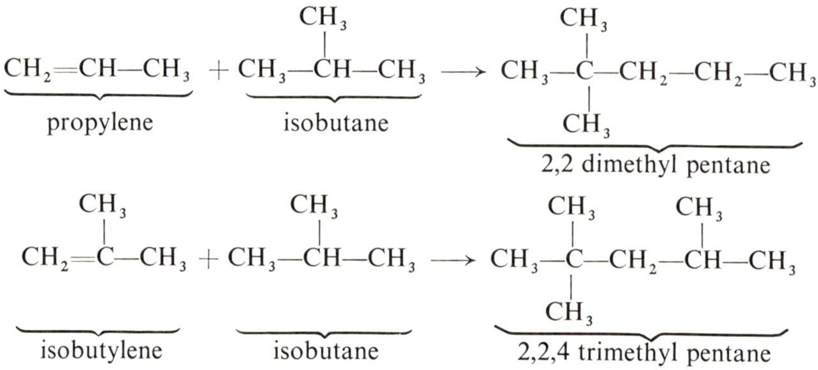

In the plant, the reaction occurs in a long, horizontal cylindrical reactor in an emulsion with a continuous acid phase, with acid strength greater than 89%. Temperatures are maintained between 7° and 13°C; the heat of reaction is removed through cooling coils.

Product/by-product purification

There are three separate purification circuits. First the acid is removed from the reactor product in a settler. Acid recycles to the reactor. Makeup acid is added to maintain acid strength because of the dilution of the acid catalyst with water from the organic feed, and because of the organic sulfates produced by side reactions. The separated organic phase from the settler is passed through a throttling valve and then through the heat exchanger coils of the reactor, where some of the heat of reaction is removed. From the reactor coils, the flow enters the product separator, which splits the organic reactor effluent into a vapor stream of propane, isobutane, and butane (for the depropanizer circuit), and a liquid stream of alkylate, butane, isobutane, and some propane (for the deisobutanizer circuit).

For the depropanizer circuit, the vapor flow from the product separator

TABLE A.1

ESTIMATED COST FOR THE PRODUCTION OF ALKYLATE FOR TWO FEEDS

Materials	Unit usage/bbl alkylate		Unit cost	Net cost/bbl alkylate		
Raw materials	Low n-paraffin feed	High n-paraffin feed		Low n-paraffin feed		High n-paraffin feed
Propylene } Olefin	0.138 bbl		$1.20/bbl	$0.17		
Olefin		1.6 bbl				$5.68
Butylene } Olefin	0.445 bbl		$3.55/bbl	1.58		
Isobutane	0.743 bbl	0.812	$3.00/bbl	2.23		2.44
Acid	0.054 bbl	0.038	$8.68/bbl	0.46		0.33
Caustic	0.57 lb	neglect	—			
n-Butane for sale	—	0.44 bbl				−1.58
Propane for sale	—	0.81 bbl				−2.88
				4.54		3.99
Utilities						
Steam	150 lb	200 lb	$0.50/1000 lb	0.08		0.10
Process water	3.3 U.S. gal		$0.015/1000 U.S. gal	—		—
Electricity	15 kw	20 kw	$0.70/100 kw	0.11		0.14
Fuel gas	0.26 MM		$0.30/MM BTU	0.08		0.08
				4.81		4.31
Labor (L): 0.025 man hour/bbl			$4.00/mh		$0.10	
Maintenance: 0.03 I_F^*					0.14	
Supervision: 0.1 L					0.01	
Depreciation: 10 I_F^*					0.45	
Indirectly attributable: $(0.75(1.1L + 0.03I_F))$					0.19	
Total manufacturing cost: (M)				$5.70		$5.20
General expense: 0.1M					0.55	
Total cost/bbl of alkylate				$6.25		$5.75

*Capital investment (I_F) in 1968 as estimated from Dryden and Furlow (1966) is $2.9 million for a 2000 bbl/day plant. It is assumed that the plant operates 320 days/year.

is liquefied, neutralized, water-washed, and fed under high pressure to the depropanizer (DeC_3) distillation column. The overhead is sold as a high-purity propane LPG. The underflow is partially cooled and then passes through the product separator heat exchanger coils. This stream contains only a small percentage of propane.

For the deisobutanizer circuit, the liquid flowing from the product separator is neutralized, water-washed, and fed to the deisobutanizer distillation column. The column feed temperature is adjusted in a bottoms-to-feed exchanger. The reboiler is gas fired. The bottoms flow is the product alkylate, and the vapor side draw is a butane-rich product. The distillate containing butane, most of the isobutane, and propane recycles to the reactor. Details of this process are given by Albright (1966), Cupit, et al. (1961), Putney (1959), and Freel (1967). An estimate of the unit costs is shown in Table A.1 which indicates the importance of the sulfuric acid makeup relative to the utilities and depreciation costs.

A.2 A Bayer Process Plant

In the Bayer process, alumina is separated from the naturally occurring mineral bauxite. The process is shown in Fig. A.2. First the alumina is dissolved in caustic solution. Dissolution is carefully controlled to minimize the amount of iron and silica in solution. The alumina-rich liquid is separated from the unwanted solids by filtration. Alumina is then recovered from the caustic by a controlled precipitation and subsequent settling to yield alumina trihy-

TABLE A.2
ESTIMATED UNIT COSTS FOR THE PRODUCTION OF ALUMINA
(FROM THE U.S. BUREAU OF MINES)

Item	Unit cost, $/ton
Raw materials	26.00
Utilities	6.25
Labor	2.90
Maintenance	3.60
Operating supplies	0.55
Indirectly attributable costs	3.25
Property and insurance costs	3.00
	45.55
Depreciation: $0.05 I_F^*$	7.10
Total manufacturing cost (M)	52.65
General expense: 0.1 M	5.26
Total estimated cost/ton of alumina	57.91

*It is assumed that there is a capital investment (I_F) of $45,700,000 for a 1000 ton/day plant and that the plant operates 320 days/year.

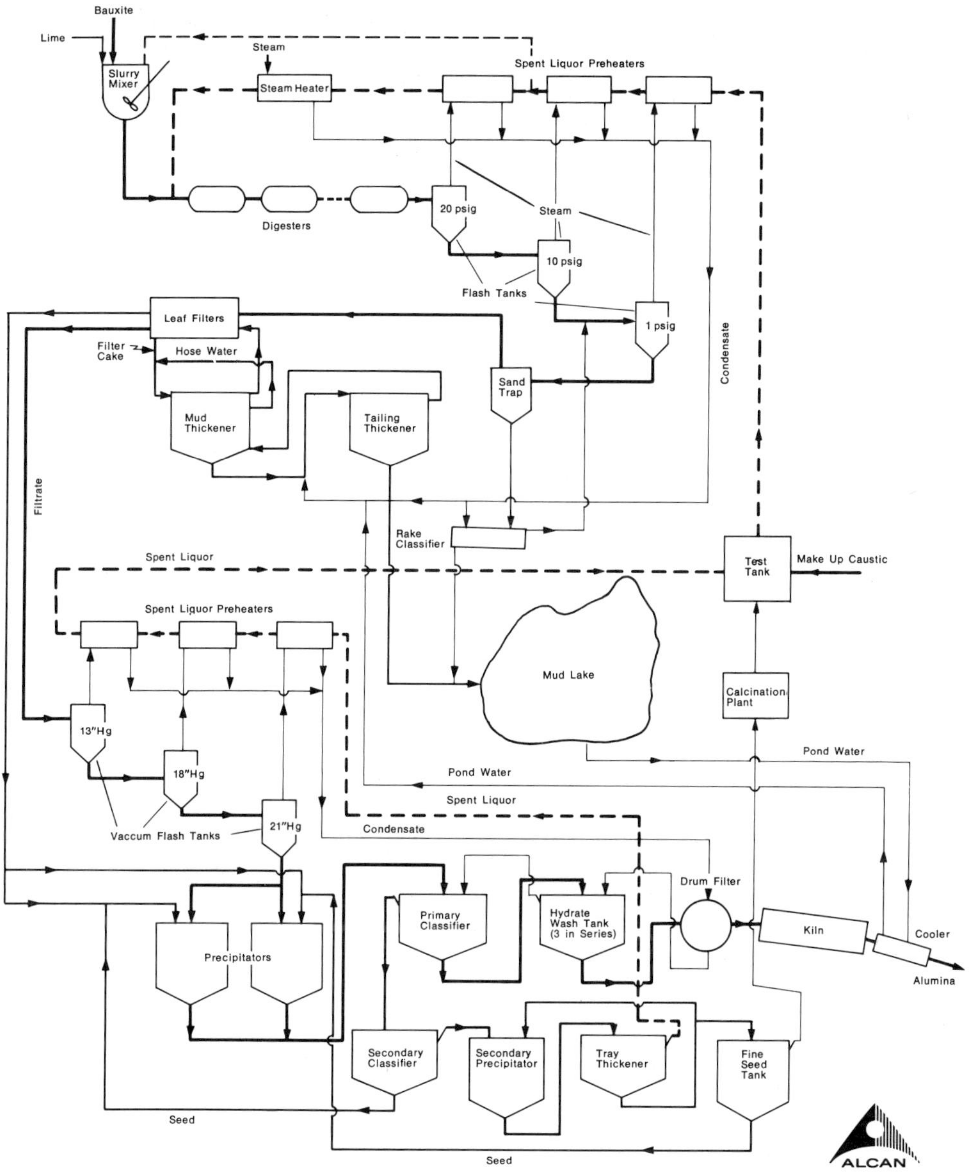

Figure A.2. Process flow diagram for the production of alumina from bauxite at the Arvida, Quebec plant of Aluminum Company of Canada, Ltd.

drate; the caustic is returned to the digestion system. The crystals are washed and calcined to give alumina. Costs are summarized in Table A.2 to indicate that the main financial problems are to minimize the caustic makeup and to conserve heat and energy.

Details of this process are given by Gerard and Stoup (1963), Pearson (1955), and Reese and Cundiff (1956).

Appendix B: Coding Sheets for MACSIM and Sample Output

Although not everyone will use PACER as an executive system, it might be helpful to illustrate the actual input data and the results of a very simple simulation. The coding sheets for MACSIM, a version of PACER, and the typed data for a run with a simple MIXER1 Unit Computation are given. This Unit Computation calculates the component and total flows. The coding sheets follow the definitions given on pages 50–53 and the format instructions of Prob. 3.1. The SN list is:

1. Stream number
2. Stream flag
3. Total flow per unit time
4. Temperature
5. Pressure
6. Total flow of water per unit time
7. Total flow of salt per unit time

B.1 Sample Coding Sheets

Coded by	T. TOONG	MACSIM CODING SHEETS	Page	1
Checked by	A. I. J.		Run No.	5
Date	Oct. 23, 1971			

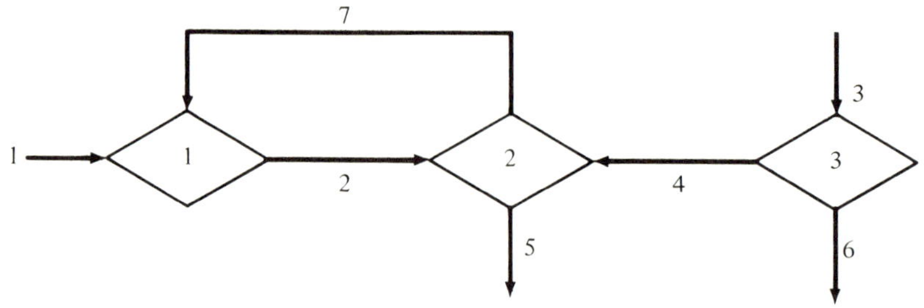

Titles

2 72

TESTING MIXER1 (MIXER—MASS BALANCE)
OCTOBER 23, 1971

Controls

5 10 15 20 25 30 35 40 45 50

KRUN	KSETS	KCLEAN	LOOPS		KCARDS	KPRINT	KONV			
5	0	0	20		0	5				

Dimensions

5 10 15 20 25 30 35 45 50

N2MAX	N3MAX	NEMAX	NSMAX	NELMAX	NECLMX	NSLMAX	NPSMAX	NOCOMP
3	5	3	7	3	0	7	0	2

Process matrix, KPM

5	7–12	14–19	25	30	35	40	45	50	55	60	65
3											
1	MIXER1		1	7	−2						
2	MIXER1		2	4	−5	−7					
3	MIXER1		3	−4	−6						

(a) Equipment lists, EN
(b) Equipment control lists, ENC
(c) Stream lists, SN

EN

5	15	30	45	60	75
3					
	1.	1.0	0.0		
	2.	0.4	0.6		
	3.	0.3	0.7		
0	ENC				
2	SN				
	1.	1.	200.	0.	0.
	50.	150.			
	3.	1.	150.		
	100.	50.			

Test list, DELS

5	15	30	45	60	75
7					
			0.001	0.	0.
	0.001	0.001			

Preferred streams, KPS

5	10	15
0		

Additional equipment lists, AEN

5
0

Note: Coding sheets are to be added here for additional equipment lists if they are entered.

List of the sequence of calculations, NELIST

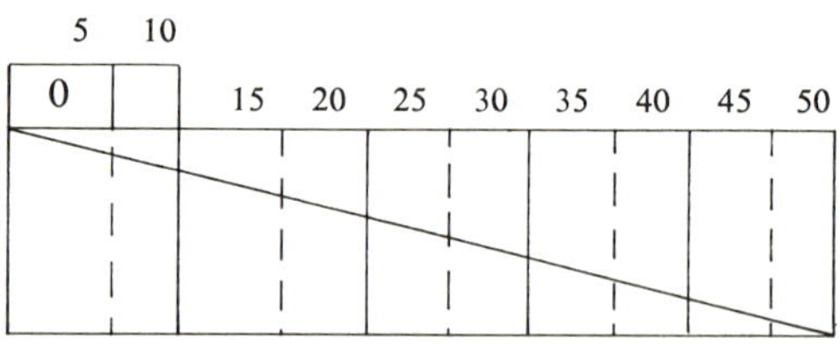

Selected streams printed out, KSPRNT

5	10	15	20	25	30	35	40	45	50
2	1	2							

B.2 Sample Printout

MACSIM—McMASTER PACER CASE STUDY

DATA OF RUN NUMBER 5

TESTING MIXER1 (MIXER—MASS BALANCE)
OCT. 23, 1971

PROCESS MATRIX

UNIT NUMBER	UNIT COMPUTATION	STREAM NUMBERS			
1	MIXER1	1	7	−2	−0
2	MIXER1	2	4	−5	−7
3	MIXER1	3	−4	−6	−0

2 STREAMS TO BE PRINTED EACH ITERATION

1	2	−0	−0	−0	−0	−0	−0	−0

STREAM CONNECTIONS

STREAM	FROM	TO
1	0	1
2	1	2
3	0	3
4	3	2
5	2	0
6	3	0
7	2	1

CONTROL VARIABLES

KSETS = 0
KCLEAN = 0
LOOPS = 20
KARDS = 0
KPRINT = 5
KONV = 0

DIMENSIONS OF THE PROBLEM

NUMBER OF ROWS IN PROCESS MATRIX	3
NUMBER OF COLUMNS IN PROCESS MATRIX	5
MAXIMUM EQUIPMENT NUMBER	3
MAXIMUM STREAM NUMBER	7
LENGTH OF STREAM PARAMETERS LIST	7
LENGTH OF EQUIPMENT PARAMETERS LIST	3
LENGTH OF EQUIPMENT CONTROL LIST	0
NUMBER OF COMPONENTS	2

STREAM VARIABLES TEST LIST, DELS

0.0000	0.0000	0.0010	0.0000	0.0000
0.0010	0.0010			

EQUIPMENT PARAMETERS MATRIX RUN 5—INITIAL

1.0000	1.0000	0.0000
2.0000	0.4000	0.6000
3.0000	0.3000	0.7000

STREAM VARIABLES MATRIX RUN 5—INITIAL

1.0000	1.0000	200.0000	0.0000	0.0000
50.0000	150.0000			
2.0000	0.0000	0.0000	0.0000	0.0000
0.0000	0.0000			
3.0000	1.0000	150.0000	0.0000	0.0000
100.0000	50.0000			
4.0000	0.0000	0.0000	0.0000	0.0000
0.0000	0.0000			
5.0000	2.0000	0.0000	0.0000	0.0000
0.0000	0.0000			
6.0000	2.0000	0.0000	0.0000	0.0000
0.0000	0.0000			
7.0000	0.0000	0.0000	0.0000	0.0000
0.0000	0.0000			

EQUIPMENT NUMBER 3 IS BEING CALCULATED DIRECTLY

SHORTEST LIST OF EQUIPMENT WILL NOW BE FOUND
AT THIS POINT

ITERATIVE CALCULATION-ASSUMED INPUT STREAMS ARE
2 0 0
2 EQUIPMENT CALCULATED IN THE ORDER
2 1

SELECTED STREAMS AT LOOP 1

1.0000	1.0000	200.0000	0.0000	0.0000
50.0000	150.0000			
2.0000	0.0000	227.0000	0.0000	0.0000
68.0000	159.0000			

SELECTED STREAMS AT LOOP 2

1.0000	1.0000	200.0000	0.0000	0.0000
50.0000	150.0000			
2.0000	0.0000	363.2000	0.0000	0.0000
108.8000	254.4000			

SELECTED STREAMS AT LOOP 3

1.0000	1.0000	200.0000	0.0000	0.0000
50.0000	150.0000			
2.0000	0.0000	444.9200	0.0000	0.0000
133.2800	311.6400			

SELECTED STREAMS AT LOOP 4

1.0000	1.0000	200.0000	0.0000	0.0000
50.0000	150.0000			
2.0000	0.0000	493.9520	0.0000	0.0000
147.9680	345.9840			

STREAM VARIABLES MATRIX AT RUN 5 LOOP 5

1.0000	1.0000	200.0000	0.0000	0.0000
50.0000	150.0000			
2.0000	0.0000	523.3712	0.0000	0.0000
156.7808	366.5904			
3.0000	1.0000	150.0000	0.0000	0.0000
100.0000	50.0000			
4.0000	0.0000	45.0000	0.0000	0.0000
30.0000	15.0000			
5.0000	2.0000	215.5808	0.0000	0.0000
71.1872	144.3936			
6.0000	2.0000	105.0000	0.0000	0.0000
70.0000	35.0000			
7.0000	0.0000	323.3712	0.0000	0.0000
106.7808	216.5904			

SELECTED STREAMS AT LOOP 5

1.0000	1.0000	200.0000	0.0000	0.0000
50.0000	150.0000			
2.0000	0.0000	523.3712	0.0000	0.0000
156.7808	366.5904			

SELECTED STREAMS AT LOOP 6

1.0000	1.0000	200.0000	0.0000	0.0000
50.0000	150.0000			
2.0000	0.0000	541.0227	0.0000	0.0000
162.0685	378.9542			

SELECTED STREAMS AT LOOP 7

1.0000	1.0000	200.0000	0.0000	0.0000
50.0000	150.0000			
2.0000	0.0000	551.6136	0.0000	0.0000
165.2411	386.3725			

SELECTED STREAMS AT LOOP 8

1.0000	1.0000	200.0000	0.0000	0.0000
50.0000	150.0000			
2.0000	0.0000	557.9682	0.0000	0.0000
167.1446	390.8235			

SELECTED STREAMS AT LOOP 9

1.0000	1.0000	200.0000	0.0000	0.0000
50.0000	150.0000			
2.0000	0.0000	561.7809	0.0000	0.0000
168.2868	393.4941			

STREAM VARIABLES MATRIX AT RUN 5 LOOP 10

1.0000	1.0000	200.0000	0.0000	0.0000
50.0000	150.0000			
2.0000	0.0000	564.0685	0.0000	0.0000
168.9721	395.0965			
3.0000	1.0000	150.0000	0.0000	0.0000
100.0000	50.0000			
4.0000	0.0000	45.0000	0.0000	0.0000
30.0000	15.0000			
5.0000	2.0000	242.7124	0.0000	0.0000
79.3147	163.3976			
6.0000	2.0000	105.0000	0.0000	0.0000
70.0000	35.0000			
7.0000	0.0000	364.0685	0.0000	0.0000
118.9721	245.0965			

SELECTED STREAMS AT LOOP 10

1.0000	1.0000	200.0000	0.0000	0.0000
50.0000	150.0000			
2.0000	0.0000	564.0685	0.0000	0.0000
168.9721	395.0965			

SELECTED STREAMS AT LOOP 11

1.0000	1.0000	200.0000	0.0000	0.0000
50.0000	150.0000			
2.0000	0.0000	565.4411	0.0000	0.0000
169.3832	396.0579			

SELECTED STREAMS AT LOOP 12

1.0000	1.0000	200.0000	0.0000	0.0000
50.0000	150.0000			
2.0000	0.0000	566.2647	0.0000	0.0000
169.6299	396.6347			

SELECTED STREAMS AT LOOP 13

1.0000	1.0000	200.0000	0.0000	0.0000
50.0000	150.0000			
2.0000	0.0000	566.7588	0.0000	0.0000
169.7780	396.9808			

SELECTED STREAMS AT LOOP 14

1.0000	1.0000	200.0000	0.0000	0.0000
50.0000	150.0000			
2.0000	0.0000	567.0553	0.0000	0.0000
169.8668	397.1885			

CONVERGENCE OBTAINED AFTER LOOP NUMBER 14

STREAM VARIABLES MATRIX RUN 5—FINAL

1.0000	1.0000	200.0000	0.0000	0.0000
50.0000	150.0000			
2.0000	0.0000	567.0553	0.0000	0.0000
169.8668	397.1885			
3.0000	1.0000	150.0000	0.0000	0.0000
100.0000	50.0000			
4.0000	0.0000	45.0000	0.0000	0.0000
30.0000	15.0000			
5.0000	2.0000	244.7035	0.0000	0.0000
79.9112	164.7923			
6.0000	2.0000	105.0000	0.0000	0.0000
70.0000	35.0000			
7.0000	0.0000	367.0553	0.0000	0.0000
119.8668	247.1885			

CALCULATION SEQUENCE
EQUIPMENT CODE

3	1
2	2
1	3

TIME IS 001 M. 19 S.

Appendix C: Symbols

Symbol	*Name*	*Typical units*
a	Surface per unit volume	sq ft/cu ft
A	Area	sq ft
c	Volumetric concentration	lb/cu ft or lb moles/cu ft
C	Specific heat	PCU/lb °C
C	Capital cost	dollars
d	Diameter	ft
D	Diffusivity	sq ft/hr
e	Material balance error	
E	Axial eddy diffusivity	
E	Effectiveness factor	
f	Friction factor	
F	Flow	lb moles/hr
F	Shape factor for radiant heat transfer	
g	Acceleration of gravity	ft/hr^2
G	Gas flow rate	lb moles/hr
h	Coefficient of heat transfer	PCU/hr sq ft °C
h	Enthalpy per unit mass	PCU/lb

H	Enthalpy per mole	PCU/lb mole
H	Henry's law constant	lb moles/cu ft atm
H	Head	ft
HTU	Height of a transfer unit	ft
I	Fixed capital investment	dollars
k	Reaction rate constant	
k	Convergence promotion factor	
k	Individual mass transfer coefficient	lb moles/hr sq ft atm
k	Thermal conductivity	PCU/hr sq ft (°C/ft)
K	Overall mass transfer coefficient	lb moles/hr sq ft atm
K	Equilibrium constant	
l	Packing depth	ft
lm	Log mean	
L	Length	ft
L	Liquid flow rate	lb moles/hr
L	Labor cost	dollars
m	Mass	lb
M	Manufacturing cost	dollars
M	Molecular weight	
n	Number in a series	
n	Loop number	
N	Number of velocity heads	
N	Flux	PCU/hr/ft^2
NTU	Number of transfer units	
p	Pressure	atm; lb/sq ft
q	Rate of heat transfer	PCU/hr
Q	Quantity of heat transferred	PCU
Q	Differential heat of solution	PCU/lb mole
Q	Volumetric flow rate	
r	Radius	ft
r	Rate of formation in a chemical reaction	
R	Gas constant	1.987 cal/gm mole °C
$\mathbf{R}$	Reachability matrix	
R	Heat flow ratio	
rpm	Revolutions per minute	min^{-1}
S	Entropy	PCU/lb mole °K
S	Arc length	ft
t	Temperature of gas	°C or °K
T	Temperature of liquid	°C or °K
u	Local velocity	ft/hr

U	Heat transfer coefficient	PCU/hr sq ft °C
v	Velocity	ft/sec
v	Specific volume	cu ft/lb
V	Vapor rate	lb moles/hr
V	Volume	cu ft
W	Mass flow rate	lb/hr
$\mathbf{W}$	Intersection matrix	
x	Distance in x direction	ft
x	Mole fraction in liquid	
X	Fractional conversion	
y	Depth	ft
y	Mole fraction in vapor	
y	Thickness of radial fin	ft
Y	Correction factor for temperature driving force	
z	Compressibility	
z	Axial position	

Greek Letters

Symbol	*Name*	*Typical units*
α	Absorptivity	
α	Activity coefficient	
α	Fraction of total input streams leaving a unit or fractional split	
β	Property group; cost allocation factor	
γ	Cost distribution factor	
Γ	Mass or molar flow rate per unit width of stream	lb/ft or lb mole/ft
δ	Partial differentiation operator; cost distribution factor	
Δ	Difference operator	
Δ	Difference, as in temperature driving force	
ϵ	Void fraction	
ϵ	Emissivity	
ϵ	Error	
η	Efficiency	
θ	Convergence factor	
θ	Time	
Θ	Convergence factor for system	
κ	Ratio of specific heats	

λ	Latent heat of vaporization	PCU/lb
μ	Absolute viscosity	lb/hr ft
ν	Stoichiometric factor	
ξ	Extent of reaction	
ρ	Density	lb/cu ft
ρ	Reflectivity	
σ	Surface tension	lb/ft
σ	Stefan-Boltzman constant	0.1713×10^{-8} Btu/ft^2 hr °R
ϕ	Angle in cylindrical coordinate system	
ω	Angular velocity	
∇	Operator grad	

Subscripts

av	Average
b	Base (of a fin)
B	Bulk
c	Critical
c	Cold phase
c	Catalyst
c	Cost
e	Exchange fluid
e	External
Eff	Effective
f	Film
G	Gas phase
h	Hot phase
hot	Hot phase
H	Heat transfer
i	Component i
i	Row number in a matrix; unit number
in	Initial
Int	Interface
j	Column number in a matrix; component number
layer	Layer
lm	Logarithmic mean
L	Liquid phase
M	Mass transfer
Mix	Mixture

n	Loop number
o	Outside
out	Outlet
p	Pressure
p	Particle
r	Radius
R	Radiation contribution
S	Surface
sat	Saturated
TP	Two-phase
V	Volume

Superscripts

0	Standard or reference state
*	Equilibrium, optimum
–	Average

Dimensionless Constants

Nu	Nusselt number
Pr	Prandtl number
Sc	Schmidt number
Sh	Sherwood number
Re_p	Particle Reynolds number

Appendix D: Glossary

Adiabatic Without transfer of heat through the bounding surface of the system.

Adjacency matrix An array which represents the connections on the information flow diagram between pairs of Unit Computations. Unity indicates a connection from the row number to the column number, while zero indicates no connection.

Algorithm A step-by-step procedure for the sequential solution of a problem.

Approximate simulation Simulation of a process to within an expected accuracy for a range of operating conditions consistent with the overall objectives.

Arguments Names of storage locations or constants indicating information transmitted between main programs, arithmetic statement functions, functions, and subroutines. The arguments appear in parentheses following the name of such functions and subroutines.

Array A set of adjacent storage places, having one or more dimensions, in a computer program.

Base case simulation Simulation of a process to within an expected accuracy for the single set of consistent operating data at which the process normally runs.

Catalyst effectiveness The ratio of the rate of reaction when limited by the diffusion into the catalyst pores to the rate of reaction without diffusion limitation.

COMMON Computer storage shared by the main program and by subroutines.

Common system A set of locations into which data are stored so as to be accessible to more than one Unit Computation. For example, physical properties and costs are usually in COMMON locations.

Compile To convert a computer program in a source language to a machine language program which is executed on a digital computer.

Convergence The state reached in an iterative calculation in which the absolute fractional change in prescribed variables between successive iterations is less than some prescribed tolerance.

Data set Information the engineer supplies to the simulation, including input conditions, arrangement of equipment, and perhaps the order in which the calculations are to be done. The data set does not include the Unit Computations.

Declaration A line in a source program which provides information necessary for the program but does not cause any mathematical operations to be carried out (cf. **Statement**).

Design Calculation of the characteristics of a device to convert inputs into specified outputs.

Diagnostic An error message concerning a source program obtained during the compilation of that program.

Dimension The size of an array, indicated by a number or numbers following the subscripted variable in a dimension declaration at the start of a source program.

Element Storage place in an array.

Equipment Actual piece of equipment; a physical device.

Equipment subroutines Type of Unit Computation that models equipment only. In this text, the only equipment subroutines referred to are Unit Computations.

Executive program A program which calls successively on different Unit Computations and passes the necessary data among them. It consists of several subroutines and may generate its own sequence of calling the Unit Computations.

Field The space on a punched card, or on a line of printing, reserved for a given number or message.

Format The form in which numbers or messages are read into or printed or punched out of a computer during the execution of a program.

FORTRAN An algebraically oriented computer source language created by IBM for the solution of scientific and engineering problems.

Function A calculation block in a computer program which may receive several items of information but returns only one item, based on these several input items. An *arithmetic* statement function is a function which can be written in one line with only one equals sign (see **Subroutine**).

GEMCS A file-oriented, interactive executive system written for time-sharing computers.

Incidence matrix An array which represents the connections on the information flow diagram between a stream number and a Unit Computation. An entering stream is given +1 and a leaving stream −1.

Information Collective name for the magnitudes of stream variables contained in a particular stream.

Information flow diagram A diagrammatic representation of the relationship among Unit Computations and streams of information for the simulation of a stagewise manufacturing process; a directed graph.

Information modifier See **Unit Computation.**

Iteration A controlled repetition of a sequence of events, almost always with a controlled change of one or more of the variables in the sequence. In PACER, iteration is the repeated calculation of a specified list of Unit Computations.

List A one-dimensional array of storage places in a computer program.

Listing A printout of a computer program, obtained either by passing the cards through a suitable printer or from the computer during compilation.

MACSIM A MAP version of PACER, with a physical properties handling system, and a library of Unit Computations.

Modeling Deciding the equations to be solved and the accuracy to be achieved in order to simulate units by Unit Computations, and preparing a suitable computer program.

Modularity Breaking the problem into portions that are to be analyzed one at a time. The portions are arranged according to the similarities in behavior. Examples of modularity include Unit Operations, Unit Processes, and, in this text, Unit Computations.

Object program A program written in machine language, resulting from the compilation of a source program.

Operating cost All directly attributable costs excluding depreciation.

PACER A digital computer executive program created by Professor Paul T. Shannon and Mr. H. A. Mosler for carrying out process calculations in any stagewise manufacturing process, regardless of the complexity of the manufacturing units. Current versions treat steady-state manufacturing processes only.

Parameter The alphanumeric name of an adjustable quantity in a simulation program which is normally fixed for each case study. The value of a parameter is part of the input data; the name is usually associated with a Unit Computation.

Plug flow The model of flow which assumes a uniform velocity profile, no axial mixing or conduction, and no radial variation in properties.

Process flow diagram A diagrammatic representation of a manufacturing process, showing the manufacturing and control equipment and the pipelines of material flow (cf. **Information flow diagram**).

Process matrix A two-dimensional array whose row has an equipment number, the name of the Unit Computation, and the numbers of the streams of information connected to the equipment unit. It is a numerical encoding of the information flow diagram.

Program The sequence of mathematical operations to be carried out on data fed to a digital computer to achieve a desired result.

Recycle A condition in which information leaving one unit affects information entering that unit.

Regression The process of reducing tabular data to equations or the process of generating an equation to represent the relationship among variables.

Serial Unit Computations In describing an information flow diagram, those Unit Computations which are not involved in a recycle calculation, that is, those that are executed only once for each calculation of the total process. Synonyms: chain, sequential.

Simulation The solution of a mathematical model corresponding to a physical process. A digital simulation solves the mathematical model using a digital computer and calculates values of the output variables, given the inputs and the behavior of the device. In this book the word is used to mean an executive-aided digital computer simulation.

Source program A program written in an algebraically oriented language.

Stage In a manufacturing process, a discrete manufacturing entity which modifies material in a desired manner (see **Unit Computation**).

Statement A line in a source program which causes mathematical or logical operations to be carried out (see **Declaration**).

Stream A carrier of information on an information flow diagram.

Subroutine A calculation block in a computer program which may receive several items of information and may return more than one item of calculated information. Items are transmitted between the subroutine and other subroutines of the main program through the arguments of the subroutine or by use of common storage.

Subscripted variable An array in a source program.

Supervisory section Unit Computations in the information flow diagram and the process matrix which handle variables and parameters concerned with any Unit Computations.

Unit or module A characteristic type of calculation. It may concern equipment or a calculational procedure.

Unit Computation In a modular simulation scheme, a mathematical model which modifies information according to an algorithm. There are several types, but the most common are analogous to manufacturing stages. Synonyms: equipment subroutine, process module, building block.

Variable The alphanumeric name of a quantity in a simulation program which varies with location and use in the program. In particular, stream variables usually change in value after passage through a Unit Computation. In this book, the name is reserved for streams.

Appendix E: References

Aizawa, M. (1966), The Simulation of the Steady-State Behaviour of an Ethylbenzene Styrene Fractionation Unit. Master of Engineering thesis, McMaster University, Hamilton, Ontario.

Albright, L. F. (1966), Alkylation: Chemical and Engineering Factors for Reactor Design. *Chem. Eng.* **73,** No. 14, 119–26.

Amelin, A. G., et al. (1949), *Zhur. Fiz. Khim.* **25,** 8.

Amundson, N. R. (1966), *Mathematical Methods in Chemical Engineering.* Prentice-Hall, Englewood Cliffs, N.J.

Andress, F. J. (1960), The Learning Curve as a Production Tool. *Harvard Business Review* **32,** No. 1, 87–97.

Aries, R. S. and Newton, R. D. (1955), *Chemical Engineering Cost Estimation.* McGraw-Hill, New York.

Aris, R. (1965), *Introduction to the Analysis of Chemical Reactors.* Prentice-Hall, Englewood Cliffs, N.J.

Ball, W. E. (1961), Computer Programs for Distillation. Paper presented at the 44th National Meeting of the American Institute of Chemical Engineers, New Orleans, Feb. 26–Mar. 1.

Bauman, H. C. (1964), *Fundamentals of Cost Engineering in the Chemical Industry.* Reinhold, New York.

Bauman, H. C. (1964a), *The Chemical Engineer's Handbook*. 4th ed. J. H. Perry, ed. McGraw-Hill, New York.

Bellman, R. (1957), *Dynamic Programming*. Princeton University Press, Princeton, N.J.

Bird, R. B., Stewart, W. E., and Lightfoot, E. N. (1960), *Transport Phenomena*. Wiley, New York.

Bobrow, D. G. and Raphael, B. (1964), A Comparison of List-Processing Computer Languages. *Comm. Association for Computing Machinery* **7,** 231.

Booth, A. D. and Morris, J. R. (1966), The Calculation of Heat and Material Balances Using a Small Digital Computer. *Proceedings of the Symposium on Cost Reductions in the Design Phase*. Symposium series 18. Fourth Congress of the European Federation of Chemical Engineering.

Boudart, M. (1956), Kinetics on Ideal and Real Surfaces. *Am. Inst. Chem. Engrs. Journal* **2,** 62.

Bowman, R. A., Mueller, A. C., and Nagle, W. M. (1940), Mean Temperature Difference in Design. *Trans. ASME* **62,** 283.

Bromley, L. A. (1949), *Calculation of Gas Viscosity as a Function of Temperature*. Lawrence Radiation Laboratory, University of California at Berkeley, Report No. UCRL 525.

Burkhardt, D. B. (1968), Increasing Conversion Efficiency. *Chem. Eng. Prog.* **64,** No. 11, 66.

Burman, L. K., Maddox, R. N., and Erbar, J. H. (1966), Evaluation of a Relaxation Procedure for Multicomponent Distillation Columns. Paper presented at the 58th National Meeting of the American Institute of Chemical Engineers, Dallas, Feb. 6–9.

Calderbank, P. H. (1953), Contact-Process Converter Design. *Chem. Eng. Prog.* **49,** 585.

Cary, J. R. (1959), Fast, Convenient Approach to Sizing Heat Exchangers. *Chem. Eng.* **66,** No. 10, 169.

Cavett, R. H. (1963), Application of Numerical Methods to the Convergence of Simulated Processes Involving Recycle Loops. *Proc. Amer. Petrol. Inst.*, Div. of Refining **43** (III), 57.

Cavett, R. H. (1964), Review and Critique of Relevant Mathematical and Computer Methods for the Analysis of Recycle Processes. Paper presented at the 53rd National Meeting of the American Institute of Chemical Engineers, Pittsburgh, May 17–20.

Chartrand, G. (1967), Optimization of the Oxidation of Sulphur Dioxide in an Existing Multi-Bed Adiabatic Reactor. Master of Engineering thesis, McMaster University, Hamilton, Ontario. See Chartrand, G. and Crowe, C. M. (1969), *Can. J. Chem. Eng.* **47,** 296.

Chemical Engineering (1967), Kellogg's Single-Train Ammonia Plants Achieve Lowest Costs. **74,** No. 20, 78; **74,** No. 24, 112; **74,** No. 25, 94.

Chemical Engineering (1968), Sulfuric Acid Process Boasts Higher Yields. **75,** No. 21, 94.

Chen, J. C. (1967), Correlation for Boiling Heat Transfer to Saturated Fluids in Convective Flow. *Ind. Eng. Chem. Fundamentals* **5,** 322.

Chilton, C. H., ed. (1960), *Cost Engineering in the Process Industries.* McGraw-Hill, New York.

CHIPS User's Manual. The Service Bureau Corporation, New York.

Christensen, J. H. (1967), Ph. D. thesis, University of Wisconsin, Madison. See Christensen, J. H. and Rudd, D. F. (1969), Structuring Design Computations. *Am. Inst. Chem. Engrs. Journal* **15,** 94.

Connor, J. M. (1968), Economics of Sulfuric Acid Manufacture. *Chem. Eng. Prog.* **64,** No. 11, 59.

Crowe, C. M., Allingham, J. P., Carpenter, G. R., Leyerzapf, V., Love, C. E., Wilson, R. G., Erme, T. E., Fitz, L., Taylor, W. K., Thibaudeau, H. A., Wisnowski, L. T., and Wong, H. W. (1966), Design report to the Department of Chemical Engineering, McMaster University, Hamilton, Ontario.

Crowe, C. M., Hamielec, A. E., Hoffman, T. W., Johnson, A. I., Woods, D. R., Anderson, R. B., and Hodgins, J. W. (1969), Teaching Experience with Design and Simulation Projects. Paper presented at the 153rd National Meeting of the American Chemical Society in 1967. Updated version submitted to *Chem. Eng. Ed.* in 1969.

Cupit, C. R., Gwyn, J. E., and Jernigan, E. C. (1962), Catalytic Alkylation. *Petro/Chem Eng.* Dec. 1961, Jan. 1962.

Dagnall, B. D. (1965), Use of Simulation for Sizing Equipment to Meet Varying Demands. *Brit. Chem. Eng.* **10,** 760–63.

Davidson, B. and Thodos, G. (1964), Kinetics of the Catalytic Oxidation of Sulfur Dioxide. *Am. Inst. Chem. Engrs. Journal* **10,** 568.

Dingman, J. C., Allen, D. L., and Moore, T. F. (1966), Minimize Corrosion in MEA Units. *Hydrocarbon Process. Petrol. Refiner* **45,** No. 9, 285.

Dryden, C. E. and Furlow, R. H. (1966), *Chemical Engineering Costs.* Ohio State University, Columbus.

Dryden, C. E. and Winfield, M. D. (1963), *Chemical Engineering Costs.* Ohio State University, Columbus.

Duecker, W. W. and West, J. R. (1959), *The Manufacture of Sulfuric Acid.* Reinhold, New York.

Dukler, A. E., Wicks, M., and Cleveland, R. G. (1964), Frictional Pressure Drop in Two-Phase Flows: B. An Approach Through Similarity Analysis. *Am. Inst. Chem. Engrs. Journal* **10,** 44.

Eckert, E. R. G. and Drake, R. M. (1959), *Introduction to the Transfer of Heat and Mass.* 2nd ed. McGraw-Hill, New York.

Eklund, R. B. (1956), Rate of Oxidation of Sulfur Dioxide with a Commercial

Vanadium Catalyst. Dissertation presented at the Royal Institute of Technology, Stockholm, Sweden. Almquist and Wiksell Publishers, Stockholm.

Elliott, D. M. and Owen, J. M. (1968), Critical Examination in Process Design. *The Chemical Engineer* (*London*) **223,** 377.

Ellis, D. S. (1947), *Elements of Hydraulic Engineering*. Van Nostrand, New York.

Ergun, S. (1952), Fluid Flow through Packed Columns. *Chem. Eng. Prog.* **48,** 89.

Evans, L. B., Steward, D. G., and Sprague, C. R. (1968), Computer-aided Chemical Process Design. *Chem. Eng. Prog.* **64,** No. 4, 39.

Fair, J. R. and Bolles, W. L. (1968), Modern Design of Distillation Columns. *Chem. Eng.* **75,** No. 9, 156.

Faith, W. L., Keyes, D. B., and Clark, R. L. (1957), *Industrial Chemicals*. rev. 1966. Wiley, New York.

Fan, L.-T. (1966), *The Continuous Maximum Principle*. Wiley, New York.

Farkas, M. D. and Dukes, R. R. (1968), Multiple Route to Sulfuric Acid. *Chem. Eng. Prog.* **64,** No. 11, 54.

Forgacs, O. L. (1963), *The Characterization of Mechanical Pulps*. part II. Pulp and Paper Research Institute of Canada, report no. 320. Montreal.

Forrester, J. W. (1961), *Industrial Dynamics*. MIT Press, Cambridge, Mass.

Fox, L. (1962), *Numerical Solution of Ordinary and Partial Differential Equations*. Addison-Wesley, Reading, Mass.

Franks, R. G. E. (1967), *Mathematical Modeling in Chemical Engineering*. Wiley, New York.

Freel, J. D. (1967), Refinery Studies of a Stratco Alkylation Reactor. A project report presented to the Department of Chemical Engineering, McMaster University, Hamilton, Ontario.

Freel, J. D. (1969), A Study of an Alkylation Reaction. Master of Engineering thesis, McMaster University, Hamilton, Ontario.

Furkert, H. (1969), Two Routes to Sulfuric Prove Tough on Pollution. *Chem. Eng.* **76,** No. 1, 70.

Gardner, K. A. (1945), Efficiency of Extended Surface. *Trans. ASME* **67,** 621.

Gary, J. H. (1967), Production of Ammonia. *Chemical Engineering Case Problems*. American Institute of Chemical Engineers, New York.

Gerard, G. and Stoup, P. T. (1963), *Extractive Metallurgy of Aluminum*. vol. 1: Alumina. Interscience, New York.

Gmitro, J. I. and Vermeulen, T. (1964), Vapor-Liquid Equilibria for Aqueous Sulfuric Acid. *Am. Inst. Chem. Engrs. Journal* **10,** 740.

Gordon, G. (1969), *System Simulation*. Prentice-Hall, Inc., Englewood Cliffs, N.J.

Grober, H., Erk, S., and Grigull, U. (1961), *Fundamentals of Heat Transfer*. 3rd ed. McGraw-Hill, New York.

Grumer, E. L. (1967), Selling Price vs. Raw Material Cost. *Chem. Eng.* **74,** No. 9, 190.

Hanson, D. N., Duffin, J. H., and Somerville, G. F. (1962), *Computation of Multistage Separation Processes.* Reinhold, New York.

Happel, J. (1958), *Chemical Process Economics.* Wiley, New York.

Harary, F., Norman, R. Z., and Cartwright, D. (1965), *Structural Models: An Introduction to the Theory of Directed Graphs.* Wiley, New York.

Himmelblau, D. M. (1966), Decomposition of Large Scale Systems. I: Systems Composed of Lumped Parameter Elements. *Chem. Eng. Sci.* **21,** 425.

Himmelblau, D. M. and Bischoff, K. B. (1968), *Process Analysis and Simulation: Deterministic Systems.* Wiley, New York.

Hirschmann, W. H. (1964), The Learning Curve. *Chem. Eng.* **71,** No. 7, 95.

Holland, C. D. (1963), *Multicomponent Distillation.* Prentice-Hall, Englewood Cliffs, N.J.

Hooke, R. and Jeeves, T. A. (1961), Direct Search Solution of Numerical and Statistical Problems. *J. Assoc. Comp. Mach.* **8,** 212.

Hottel, H. C. and Sarofim, A. F. (1967), *Radiative Transfer.* McGraw-Hill, New York.

Hougen, O. A., Watson, K. M., and Ragatz, R. A. (1954), *Chemical Process Principles.* part I: Material and Energy Balances. 2nd ed. Wiley, New York.

Hougen, O. A., Watson, K. M., and Ragatz, R. A. (1959), *Chemical Process Principles.* part II: Thermodynamics. 2nd ed. Wiley, New York.

Hughmark, G. A. (1962), Holdup in Gas-Liquid Flow. *Chem. Eng. Prog.* **58,** No. 4, 62.

Hui, A. W. T. and Hamielec, A. E. (1969), Polymer Reactors and Molecular Weight Distribution: Role of Viscosity and Recycle in Reactor Systems. *Ind. Eng. Chem., Process Design and Development* **8,** 105.

Hydrocarbon Processing and Petroleum Refiner (1967), Ammonia Flow Diagrams for Bran, Fluor, Foster Wheeler, and Kellogg. **46,** No. 11, 145–48.

Industrial and Engineering Chemistry (1950), Modern Chemical Processes. vol. 1. Reinhold, New York.

International Critical Tables (1930), C. J. West, compiler. McGraw-Hill, New York.

Jackson, R. E., Klomparens, A. J., and Westbrook, G T. (1965), Long Range Planning. *Chem. Eng. Prog.* **61,** No. 1, 83–87.

James, J. L., Gardner, N. F., Reinhart, L. R., and Hellenack, L. J. (1966), Economic Design by Flexible Flowsheet Analysis. *Proceedings of the Symposium on Cost Reductions in the Design Phase.* Symposium series 18. Fourth Congress of the European Federation of Chemical Engineering. Institution of Chemical Engineers, London.

Johnson, A. I. and Associates (1970), *GEMCS/6400: A General Engineering and Management Computation System (CDC 6400 version)*. Department of Chemical Engineering, McMaster University, Hamilton, Ontario.

Johnson, A. I., Aizawa, M., and Petryschuk, W. F. (1968), Simulation of a Synthetic Rubber Plant. *Brit. Chem. Eng.* **13**, 1432.

Johnson, A. I., Baird, M. H. I., Crowe, C. M., Hamielec, A. E., Hoffman, T. W., Pollock, K. G., Woods, D. R., Chappel, J. F., Gilmour, R. H., Gleave, A. F., Haney, D. C., Kitcher, E. D., Powell, J. A., Shaw, I. D., Sutton, T. L., Brown, A. G., Canning, H. I., Gates, J. E., Melnyk, P. B., Orlickas, A. J., Williams, D. E., Chartrand, G., Freel, J. D., LeClair, B. P., Li, M. C., Milburn, C. R., Orbach, O., Pearce, S. L., Prowse, D. R., Yau, A. Y., and Toong, T. (1968), *A Comprehensive Report on a Digital Simulation of an Alkylation Unit*. Department of Chemical Engineering, McMaster University, Hamilton, Ontario.

Johnson, A. I., Bobrow, S., and Ponton, J. W. (1970), *DYNSYS: Manual and Application Studies*. Department of Chemical Engineering, McMaster University, Hamilton, Ontario.

Johnson, A. I., Bobrow, S., Prowse, D. R., Leather, A., and Ponton, J. (1969), *DYNSYS: Digital Computer Programs for Studying the Transient Behaviour of Systems Using a Modular Approach*. Department of Chemical Engineering, McMaster University, Hamilton, Ontario.

Johnson, A. I. and Toong, T. (1968) [See Johnson and Associates (1970)].

Johnstone, R. E. and Thring, M. W. (1957), *Pilot Plants Models and Scale-up Methods in Chemical Engineering*. McGraw-Hill, New York.

Johri, H. P. (1968), Master of Applied Science thesis, University of Waterloo, Waterloo, Ontario. To appear as project paper in Stochastic Simulation Model of a Chemical Plant coauthored by Broekhoven, L. and Reilly, P. M.

Kaufmann, W. (1936), *Fluid Mechanics*. trans. E. G. Chilton. McGraw-Hill, New York.

Kays, W. M. and London, A. L. (1955), *Compact Heat Exchangers*. 2nd ed. McGraw-Hill, New York.

Keenan, J. H. and Keyes, F. G. (1963), *Thermodynamic Properties of Steam*. Wiley, New York.

Kelley, K. K. (1949), *Contributions to the Data on Theoretical Metallurgy*. X: High-Temperature Heat-Content, Heat Capacity, and Entropy Data for Inorganic Compounds. U.S. Department of the Interior, Bureau of Mines Bulletin 476.

Kelley, K. K. (1960), *Contributions to the Data on Theoretical Metallurgy*. XIII: High-Temperature, Heat Capacity, and Entropy Data for the Elements and Inorganic Compounds. U.S. Department of the Interior, Bureau of Mines Bulletin 584.

King, C. J. (1967), *Chemical Engineering Case Problems*. chap. 5. American Institute of Chemical Engineers, New York.

Kline, S. J. (1965), *Similitude and Approximation Theory*. McGraw-Hill, New York.

Kreith, F. (1965), *Principles of Heat Transfer.* 2nd ed. International Textbook Co., Scranton, Pa.

Krichevskaya, E. L. (1947), Kinetics of the Oxidation of Sulfur Dioxide on Vanadium Pentoxide. *J. Phys. Chem. USSR* **21,** 287.

Kronseder, J. G. (1968), Cost of Reducing Sulfur Dioxide Emissions. *Chem. Eng. Prog.* **64,** No. 11, 71.

Kubota, H., Ishizawa, M., and Shindo, M. (1959), The Rate of Catalytic Oxidation of Gaseous Sulfur Dioxide with Vanadium Catalyst. *Sulfuric Acid Japan* **12,** 243.

Lapidus, L. (1962), *Digital Computation for Chemical Engineers.* McGraw-Hill, New York.

Lapidus, L. (1968), Annual Review of Applied Mathematics. *Ind. Eng. Chem.* **60,** No. 12, 44.

Lawrence, A. E., et al. (1954), *Chemical Business Handbook.* J. H. Perry, ed. McGraw-Hill, New York.

Lederman, P. B. (1968), Flowsheet Simulation and Beyond. *Chem. Eng.* **75,** No. 25, 127.

Lee, K. Y. and Aris, R. (1963), *Ind. Eng. Chem. Process Design and Development* **2,** 300, 306.

Lee, W., Christensen, J. H., and Rudd, D. F. (1966), Design Variable Selection to Simplify Process Calculations. *Am. Inst. Chem. Engrs. Journal* **12,** 1104.

Levenspiel, O. (1962), *Chemical Reaction Engineering.* Wiley, New York.

Lowry, I. and Pike, R. W. (1967), A Dynamic Programming Study of the Contact Process. Paper presented at the 61st National Meeting of the American Institute of Chemical Engineers, Houston, Feb. 19–23.

Mandelik, B. G. and Pierson, C. U. (1968), New Sources for Sulfur. *Chem. Eng. Prog.* **64,** No. 11, 75.

Markowitz, H. M., Hausner, B., and Karr, H. W. (1964), *SIMSCRIPT: A Simulation Programming Language.* Prentice-Hall, Englewood Cliffs, N.J.

Martin, T. W. (1954), Journal of Heat Transfer. *Trans. ASME* **76,** 515.

McAdams, W. H. (1954), *Heat Transmission.* 3rd ed. McGraw-Hill, New York.

Meier, R. C., Newell, W. T., and Pazer, H. L. (1969), *Simulation in Business and Economics.* Prentice-Hall, Englewood Cliffs, N.J.

Mezaki, R. and Kittrell, J. R. (1967), Parametric Sensitivity in Fitting Nonlinear Kinetic Models. *Ind. Eng. Chem.* **59,** No. 5, 63.

Miller, R. L. (1967), Solving Practical Problems in Big Ammonia Plants. *Chem. Eng.* **74,** No. 12, 125.

Moder, J. J. and Phillips, C. R. (1964), *Project Management with CPM and PERT.* Reinhold, New York.

Modern Computation Methods (1957), Notes on Applied Science no. 16, National Physical Laboratory. Her Majesty's Stationery Office, London.

Modrey, J. (1960), On the Incompatibility of Present Graduate Schools and Engineering Design Education. Proceedings of the First Conference on Engineering Design Education. Case Institute of Technology, Cleveland.

Morris, G. A. and Jackson, J. (1953), *Absorption Towers.* Butterworth, London.

Mosler, H. A. (1964), PACER—A Digital Computer Executive Program for Process Simulation and Design. Master of Science thesis, Purdue University, Lafayette, Ind.

Motard, R. L., Lee, H. M., Barkley, R. W., and Ingels, D. M. (1968), *CHESS, Chemical Engineering Simulation System, System Guide.* Technical Publishing Co., Houston.

Nagiev, M. F. (1957), Material Balance in Complex and Multistage Recycle Chemical Processes. *Chem. Eng. Prog.* **53,** No. 6, 297.

Nagiev, M. F. (1964), *The Theory of Recycle Processes in Chemical Engineering.* Macmillan, New York.

Naphtali, L. M. (1964), Process Heat and Material Balances. *Chem. Eng. Prog.* **60,** No. 9, 70.

Nemhauser, G. L. (1966), *Introduction to Dynamic Programming.* Wiley, New York.

Nenniger, E. jr. (1966), The Engineering Design Process. *Chemistry in Canada* **18,** No. 9, 68.

Norman, R. L. (1965), A Matrix Method for Location of Cycles in a Directed Graph. *Am. Inst. Chem. Engrs. Journal* **11,** 450.

Norman, R. L. and Ravicz, A. E. (1964), *User Manual, Calresearch Generalized Heat and Material Balancing Program.* California Research Corporation, Richmond, Calif.

Northcott, T. H. (1967), The Numerical Solution of Chemical Engineering Problems Using MIMIC. Part A of a major report to the Department of Chemical Engineering, McMaster University, Hamilton, Ontario.

Orbach, O. (1969), A Study of the Promotion of Convergence in the Calculation of Complex Chemical Plants. Ph.D. thesis, McMaster University, Hamilton, Ontario.

Pearson, T. G. (1955), *The Chemical Background of the Aluminum Industry.* Lectures, Monographs, and Reports no. 3, Royal Institute of Chemistry, London.

Perry, J. H., ed. (1950), *Chemical Engineers' Handbook.* 3rd ed. McGraw-Hill, New York.

Perry, J. H., ed. (1963), *Chemical Engineers' Handbook.* 4th ed. McGraw-Hill, New York.

Peters, M. S. (1958), *Plant Design and Economics for Chemical Engineers.* McGraw-Hill, New York.

Peters, M. S. and Timmerhaus, K. (1968), *Plant Design and Economics for Chemical Engineers.* 2nd ed. McGraw-Hill, New York.

Petersen, E. E. (1965), *Chemical Reaction Analysis.* Prentice-Hall, Englewood Cliffs, N.J.

Petryschuk, W. F. (1967), The Mathematical Representation and Analysis of a Light Hydrocarbon Refining Network. Ph.D. thesis, McMaster University, Hamilton, Ontario.

Petryschuk, W. F. and Johnson, A. I. (1965), Simulation of the Steady-State Behaviour of a Multicomponent, Multifeed Reboiled Absorber. *Can. J. Chem. Eng.* **43,** 209.

Petryschuk, W. F. and Johnson, A. I. (1966), A Simulation and Parametric Study of Four Existing Multi-Component Distillation Columns. *Can. J. Chem. Eng.* **44,** 241.

Polya, G. (1965), *How To Solve It.* 2nd ed. Anchor Books, Doubleday, Garden City, New York.

Pontryagin, L. S., Boltyanskii, V. G., Gamkrelidze, R. V., and Mishchenko, E. F. (1962), *The Mathematical Theory of Optimal Processes.* Interscience, New York.

Porter, A. W. (1917), The Thermal Properties of Sulphuric Acid and Oleum. *Trans. Faraday Soc.* **13,** 373.

Putney, D. H. (1959), Sulfuric Acid Alkylation of Paraffins. *Advances in Petroleum Chemistry and Refining.* vol. 2. Kobe, K. A. and McKetta, J. J., jr., eds. Interscience, New York.

Ramsay, W. and Young, S. (1886), Evaporation and Dissociation. IV: Acetic Acid. *J. Chem. Soc.* **49,** 790.

Rase, H. F. and Barrow, M. H. (1957), *Project Engineering of Process Plants.* Wiley, New York.

Ravicz, A. E. and Norman, R. L. (1964), Heat and Mass Balancing on a Digital Computer. *Chem. Eng. Prog.* **60,** No. 5, 71.

Reese, K. M. and Cundiff, W. H. (1956), *Modern Chemical Processes.* vol. IV, 180–88. Reinhold, New York.

Reid, R. C. (1965), Engineering Properties of Fluids. *Chem. Eng. Prog.* **61,** No. 12, 58–64.

Reid, R. C. and Sherwood, T. K. (1958), *Properties of Gases and Liquids.* McGraw-Hill, New York.

Reid, R. C. and Sherwood, T. K. (1966), *Properties of Gases and Liquids.* 2nd ed. McGraw-Hill, New York.

Reilly, P. M. and Greene, A. L. (1966), The Application of Adaptive Control to a Continuous SBR Plant. Paper presented at the 16th Chemical Engineering Conference of the Chemical Institute of Canada, Windsor, Ontario, October 17–19.

Rich, L. G. (1961), *Unit Operations of Sanitary Engineering.* Wiley, New York.

Rosen, E. M. (1962), A Machine Computation Method for Performing Material Balances. *Chem. Eng. Prog.* **58,** No. 10, 69.

Rosenbrock, H. H. and Storey, C. (1966), *Computational Techniques for Chemical Engineers*. Pergamon, New York.

Rosenstein, A. B. and English, J. M. (1960), Design as a Basis for a Unified Engineering Curriculum. *Proceedings of the First Conference on Engineering Design Education*. Case Institute of Technology, Cleveland.

Rowe, P. N. (1963), The Correlation of Engineering Data. *The Chemical Engineer* (*London*) **166,** 69. Published with *Trans. Inst. Chem. Engrs.* **41,** 2.

Rubin, D. I. (1962), Generalized Material Balance. *Chem. Eng. Prog. Symp. Ser.* 58, nos. 37 and 54.

Rudd, D. F. and Watson, C. C. (1968), *Strategy in Process Design*. Wiley, New York.

Sanson, F. J. and Petersen, H. E. (1967), *MIMIC Programming Manual*. Technical report SEG-TR-67-31, Wright-Patterson Air Force Base, Ohio.

Sargent, R. W. H. (1967), Integrated Design and Optimization of Processes. *Chem. Eng. Prog.* **63,** 9, 71.

Sargent, R. W. H. and Westerberg, (1964), Speed-up in Chemical Engineering Design. *Trans. Inst. Chem. Engrs.* **42,** 190.

Sawistowski, H. and Smith, W. (1963), *Mass Transfer Process Calculations*. Interscience, New York.

Scandrett, H. F. (1963), Equations for Calculating Recovery of Soluble Values in a Countercurrent Decantation Washing System. *Extractive Metallurgy of Aluminum*, p. 83. [See Gerard and Stoup (1963).]

Schweyer, H. E. (1955), *Process Engineering Economics*. McGraw-Hill, New York.

Shannon, P. T. (1965), Notes of introduction to ES 21. Dartmouth College, Hanover, N.H.

Shannon, P. T., Johnson, A. I., Crowe, C. M., Hoffman, T. W., Hamielec, A. E., and Woods, D. R. (1966), Computer Simulation of a Sulfuric Acid Plant. *Chem. Eng. Prog.* **62,** No. 6, 49.

Shaw, I. D. (1969), An Appraisal of the MACSIM Simulation Routine in its Application to an Alkylation Plant. Master of Engineering thesis, McMaster University, Hamilton, Ontario.

Sherwood, T. K. (1963), *A Course in Process Design*. MIT Press, Cambridge.

Sherwood, T. K. and Pigford, R. L. (1952), *Absorption and Extraction*. McGraw-Hill, New York.

Shreve, R. N. (1967), *Chemical Process Industries*. 3rd ed. McGraw-Hill, New York.

Simon-Carves Ltd. (1958), *Sulfuric Acid Plants*. Cheadle Heath, Stockport, England.

Smalter, D. J. (1964), Management by Priority of Challenge a Procedure. Paper presented at the 53rd National Meeting of the American Institute of Chemical Engineers, Pittsburgh, May 17–20.

Smith, R., Eilers, R. G., and Hall, E. D. (1968), Executive Digital Computer Program for Preliminary Design of Wastewater Treatment Systems. Report no. WP-20-14, U.S. Department of the Interior, Advanced Water Treatment Branch, Division of Research, Cincinnati Water Research Laboratory, Cincinnati.

Sparrow, E. M. and Cess, R. D. (1966), *Radiation Heat Transfer*. Brooks/Cole Publishing Co., Belmont, Calif.

Sparrow, E. M., Hallman, T. M., and Siegel, R. (1955), Turbulent Heat Transfer in the Thermal Entrance Region of a Pipe with Uniform Heat Flux. *App. Sci. Res.* **A7,** 37.

Spencer, H. M. (1954), Empirical Heat Capacity Equations of Various Gases. *J. Am. Chem. Soc.* **67,** 1859.

Steltz, W. G. and Silvestri, G. J. (1958), The Formulation of Steam Properties for Digital Computer Application. *Trans. ASME* **80,** 967.

Steward, D. G. (1967), A Survey of Computer-Aided Chemical Process Design. Report ESL-R-304, MIT project DSR 74994, Electronic Systems Laboratory, MIT, Cambridge.

Tate, G. E. and Cartinhour, J. (1945), Disk Extended Surface for High Heat Absorption Rates. *Trans. ASME* **67,** 687.

Taylor, E. S., et al. (1961), Report on Engineering Design. *J. Eng. Educ.* **51,** 645.

Tayyabkhan, M. and Richardson, T. C. (1965), Monte Carlo Methods. *Chem. Eng. Prog.* **61,** No. 1, 78–82.

Ten States Standards (1960), Recommended Standards for Sewage Works. Great Lakes/Upper Mississippi River Board of State Sanitary Engineers. Available from the State Department of Health of any of the member states.

Tocher, K. D. (1963), *The Art of Simulation*. Van Nostrand, Princeton, N. J. or The English Universities Press Ltd., London.

Tong, L. S. (1965), *Boiling Heat Transfer and Two-Phase Flow*. Wiley, New York.

Vilbrandt, F. C. and Dryden, C. E. (1959), *Chemical Engineering Plant Design*. 4th ed. McGraw-Hill, New York.

Water Pollution Control Federation (1967), *Sewage Treatment Plant Design*. Manual No. 8, p. 90.

Wegstein, J. H. (1958), Accelerating Convergence of Iterative Processes. *Comm. Association for Computing Machinery* **1,** 9.

Wei, J. (1966), Disguised Kinetics. *Ind. Eng. Chem.* **58,** No. 9, 38.

Weller, S. (1956), Analysis of Kinetic Data for Heterogeneous Reactions. *Am. Inst. Chem. Engrs. Journal* **2,** 59.

Weltner, W., jr., (1955), The Vibrational Spectrum, Associative and Thermodynamic Properties of Acetic Acid Vapor. *J. Am. Chem. Soc.* **77,** 3941–50.

Wheelock, T. D. and Boylan, D. R. (1968), Sulfuric Acid from Calcium Sulfate. *Chem. Eng. Prog.* **64,** No. 11, 87.

Whitney, R. P. and Vivian, J. E. (1963). [See Sawistowski and Smith (1963) or *Chem. Eng. Prog.* **45,** 323 (1949).]

Wilde, D. J. (1964), *Optimum Seeking Methods.* Prentice-Hall, Englewood Cliffs, N.J.

Wilde, D. J. and Beightler, C. S. (1967), *Foundations of Optimization.* Prentice-Hall, Englewood Cliffs, N.J.

Williams, T. J. (1969), Computer and Process Control, Annual Review of Fundamentals. *Ind. Eng. Chem.* **61,** No. 1, 76.

Woods, D. R. (1968), Innovations in a Process Design and Development Course. *Chem. Eng. Educ.* **2,** 162.

Woods, D. R. (1969), *A Chemical Engineer in Plant Operations, Design and Research and Development.* Department of Chemical Engineering, McMaster University, Hamilton, Ontario.

Woodson, W. S. (1964), Creative Techniques—A Comparative Analysis. *J. Ind. Eng.* **15,** 60.

Wright, L. T. (1964), Analog Solves Fire Water Problems Fast. *Hydrocarbon Process. Petrol. Refiner* **41,** No. 9, 274.

Author Index

Aizawa, M., 236, 283, 296, 348
Albright, L.F., 322
Allen, D.L., 301, 345
Allingham, J.P., 308, 345
Amelin, A.G., 126
American Petroleum Institute Project 44, 129, 190
Anderson, R.B., 315, 345
Andress, F.J., 309
Aries, R.S., 110
Aris, R., 199, 268, 349
Armett, R.L., 129

Baird, M.H.I., 293, 348
Ball, W.E., 280
Barkley, R.W., 67, 316, 350
Barrow, M.H., 77, 351
Bauman, H.C., 110, 241, 245
Beightler, C.S., 257, 273, 354
Beilstein, 129
Bellman, R., 266
Bird, R.B., 132, 216
Bischoff, K.B., 25, 233, 347
Bobrow, D.G., 25
Bobrow, S., 310, 348
Bolles, W.L., 278, 306, 346
Boltyanskii, V.G., 257, 351
Booth, A.D., 236
Bornstein, R., 128
Bottle, R.T., 129, 131
Boudart, M., 189
Bowman, R.A., 156, 167
Boylan, D.R., 256, 353
Braun, R.M., 129
Braun, W.G., 129
Bromley, L.A., 122, 126
Burkhardt, D.B., 74, 95, 226, 255
Burman, L.K., 280

Calderbank, P.H., 189
Canjar, L.N., 130
Carnegie Institute, 128
Carpenter, G.R., 308, 345
Cartinhour, J., 158, 166, 353
Cary, J.R., 157–58
Cavett, R.H., 25
Cess, R.D., 161, 353
Chartrand, G., 259, 265, 270–71
Chemical Engineering, 209, 255, 301
Chen, J.C., 158, 173
Chilton, C.H., 241
Christensen, J.H., 32, 33, 35, 349
Chu, J.C., 129
Chueh, P.L., 131
Clark, R.L., 75, 301, 346
Cleveland, R.G., 173, 345
Coffey, S., 130
Connor, J.M., 74, 96, 255
Crowe, C.M., 32, 293, 308, 315, 348, 352
Cundiff, W.H., 324, 351
Cupit, C.R., 322
Curtiss, C.F., 132

Dagnall, B.D., 310
Davidson, B., 189
Din, F., 128
Dingman, J.C., 301
Dow Chemical Company, 129
Drake, R.M., 122, 127, 155, 345
Dreisbach, R.R., 128, 130

Dryden, C.E., 75, 77, 110, 353
Duecker, W.W., 74, 95–97, 209, 259
Dukes, R.R., 74, 255, 346
Dukler, A.E., 173

Eckert, C.A., 131
Eckert, E.R.G., 122, 127, 155
Edmister, W.C., 130
Egloff, G., 130
Eilers, R.G., 67, 353
Eklund, R.B., 189
Elliott, D.M., 97
Ellis, D.S., 142
Emmett, P.H., 128
English, J.M., 77, 352
Erbar, J.H., 280, 344
Ergun, S., 186, 196
Erk, S., 155, 346
Erme, T.E., 308, 345
Eucken, A., 122
Evans, L.B., 7, 236
Evans, W.H., 129

Fair, J.R., 278, 306
Faith, W.L., 75, 301
Fan, L.T., 257
Farkas, M.D., 74, 255
Fenske, M.R., 129
Fitz, L., 308, 345
Forgacs, O.L., 291
Forker, G.M., 128
Forrester, J.W., 310
Fox, L., 188, 191
Francis, A.W., 130
Franks, R.G.E., 311
Freel, J.D., 293, 322
Fried, V., 129
Furkert, H., 226, 255
Furlow, R.H., 110, 345

Gallant, R.W., 131
Gambill, W.R., 132
Gamkrelidze, R.V., 257, 351
Gardner, K.A., 160, 166
Gardner, N.F., 245, 278, 347
Gary, J.H., 301
Gerard, G., 324
Gmelin, L., 131
Gmitro, J.I., 120
Gordon, G., 311
Greene, A.L., 311, 351
Grigull, U., 155, 346
Grober, H., 155
Grumer, E.L., 301
Gwyn, J.E., 322, 345

Hala, F., 129
Hall, E.D., 67, 353
Hallman, T.M., 159, 353
Hamielec, A.E., 32, 103, 293–94, 315, 345, 347, 348, 352
Happel, J., 110, 245
Hausner, B., 310, 349
Hellenack, L.J., 245, 278, 347
Himmelblau, D.M., 26, 233
Hirschfelder, J.O., 132
Hirschmann, W.H., 309
Hodgins, J.W., 315, 345
Hoffman, T.W., 32, 293, 315, 345, 348, 352
Holland, C.D., 131, 236, 280
Hooke, R., 260, 263–65, 269, 272
Horsley, L.H., 129
Hottel, H.C., 161
Hougen, O.A., 96, 130, 185
Hughmark, G.A., 173
Hui, A.W.T., 101, 103, 294
Hydrocarbon Processing, 38, 227, 301

Industrial and Engineering Chemistry, 74–75, 96–97, 301
Ingels, D.M., 67, 316, 350
Ishizawa, M., 189, 349

Jackson, J., 220, 350
Jackson, R.E., 310
Jacobs, M.B., 130
Jaffe, I., 129
James, J.L., 245, 278
Janaf, 131
Janz, G.J., 132
Jeeves, T.A., 260, 263–65, 269, 272, 347
Jernigan, E.C., 322, 345
Johnson, A.I., 32, 67, 236, 280, 283, 293, 310, 312, 315, 326, 345, 351, 352
Johnstone, R.E., 304
Johri, H.P., 311
Jordan, T.E., 129

Karr, H.W., 310, 349
Kauffmann, W., 174
Kaye, G.W.C., 128
Kays, W.M., 8, 157–58
Keenan, J.H., 120
Kelley, K.K., 122, 132, 190–91
Keyes, D.B., 75, 301, 346
Keyes, F.G., 120, 348
Kittrell, J.R., 189, 349
Kline, S.J., 304
Klomparens, A.J., 310, 347
Kreith, F., 155–56, 158
Krichevskaya, E.L., 189
Kronseder, J.G., 226, 255
Kubota, H., 189

Laby, T.H., 128
Landolt, H.H., 128
Lapidus, L., 292
Lawrence, A.E., 110
Lederman, P.B., 236
Lee, H.M., 67, 316, 350
Lee, K.Y., 268
Lee, W., 32–35, 232
Lenoir, J.M., 130
Levenspiel, O., 103
Levine, S., 129
Leyerzapf, V., 308, 345
Lightfoot, E.N., 216, 344
London, A.L., 8, 157–58, 348
Love, C.E., 308, 345
Lowry, I., 267–68

McAdams, W.H., 156, 158, 162
Maddox, R.N., 280, 344
Mandelik, B.G., 256
Markowitz, H.W., 310
Marsden, C., 130
Martin, T.W., 120
Matheson Company, 128
Maxwell, J.B., 129
Meier, R.C., 310
Mellor, J.W., 132
Mezaki, R., 189
Miller, R.L., 301
Mishchenko, E.F., 257, 351
Moder, J.J., 316
Modrey, J., 77
Moore, T.F., 301, 345
Morris, G.A., 220
Morris, J.R., 236, 344
Mosler, H.A., 40
Motard, R.L., 67, 316
Mueller, A.C., 167, 344

Nagiev, M.F., 35, 317
Nagle, W.M., 167, 344
Naphtali, L.M., 35, 236
Natural Gas Processors, 129
Natural Gasoline Association, 129
Nemhauser, G.L., 257
Newell, W.T., 310, 349
Newton, R.D., 110, 343
Nenniger, E., Jr., 77
Norman, R.L., 25, 29, 35, 351
Northcott, T.H., 311

O'Connell, J.P., 131
Orbach, O., 233, 235–36, 298
Orye, R.V., 131
Owen, J.M., 97, 346

Pazer, H.L., 310, 349
Pearson, T.G., 324
Perry, J.H., 105, 127–28, 155, 168, 173
Peters, M.S., 77, 110, 241, 245
Petersen, E.E., 181–82, 184
Petersen, H.E., 311, 352
Petryschuk, W.F., 236, 280, 283, 298, 348
Phillips, C.R., 316, 349
Pick, J., 129
Pierson, C.U., 256, 349
Pigford, R.L., 218, 352
Pike, R.W., 267–68, 349
Pimentel, G.C., 129
Pitzer, K.S., 129
Pollock, K.G., 293, 348
Polya, G., 76–77
Ponton, J.W., 310, 348
Pontryagin, L.S., 257
Porter, A.W., 150
Prausnitz, J.M., 131
Putney, D.H., 322

Ragatz, R.A., 96, 130, 185, 347
Ramsay, W., 105
Raphael, B., 25, 344
Rase, H.F., 77
Ravicz, A.E., 35, 350
Reese, K.M., 324
Reid, R.C., 106–7, 108, 122, 126, 132
Reilly, P.M., 311
Reinhart, L.R., 245, 278, 347
Rich, L.G., 304
Richardson, T.C., 310, 353
Rodd, E.H., 130
Rosen, E.M., 35
Rosenstein, A.B., 77
Rossini, F.D., 129
Rowe, P.N., 109
Rubin, D.I., 32, 232
Rudd, D.F., 25, 32–35, 127, 232–33, 241, 257, 278, 304, 310, 314, 345, 349

Sansom, F.J., 311
Sargent, R.W.H., 25, 32
Sarofim, A.F., 161, 347
Sawistowski, H., 217, 219–20
Sax, N.I., 128
Scandrett, H.F., 152
Scheflan, L., 130
Schutt, H.C., 196
Schweyer, H.E., 241
Seidell, A., 128
Shannon, P.T., 32, 40, 55, 77
Shaw, I.D., 243, 246
Sherwood, T.K., 77, 108, 122, 126, 132, 218, 351
Shindo, M., 189, 349
Shreve, R.N., 96, 301
Siegel, R., 159, 353
Silvestri, G.J., 120, 353
Simon-Carves, 255–56
Smalter, D.J., 77
Smith, R., 67
Smith, W., 217, 219–20, 352
Snow, R.H., 196
Sparrow, E.M., 159, 161
Spencer, H.M., 122
Spiers, H.M., 130
Sprague, C.R., 7, 236, 346
Steltz, W.G., 120
Stephen, H., 128
Stephen, T., 128
Steward, D.G., 7, 236, 346
Stewart, W.E., 216, 344
Stoup, P.T., 324, 346

Tate, G.E., 158, 166
Taylor, E.S., 77
Taylor, W.K., 308, 345
Tayyabkhan, M., 310
Ten States Standards, 305
Thibaudeau, H.A., 308, 345
Thodos, G., 189, 345
Thompson, W.H., 129
Thring, M. W., 304, 348
Timmerhaus, K., 241, 245, 350
Timmermans, J., 130
Tong, L.S., 173
Toong, T., 139–41, 312, 326, 348
Touloukian, Y.S., 128

Vermeulen, T., 120, 346
Vilbrandt, F.C., 77
Vilim, O., 129
Vivian, J.E., 220, 354

Wagman, W.H., 129
Wang, C.-S., 257, 346
Water Pollution Control Federation, 306
Watson, C.C., 127, 232–33, 241, 257, 278, 304, 310, 314, 352
Watson, K.M., 96, 130, 185, 347
Weast, R.C., 128
Wegstein, J.H., 235
Wei, J., 299
Weller, S., 189
Weltner, W., 105
West, J.R., 74, 95–97, 209, 259, 345
Westbrook, G.T., 310, 347
Westerberg, A.W., 25, 32, 352
Wheelock, T.D., 256
Whitney, R.P., 220
Wicks, M., 173, 345
Wilde, D.J., 257, 260, 263–64, 273
Wilke, C.R., 122
Williams, T.J., 310
Wilson, R.G., 308, 345
Winfield, M.D., 75, 345
Wisnowski, L.T., 308, 345
Wong, H.W., 308, 345
Woods, D.R., 32, 97, 293, 315, 345, 348, 352
Woodson, W.S., 77
Wright, L.T., 310

Young, S., 105, 351

Subject Index

Absorber:
design by simulation, 308
details, of acid and oleum, 74, 213
simulation, 212–25:
accuracy, 214–15, 223
algorithm for, 222
data for, 220
heat and mass balance, 218
heat transfer in, 217
mass transfer in, 213, 216–17
pressure drop in, 220
programming, 220–21
Absorber section:
description, 74
information flow diagram, 212
process matrix, 224
simulation of, 223–25
Accuracy:
of complete simulation, 111
excessive, 177, 315
importance for SN list, 282
needed, from sensitivity analysis, 113, 291
of simulation and time required, 5
Acid Plant (*See* Sulfuric acid plant)
Additional equipment lists (*See* AEN *and* ENC)
Adjacency matrix, 23, 25, 27–29
AEN, auxiliary equipment list, PACER term:
dimensioning, 52
format, 65, 67
Air addition, optimum, for reactor system, 206, 208
Air drier, simulation in acid plant, 223
Algebraic equations, solution of sets of, 35–36
Algorithm for:
BLOW1, 143
COOL7, 169
dynamic programming, 267
primary loop of PACER, 60
REACT4, 192, 194
recycle calculations in PACER, 61
SORB5, 222
strategy of simulation, 276–77
Alkylation plant:
brief reference to, 111, 114, 150–51, 243
description of, 318–22:
costs for, 247, 321
process flow diagram for, 319
reactions in, 320
information flow diagrams of, 287–90
simulation of, 239, 240, 246–47, 280, 281, 285, 287–89, 293, 298
Analog computers, 311
Auxiliary equipment lists (*See* AEN)
Axial dispersion, in reactor models, 181

Base case simulation, 291:
developing, 81–93, 275
Batch processes, simulation of, 309
Bayer process, 236, 237, 285:
costs of, 322
flow diagram for, 323
process description of, 322–24
simulation of,
collection of data for, 282–83
sample problem in filtration in, 99–102, 151–52
SN list for, 98–99, 281

Block relaxation for convergence promotion, 296
Blower, description of, in acid plant, 70
BLOW, Unit Computation for blower or pump, 118, 135–43
BLOW1:
  algorithm, 143
  development of Unit Computation, 137–38, 142–43
BLOW2, 150
BLOW3, 150
Boilers:
  description for acid plant, 73, 171–72
  simulation of, 173–75
BOILS1, Unit Computation, use of, 201, 202
BOILS2, Unit Computation:
  development of, 173–74
  use of, 205
Boolean Arithmetic, 29
Business analysis, executive programs for, 311
Butane, convergence promotion in, process, 296

Calculation order list:
  dimensioning, 52
  format, 67
Canadian General Electric, 312
Canadian Industries Limited, 11, 69, 72, 74, 86–87, 88
Catalyst bed, void volume for, in simulation, 205
Catalyst data, for acid plant, 200
Catalyst effectiveness, 182
CHEOPS, an executive system, 7
CHESS, an executive system, 7, 67, 316
CHEVRON, an executive system, 7
CHIPS, an executive system, 7
Coding sheets for executive programs, examples of, 66–67, 326–28
Comment cards, use in Unit Computations, 137, 139–41, 295
COMMON:
  example of, 49
  use of, 49, 58
Communicating (*See* Print out)
Communication, organization of, for project, 9, 10
Complete plant studies:
  general 228–46, 295–99
  of the acid plant, 247–53
Compressor-condenser system, description, 151
Computer, impact of hardware developments of, 311–12 (*See also* Time)
Conduction, role in heat transfer, 160
Connection matrix:
  description, 21–22
  example, 62
CONPO1, Unit Computation for convergence promotion, use, 239
Constraints:
  choice of, for optimization, 259
  handling, in OPTIM1, 262
  for production of sulfuric acid, 74
Constraints (*cont.*)
  for simulations, 84
Contact sulfuric acid plant (*See* Sulfuric acid plant)
CONTØ1 Unit Computation for convergence promotion, 250
Control, interplay with a simulation:
  in acid plant, 252
  in alkylation, 280
Control instructions:
  coding of, 326
  list of, 50–51
  for PACER, 46
Control scheme, to ensure mass balance, 237, 250
Convergence:
  defining problem of, 229
  effect of recycle on rate of, 37, 224, 230
  effect of starting point of calculations on rate of, 231–32
  example of, 225
  oscillating answers in, 231
  slow, description of, 229–30
Convergence promotion:
  description of techniques for, 231–36
  for distillation columns, 236
  examples of difficulties in, 295–96
  fitting into PACER, 238–39
  methods of, for acid plant, 247–50
  rate of, for acid plant, 232
  use of:
    block relaxation technique, 296
    dominant eigenvalue method of, 298
    examination of information flow for, 233
    by extrapolation, 233–35, 254
    Newton-Raphson for, 296
    sequence of calculations for, 233
    stream variables for, 231
    system theta method for, example, 296
Conversion, air addition and catalyst activity, effect on, 206
Coolers, acid:
  in the acid plant, 74, 175–76
  simulation of, 175–76
COOL, general model and EN list for, 118, 119
COOL3, use of, 224
COOL7:
  algorithm, 169
  checking accuracy of, 170–71
  development, 168–70
  modification to include boiling, 174
  use in system simulation, 201
Correlating physical properties, 108–9
Correlations, used to replace rigorous Unit Computations, 315
Corroborating data, need and collection, 92 (*See also* Data)
Costs:
  for acid plant, 75, 252
  for alkylation plant, 321
  allocation of, among products, 243, 244
  for alumina plant, 322

Costs (*cont.*)
  associated with equipment, 114, 118, 243, 246
  classification, 242
  collection, 109, 110
  and ENC, 118, 246
  indices for updating, 110
  reporting, 246, 247
  as stream variables, 243
Cost flow, determination of, 241
Criteria, for acid plant simulation, 80–81
Critical path method, 316
CSMP, digital simulator of analog computer, 311
Cut streams, use in convergence promotion, 231
Cycle matrix, 34

Data:
  consistency of, 105, 122
  errors in extrapolation, 105
  inconsistencies in, 94, 110, 275
  lack of, impact on analysis, 315
  need for, 92
  needed for base case, 283
  summary of operating, for acid plant, 86–87
Data collection:
  difficulties, 291, 293
  examples of:
    in alkylation process, 293
    in polymer systems, 294
    in pulp and paper systems, 293
    for simulation, 10, 283
    suggestions for, 284
    use of statistics in, 292
Data set, required for simulation, 7
Decisions:
  needed for simulation, 79
  required for optimization, 257
  when to simulate, 6, 278
Decision variables, for optimization, 257–58
Defining the problem, 274: for the acid plant case, 80–95
Dehydrogenation of ethylbenzene, 295
DELS, PACER, term:
  dimensioning and definition, 52
  use, 48, 55, 230
Depreciation costs, when to calculate, 241, 243
Design:
  comparison with simulation, 303
  definition of, 303
  errors in, 304
  fitting, into PACER, 307
  procedure for settler, 304
  relation of, to simulation and research, 308–9
  of a smelter off-gas sulfuric acid plant, 307–8
  strategy for, 76, 307
  use of:
    executive programs for, 302–3
    REACT for, 308
    SORB5 for, 221

Diagram:
  information flow (*See* Information flow diagram)
  process flow (*See* Process flow diagram)
Differential equations, method of solution of, 191–92
Diffusion and heat conduction, effect on reaction rate of, 182
Dilution tank, in the acid plant, 74
Dimension deck, example, 49
Dimension variables for PACER, 51–53
Direct search optimization, 263
Direct solution of recycle set, 37
Discontinuities, in Unit Computations, 285
Discounted cash flow, 245
Distillation columns:
  convergence promotion in, 236, 295, 296
  simulating, 105, 114, 315
Dollar flow:
  determination of, 241
  effect of recycle on, 243
  use, 245 (*See also* Costs)
DPRINT, PACER term, definition, 56
DREAD, PACER term:
  definition, 55
  use, 62
Drier, process details of, for acid plant, 73, 212–13
DRUM1, Unit Computation for steam drum in acid plant, 201
Dynamic Programming:
  algorithm for, 267
  example of, 270–71
  fitting, into PACER, 268
  use of, 266
DYNAMO, 310
DYNSYS, 310

Economizer, description of, for acid plant, 74
Effectiveness factor, for catalyst:
  definition of, 182
  matched to data, 204
Effectiveness of heat exchangers:
  in the acid plant, 167
  definition of, 8, 157
Emissivity, for surfaces and gases, 161, 162
ENC, auxiliary equipment parameter list:
  dimensions, 52
  examples of costs stored in, 118
  format, 67, 327
  use of:
    for cost information, 246
    for physical properties, 316
Encoding the information flow diagram, 20–23 (*See also* Information flow diagram, adjacency, stream connection, incidence and process matrices)
Energy balance, for reactor Unit Computation, 184
EN list (*See also* Equipment parameter list, ENC, AEN)
  dimensions, 52
  format, 67, 327

EN list (*cont.*)
  example of:
    for COOL, 118
    with costs, 114–15
    how PACER uses it, 42–44
    input to PACER, 64
    for MIXER, 43
    in optimization, 261, 265, 268
  generality of Unit Computations, 112, 114–15
EQCAL, PACER term, definition, 56
Equations, algebraic, solution of, 35–36
Equilibrium model versus rate model of absorber, 211
EQUIP, PACER term, definition, 56
Equipment:
  and associated costs, 243
  collecting information about, 109
  as Unit Computations, examples of, 19
Equipment parameter information, supplied to Unit Computation, 65 (*See also* EN, ENC, *and* AEN)
Equipment Unit, as opposed to Unit Computation, 42
Error messages:
  to test technical feasibility, 239, 240
  in Unit Computations, 239
Errors:
  in design, 304
  typographical, 105
  uncovering, through simulation, 283
Estimation of physical properties, 105
Ethylbenzene, dehydrogenation of, 295
Ethylene:
  convergence promotion in process for, 296
  cracking furnace, data required for, 283
  problem on, 196
EXCHØ1, Unit Computation for heat exchange, description of, 78
Executive program (*See also* PACER, CHESS):
  characteristics, 13
  general description of, 7
  names, 7
  need for, 2, 4
  as part of simulation, 7
  specifications for, 41
  useful characteristics and extensions, 302–15
EXPRNT, PACER term, definition, 56
EXREAD, PACER term, definition, 56
Extrapolation:
  for convergence promotion, 233
  of Unit Computations, 113

Factor (*See* Effectiveness factor)
Feasibility:
  checking technical, for complete plant, 239
  financial, 239, 241
Feed streams, flag for in PACER, 62
Fiber length distribution, use in SN list, 291
Fibonacci optimization, 260
Filter press, problem of simulating, 151
Filtration, problem in, 99
Fin:
  details of, 163, 166
  efficiency, 160
Financial attractiveness of investments:
  calculation of, 239
  fitting into PACER, 245–46
Flag:
  example of, 62
  special, use of, 82
  stream, 44
FLEXIFLOW, an executive system, 7, 245
Flow diagram:
  information (*See* Information flow diagram)
  process (*See* Process flow diagram)
Format:
  for calculation order list, 67
  for control variables, 67
  for EN list of MIXER, 43
  for equipment information, 67
  for KSPRNT, 67
  for preferred streams, 67
  for process matrix, 67
  for stream information, 67
  for TEST lists, 67 (*See also* Coding sheets)
Future developments, 302

Gas absorption, 213, 216–67
Geometric progression, in convergence promotion, 235, 254
GEMCS, an executive program, 67, 312
GPSS, an executive program, 310
Golden section:
  derivation of, 260
  example of, 260–63
  fitting, into PACER, 261–62
  use of, in dynamic programming, 268
GUESS1, PACER, term:
  definition of, 56
  use of, 63
GUESS2, PACER term:
  definition of, 56
  use of, 63, 64

Hardware developments, and impact on analysis, 311–12
Heat and mass transfer, effect on reaction rate, 182
Heat conduction, effect on reaction rate, 182
Heat exchange, equations for, 155
Heat exchanger effectiveness for acid plant, 167
Heat exchangers:
  simulation of, 8, 153–78 (*See also* COOL, BOILS, EXCHØ1)
  simulation of economizer or superheater, 168
Heat transfer:
  accounting for, in absorption, 217
  similarity of, to mass transfer, 218
  summary equations for, 158–59
Heat transfer coefficient:
  adjusting, 201
  definition of, 155

Heat transfer coefficient (*cont.*)
role of conduction in, 160
role of convection and radiation in, 159, 160
Height of transfer unit, for absorption, 219
Hooke-Jeeves:
direct search, 263
example of, 265
fitting, into PACER, 265
Hybrid computers, 311

Imbalance in converging solutions, 230
Improvement:
check list of sources for process, 300–301
strategy for process, 77
Incidence matrix, 22
Index, cost, 110
Information:
definition of, 13
in a stream, 13
sources of, for Unit Computations, 64
Information flow, used to promote convergence, 233
Information flow diagram:
of absorber section, 212
for acid plant, 90, 91, 249
classification of, into recycle or serial sets, 25
coding, numerically, 20–23
development of, for alkylation plant, 287–90
methods for developing, 19, 84, 88–89, 284–85, 287–90
need for simplicity in, 276
of PACER, 57, 60–61
of reactor system, 202
Input data:
collection of, 104
required for PACER, 64
Inputs to a simulation, example of, 4, 83, 325–28
Instability, in SORB5, 221
Interaction between simulation, design and operation, 308
Inventory-dominated scheduling, use of executive in, 310
Iterative calculations, printout during, 328–33

JUNCO, Unit Computation (*See* MIXER)

Kachkaroff process, 256
KBACK2, 262, 265
KCLEAN, PACER term, definition of, 50
KEFLAG, PACER term, use of, 58, 59
Kellogg flexible flow sheet, 7, 245
Kinetics for reactor Unit Computation, 186
KK, PACER term, definition of, 52
KONV, PACER term, definition of, 51
KPM, PACER term, definition of, 52
KPRINT, PACER term, definition of, 51
KRUN, PACER term, definition of, 50
KSETS, PACER term, definition of, 50
KSFLAG, PACER term, use, 58
KSPRINT, PACER term:
coding of, 328
definition of, 53
format of, 67

Langmuir-Hinshelwood Adsorption, 189
Learning Curve, use of, 308
Leonard Construction Co., 70
Light ends recovery unit, convergence promotion in, 296
Linear equations, solution of sets of, 35, 317
Linear models, acceptable forms of, 285
LINKS, use of, 58
LISTT, PACER or MACSIM term, 67
Logic (*See also* Algorithm):
for primary loop of PACER, 60–61
LOOPS, PACER term, definition of, 50

MACSIM, an executive system, 66–67, 313, 316:
coding of, 325
control data for, 66–67
printout for, 328
Maintenance costs, calculation of, 75, 241
MAIN3, PACER term, 56, 57
MAJOR, PACER term:
definition of, 56
function of, in PACER, 57
use of, 62, 63
Mass balance:
and cost calculation, 243
solution of, 6
Mass and heat transfer, and reaction rate, 182–83
Mass transfer:
in absorber simulation, 211
in liquid phase, 217
similarity with heat transfer, 218
Material balance (*See* Mass balance)
Mathematical model of equipment (*See* Unit Computations)
Matrices:
intersection of, 30
relation between, within PACER, 47
Matrix (*See specific title* Adjacency, incidence, process, stream connection)
McMaster University (*See* DYNSYS, GEMCS *and* MACSIM)
MIMIC, 311
MIXER, Unit Computations, 16, 17, 42–43, 135:
calculations in, 36
diagram of equation for, 42
EN list for, 43
example of calculations in, 36
operation of, within PACER, 47–48
MIXER1:
description, 78
use, 325
MIXER3, 201:
use of, 224
MIXER4, use of, 224
MIXER5, 150

MIXER6:
algorithm for, 146–47
development of, 143–48
use of, 201, 224
MIXER7, 150
MIXER8, 150
Mixing, unit operation, 16
Modeling, precautions for correct answer in, 224
Models, analytical, 116
Modular executive:
evaluation of, 253
use of, in unsteady-state problems, 310
Modularity, 13, 18
Molar flows, use in stream list, 281
Monsanto Company, 70
Monte Carlo:
experimentation, 314
simulation, example of, 310

N2MAX, PACER term, definition of, 51
N3MAX, PACER term, definition of, 52
NDELS, PACER term, definition of, 52
NE, PACER term, definition of, 43, 44
NECLMX, PACER term, definition of, 52
Need (*See also* Questions):
general, for simulation, 5
identification of, 80
for recycle in chemical processing, 14
satisfying an engineering, by simulation, 1
summary of, for creating a simulation, 270
NELMAX, PACER term, definition of, 52
NEMAX, PACER term, definition of, 52
Newton-Raphson convergence promotion, 236, 296
NIN, PACER term, definition of, 44, 46, 47
NOAEN, PACER term, definition of, 52
NOCOMP, PACER term, definition of, 52
NOEN, PACER term, definition of, 52
NOENC, PACER term, definition of, 52
NOKPM, PACER term, definition of, 52
Nomenclature:
consistency of, 315
used in acid plant simulation, 121
NOPP, term for physical properties system, 313
NOSN, PACER term, definition of, 52
NOUT, PACER term, definition of, 44, 46, 47, 48
NPS, PACER term, definition of, 52
NPSMAX, PACER term, definition of, 52
NSLMAX, PACER term, definition of, 52
NSMAX, PACER term, definition of, 52

Objectives (*See* Questions)
Objective function, for optimization, 257, 258
Oleum absorber, description, 74, 213
Optimization:
definition of problem of, 257–58
fitting, into PACER, 261, 265, 268
four major decisions in, 257
methods of:
comparison of Hooke-Jeeves and dynamic programming, 269, 272
dynamic programming, 266
Optimization (*cont.*)
methods of (*cont.*)
golden section, 260
Hooke-Jeeves search, 263
review of, 257
selecting, 260
OPTIM1, Unit Computation for Optimization, 261–62
OPTIM2, Unit Computation for Optimization, 265
OPTIM3, Unit Computation for Optimization, 268
Order:
of calculations (*See* Sequence of calculations)
of recycle, 39, 201
of streams in process matrix, 21
Organization needed:
in creating a simulation, 9
of main communication, 10
of personnel, 10
of resources, 79:
example of, for acid plant study, 123
Oscillations:
from convergence promoters, 239
for converging solutions, 231
Output from a simulation, examples, 83, 328–33

PACER, 40–66:
algorithm of primary loop of, 60
algorithm for recycle calculations in, 61
availability of, 55
control information for, 50
data required for, 64
description of, 41, 49:
information flow diagram of, 57
list of subroutines in, 55
relation between information blocks in, 47
structure of, 41
evaluation of, 253
example of:
coding sheets for input information to, 325
of printout from case, 328
fitting:
convergence promotion into, 238
design into, 307
dynamic programming into, 268
financial calculations into, 246
golden section optimization into, 261
Hooke-Jeeves optimization method into, 265
how it:
executes a recycle calculation, 53
finds the sequence of calculation, 58
operates, 42, 47
uses:
the EN list, 44
flags, 62
the SN list, 43
STRMI and STRMO, 44
how to use:
coding sheets for, 325

PACER (*cont.*)
how to use (*cont.*)
deck arrangement for, 66
input data to, 64
summary of:
arrays in, 46
data requirements for, 64
organization of Unit Computation for, 45
Packed beds, pressure drop through, 186
Particle size distribution, use of SN list for, 282
People, organization of, for simulation projects, 10, 123
Petrochemical processing, convergence promotion in, 295
Physical properties:
collection of, 105, 282
estimation of, 105, 108
method of handling, 115–17, 118–23, 312
within PACER, 115
references for locating, 128–32
return of investment, 240
storage of, 108
Plant, studies of complete, 228–53, 295
Plant data (*See also* Data):
need for, in simulation, 92
suggestions for collection of, 284
for sulfuric acid plant, 86–87
Plant operation, importance of, in simulation, 275, 280, 284, 308
PLINE1, Unit Computation for pressure drop, 150
Polymer Corporation Limited, Sarnia, Ontario, 295, 296
Polymerization, collection of data for, 294
POWERFACTS, an executive system, 7
Predictor-Corrector method, 188, 220
Preferred streams:
dimensioning of, 52
format of, 67
PRESEQ, a Unit Computation, for convergence of acid plant, 252
Pressure drop:
in gas absorber, problem for, 220
through packed beds of catalyst, 186
Primary settlers, design of, and errors, 304–5
Print out:
control of, 51
of cost report, 246, 247
of simulation, 328
of technical feasibility, 239, 240
Problems:
characteristics of solving, 2, 76–77
how engineer solves, 6
solving complex, 274
strategy for solving, 274–99
Process analysis, strategy for, 274–99
Process flow diagram:
for acid plant, 71
for alkylation plant, 319
for alumina plant, 323
and information flow diagram, 18 (*See also* Information flow diagram)
Process matrix:
coding sheets for, 327
description of, 20
dimensioning of, 52
examples of, 21, 59, 64, 85:
for the acid plant, 88
for absorber section, 224
format of, 67
as input to PACER, 64
ordering streams in, 21
use of, 25
Product streams, flagging, 62
Production rate, and costs, 242
Program executive (*See* Executive program)
Programmer, information supplied by:
for PACER, 64
for Unit Computations, 65
Promotion of convergence (*See* Convergence promotion)
PTRACE, a Unit Computation, use for convergence promotion in acid plant, 252
Pulp and paper process:
example of data collection for, 293
problems in simulating, 291
PUMP, a Unit Computation, 113, 135
PUMPS1, Unit Computation of a pump, use of, 136, 201, 224
PUMPS2, 150
PUMPS3, 150
PUMPS4, 150

Questions, interplay of, 275
Questions:
answering those posed for the simulation, 80, 252, 279, 295
posed for acid plant, 80

Radial dispersion, 181
Radiation, role in heat transfer, 160
Reachability matrix, 29
Reaction rate:
effect of bulk transport on, 181
effect of diffusion and heat conduction on, 182
effect of mass and heat transfer on, 182
Reactions:
for acid plant, 70
for alkylation plant, 320
need for adequate data for, 282
Reactor:
description of, for the acid plant, 73, 198
problem of collection of data for, 293
selecting the rate processes to simulate, 180
simulation of, 112, 179–95:
with heat exchange, 197–208
studies of, in PACER, 200
use of Unit Computation for design of, 308
Reactor section:
information flow diagram for, 202
plant data for, 199
testing models for, 204, 205
REACT, Unit Computation:
brief reference to, 116

REACT (*cont.*)
data required for, 188
development of, 179–95
energy balance in, 184
heat transfer data for, 191
kinetic data for, 189
kinetics and stoichiometry in, 186
mass balance in, 183
momentum balance in, 185
summary of equations for, 187
thermodynamic data for, 190
REACT3, 193
REACT4, Unit Computation for reactor:
algorithm for, 192, 194
development of, 191–95
matching, to plant data, 193
use of, in reactor section, 201–3
Recycle:
definition of, for a process, 24
effect of:
on calculations, 1, 14
on convergence, 37
on convergence promotion, 224
on dollar flow, 243
examples of, 2, 5, 14, 15, 53:
identifying, 23–30:
by adjacency matrix, 29
by reachability matrix, 29
PACER algorithm, with, 61
need for, 14
need to recognize, 2
planning calculations when present, 31, 53, 61
sequential and simultaneous solution of problems with, 36
Recycle set:
definition of, 24
separation of, from serial set, 25
Recycle streams:
effect of, on convergence, 285
importance of information in, for SN list, 281
Regression analysis, 109
Reliability, incorporation of, in executive program, 314
Reporting (*See* Print out)
Research:
needed, in pulp and paper process, 291
relation with design in simulation, 308
Resources:
analysis of, for simulation, 292
need for analysis of, 275
organization of, 79:
for acid plant study, 123
required for the acid plant study, 252
RPRINT, PACER term, definition of, 58
Runge-Kutta, use of, 188, 191–92, 220

Scale up:
difficulties in, 304
of plant costs, 110
SEFLAG, PACER term, definition of, 56
Sensitivity analysis:
description of, and need for, 84, 275
use of, 114, 291
SEPAØ1, Unit Computation for a separator, 78, 101, 284–85
SEPAØ2, 284–85
Sequence of calculations:
coding of, 328
in data set, 7
determination of, 31
how PACER determines, 58
method of selecting, 32
PACER operation, when specified, 46
use of, to promote convergence, 233
Sequential and simultaneous calculations, comparison of, 36
Serial set:
definition of, 24
separating from recycle sets, 25
SETSP1, Unit Computation, 201, 202, 208, 229, 247:
description of, 208
use of, for convergence promotion, 208, 233, 248
SETST1, Unit Computation:
problem for, 67
use of, 89
Settler, design of, 304–5
Settling and washing, example of SN list for, 282
SETUP, PACER term definition, 56
Significant variables, identification by sensitivity analysis, 84, 291
SIMSCRIPT, 310
Simulation:
analogy with real process, 6
approximate, 85, 286, 291
comparison of, with design, 303
deciding when needed, 5, 275, 278
definition of, 2, 6, 303
design with, 307
how to create, general, 8, 76–79, 274–99
parts of, 7
printout of, 225, 328
relation to design and research, 308
significant variables in, 81, 84
uses of, 279, 299, 300, 309–11
SN list, PACER term, 41 (*See also* Stream variables):
coding of, 327
definition of, 41
dimensioning of, 52
example of, 44:
for acid plant, 225
example problems in selection of, 97–99
example values in, 54–55
format of, 67
as input to PACER, 64
printout from, 330
selecting the variables for, 81–83, 281–82
units of measurement for, 81, 102, 121, 281
use of:
in PACER, 43, 47–48
in optimization, 261–62, 265, 268
Solving problems, strategy for, 76, 274–99
SORB5:
algorithm for, 222
instability in, 221

SORB5 (*cont.*)
matching, with plant data, 223
Special flag, use of, 82, 208
Specification:
of criteria for acid plant simulation, 80
for executive systems, 41
of sequence of calculation, for PACER, 46
STABLE, physical property program, 120
Starting point of calculations and rate of convergence, 231–32
Statistical:
correlations, to replace approximate models, 315
models for equipment, 113
Statistics, use to collect data, 292
Stoichiometry for reactor Unit Computation, 186
Storage of stream information, temporary, 44 (*See also* SN list)
Strategy:
in creating a simulation, 8, 76, 274
in developing Unit Computations, 64, 133, 295
for plant improvement, 77
for process design, 76
for solving problems, 76
Stream (*See also* SN list, stream information):
definition of, 19
flag in, 43
flagging, by PACER, 62
introduction of term, 6
information in, 13
numbering of, 43
Stream connection matrix:
description of, 21
example of, 21, 62, 329
Stream information:
dimensioning of, 52
format of, 67
and SN list, 43
as supplied to Unit Computation, 64
use of previous, for convergence promotion, 232
Stream matrices, sample values in, 54–55
Stream variables (*See also* SN list):
definition of, 19
examples of, 19
selection of, for convergence promotion, 231
use of costs as, 243
Stream variables list (*See* SN list)
Stream variables test lists:
dimensioning of, 52
format of, 67
STRM matrices (*See* STRMI *and* STRMO)
STRMI, PACER term:
definition of, 44
sample values in, 54–55
use of, 47–48
STRMO, PACER, term:
definition of, 44
sample values in, 54–55
use of, 48
Styrene, production of, 295
Subroutines of PACER, 55
Successive substitution, method of, 36
Sulfur burner, description of, 73
Sulfur dioxide reactor, constraints for, 259 (*See also* Reactor)
Sulfuric acid plant:
alternative processing conditions for:
ammonia scrubber, 255
double absorption for, 226, 255
produced from off-gas, 255, 298
description of, 70–76:
constraints on, 74
costs for, 75
equations for reaction in, 70
operating data for, 86–87
photograph of, 72
process flow diagram of, 71
design, simulation models for, 307
optimum conditions for, 258
simulation of:
additional data required for, 92, 283, 284
build up of components in, 286
COMMON listing, 49
computer time required, 252
conclusions from, 252
convergence promotion of, 247
costs included in, 252
information flow diagrams for, 90, 91
man-hours required for, 252
process matrix for, 88
questions asked for, 279
SUPER, physical property program, 120
Superheater and economizer, description of, 162

Technical units dominate unsteady-state, 310
Temperature measurements, effect of errors in, 283
TEST, PACER term:
definition of, 58
use of, 64
TEST for convergence, precautions with recycle, 230
TEST lists, PACER term:
coding of, 237
dimensioning of, 52
Theta method of convergence promotion, 236
use of, 296
Time, computer:
for acid plant case study, 252
distribution of, for simulation, 123
versus accuracy, 5
Time-sharing executive program, 312 (*See also* GEMCS)
Tolerance in checking convergence, 229
Transfer unit, correlation for height of, 219
Transient behavior, simulation of, 309
Trial and error calculations (*See* Recycle)
Two-phase flow:
in boiling models, 174
problem for, 152
review of, 173
Typographical errors in data, 105

Uncertainty, use of, in executive programs, 314
Unimodality, definition of, 261
Unit Computations (*See also specific models*: BLOW1, BOILS1, CONPØ1, COSTØ1, DRUM1, CONTØ1, COOL, EXCHØ1, MIXER, OPTIM, PLINE, PRESEQ, PTRACE, PUMP, REACT, SEPA, SETSP, SETST, SORB):
  example of, 19:
    for absorbers, 210–21
    for costing, 246
    for fluid handling equipment, 133–49
    for heat exchangers, 8, 153–77
    for mixers, 17
    for optimization, 261, 265, 268
    to promote convergence, 238–39, 247
    for reactors, 179–97
    simple models for, 78
  general discussion of:
    analogy of, with Unit Operation, 7
    compared to equipment unit, 42
    control of calculations within, 48
    definition of, 6, 7
    as input to PACER, 65
    MIXER, operation within PACER, 47
    as part of a simulation, 6
    use of EN list to generalize, 42–43
  principles in preparing:
    accuracy required for, 111, 284
    decisions about extrapolation in, 113
    generality for, 42, 112
    handling constraints within, 295
    how to prepare, 64, 133, 294
    need for simplicity in, 111, 177, 278, 284
    selection of, for information flow diagram, 19
Unit computations (*cont.*)
  principles in preparing (*cont.*)
    similarity with plant operation, 280, 284
    source of information for, 64
    use of comment cards for, 295
  use of:
    group of models for, 284
    linear models for, 285
    MIXER as example of, 325
    simple models for, 100, 284
Unit of equipment, versus Unit Computation, 42
Units of measurement:
  for acid plant, 121
  selecting, for stream list, 81, 281
Unit operation (*See under specific heading*)
Unsteady-state, executive programs for, 309–11
Utility costs, calculation of, 243
UWOPS, an executive program, 7

Variable (*See also* Stream variables, SN list):
  decision, definition of, 257
Venture worth, 245
Void fraction, selection of, for catalyst bed, 205

Waste water treatment, design of primary settler for, 304

Yield, air addition and catalyst activity on, 206